Bilingual Textbook of Forage Cultivation

牧草栽培学双语辑要

主编　龙明秀　呼天明
编者　龙明秀　呼天明　杨培志
　　　何学青　何树斌　许岳飞

高等教育出版社·北京

内容简介

本书共10章，绪论主要从牧草产业发展的角度阐释牧草的重要性，世界牧草生产格局与供求态势，中国牧草产业现状与前景，本课程的性质、任务及内容等。第1章至第4章为总论部分，主要介绍牧草的分类，牧草的生长繁殖特性与规律，牧草生长发育与光、温、水、土等气候及环境因子的关系，人工草地的建植与管护，牧草混播及草田轮作技术等。第5章至第9章为各论部分，主要介绍主要豆科牧草、禾本科牧草、禾谷类饲料作物、根茎瓜类饲料作物等的植物学特征、生物学特性、栽培技术、饲用价值与利用方式等。第10章主要介绍牧草的起源、分布及区划的原则与依据，并概述了我国牧草的九大分区。

本书主要作为草业科学专业本科生教材使用，同时也适用于动物科学、水土保持等专业，也可以作为研究生或草业科学相关教学和科研人员的参考书。

图书在版编目（CIP）数据

牧草栽培学双语辑要 / 龙明秀，呼天明主编. -- 北京 : 高等教育出版社，2019.3
ISBN 978-7-04-047040-6

Ⅰ. ①牧… Ⅱ. ①龙… ②呼… Ⅲ. ①牧草－栽培技术－高等学校－教材 Ⅳ. ①S54

中国版本图书馆CIP数据核字(2017)第031540号

MUCAO ZAIPEIXUE SHUANGYU JIYAO

策划编辑 李光跃　责任编辑 李光跃　封面设计 王 鹏　版式设计 马 云
责任印制 田 甜

出版发行 高等教育出版社
社　　址 北京市西城区德外大街4号
邮政编码 100120
印　　刷 北京信彩瑞禾印刷厂
开　　本 850mm×1168mm 1/16
印　　张 20
字　　数 510千字
购书热线 010-58581118
咨询电话 400-810-0598
网　　址 http://www.hep.edu.cn
　　　　 http://www.hep.com.cn
网上订购 http://www.hepmall.com.cn
　　　　 http://www.hepmall.com
　　　　 http://www.hepmall.cn
版　　次 2019年3月第1版
印　　次 2019年3月第1次印刷
定　　价 68.00元

物 料 号 47040-00

本书作者，左起依次为龙明秀、何树斌、许岳飞、呼天明、杨培志、何学青

Preface 前言

Grass, as a producer of the most extensive coverage of the Earth, is able to maintain the life circle of the entire planet along with other producers by photosynthesis which allows solar radiant energy into biosystem and feeding herbivores in the form of organic matter with bioenergy, by which means energy is passed on from one species to another.

Grass is the forgiveness of nature-her constant benediction. It has immortal vigor and aggression to protect the ecological safety year after year. It bears no blazonry or bloom to charm the senses with fragrance or splendor but its homely hue is more enchanting than the lily or the rose. Forage production directly affect the supplication of animal food and products.

Accounting for more than 40% of national territory area and dominated by perennial herbal plants, rangeland determines the quality of ecological environment to a great extent. Lawn developed in high-cost urban land makes a great difference to the urban air quality and people's feelings and mood. Cultivated grassland can develop into an independent industry and its amount and quality is also an important indicator of animal husbandry modernization for a region and even for a country. Accordingly, forage cultivation is a systematic technique connecting the restoration improvement of rangeland vegetation, the establishment of cultivated grassland and the establishment of lawn in urban area.

Since 1942 when professor Wang Dong, the founder of modern grassland science in China, pioneered to teach and study forage science, the major of grassland science has been developed with great speed in Northwest A&F University. To be specific, postgraduates have been cultivated to study forage science from 1986; undergraduates and postgraduates were enrolled in 2000 and 2003 separately to study the major of grassland science; and then doctoral students were recruited in 2006 when the major was the secondary discipline of animal hus-

草——作为覆盖地球最为广阔的生产者，通过光合作用使太阳辐射能进入生命系统，并连同其他生产者使太阳辐射能以含能有机物的形式，经由草食动物，传递给一个又一个生物种群，进而维系整个地球生命界。

草是自然界的宽宥，不断地赐福于我们。它虽不如鲜花娇艳，更不像树木结果，然而年年岁岁、岁岁年年给予我们生态庇护，牧草生产直接影响动物性产品的供给。

占国土面积 40% 以上，以多年生草本植物建群的天然草地很大程度上决定着我国生态环境的优劣。寸土寸金的城市土地上的草坪则关乎城市的空气质量和人们的心情状态。人工草地不仅可以形成独立的产业，其多寡优劣更是一个地区乃至一个国家畜牧业现代化的重要标志。而牧草栽培则是关联天然草地植被修复改良、人工草地建植、城市草坪建设的系统技术。

自 1942 年我国现代草业科学奠基人王栋教授率先开展牧草学教学和研究以来，西北农林科技大学的草业科学学科经长期快速发展，1986 年在饲料科学硕士点开始培养牧草学方向研究生，2000 年招收草业科学专业本科生，2003 年招收草业科学硕士生，2006 年作为畜牧学二级学科招收草业科学博士生，2011 年晋升为草业科

bandry science; until 2011, grassland science was successfully promoted to the first-level discipline and approved to be the key discipline of Shaanxi Province. After decades of unremitting efforts of several generations, a young and vibrant team with more than twenty people for teaching and research has been gradually formed in the department of grassland science in Northwest A&F University. "Forage cultivation" was approved as the national excellent course in 2010 and the project of national excellent resource sharing course in 2013. In the respect of scientific research, our team has undertook a number of national and provincial major scientific research projects and international cooperation projects, focusing on the collection, evaluation, cultivation and utilization of forage resource and the forage adversity biology.

As modern education tends to be more and more international, we have been exploring and trying to bilingually teach the course of forage cultivation with the textbook *Forages* of original edition from America. However, for this course is characterized by extremely practical application and distinct regional feature, during the course of bilingual teaching we found that the system and content of the original textbook from America cannot be well applied to the practical forage production in China. Coupled with the reason that the original textbook is too expensive for students to afford, it is urgent to compile a bilingual textbook of forage cultivation which is well matched with the regional feature of our country. Therefore, the members of our team with years of teaching and practical experience and insights organized and compiled the *Bilingual Text of Forage Cultivation* based on existing domestic outstanding textbooks and teaching materials of original edition from America.

This text is divided into two main parts, general theories and special forages. Part 1 generally elaborated the basic knowledge of forage agronomy and regular techniques and principles for forage cultivation including the growth and reproduction characters of forages, the relationship between forage and the environment, the establishment and management of pastures etc. Accordingly, part 2 mainly illustrated the botanical and biological characteristics and the techniques of cultivation and utilization of the most common but very important cultivated forages and forage crops in practical production in China.

学一级学科并获批陕西省重点学科。历经几代人几十年的不懈努力，西北农林科技大学草业科学系已形成了一支具有20余人的年轻并充满活力的教学科研团队。“牧草栽培学”课程于2010年获批国家精品课程，2013年获批国家精品资源共享课。科学研究方面，围绕牧草资源收集评价、培育利用和牧草逆境生物学，承担了多项国家及省部级重大科研项目及国际合作项目。

在教育越来越国际化的大趋势下，我校早在2006年就已经开始了“牧草栽培学”双语教学的探索与尝试，并选用美国原版教材*Forages*。然而，因本课程是一门实践性极强的应用型课程，具有明显的地域性。双语教学过程中，我们深感原版教材的体系和内容与我国的饲草生产实际存在较大出入，加之原版教材价格昂贵，学生无力承担。因此亟待编写一本适宜于我国地域特色的双语版教材。鉴于此，本课程教学团队结合多年的教学经验和感悟，根据国内现有的优秀中文教材和美国原版英文教材，组织编写了这本《牧草栽培学双语辑要》。

《牧草栽培学双语辑要》分为总论和各论两大部分。总论包括牧草的生长繁殖特性、生长发育与环境、人工草地的建植与管护等，从总体上阐述牧草栽培的农艺学基础、常规技术和原则。各论部分主要介绍我国生产上常见的重要栽培牧草及主要饲料作物的植物学特征、生物学特性及栽培利用技术等。

本教材突出的特色是，形式上文字简练、图文并茂；语言上，以英文

The prominent features of this work are concise and well-illustrated in form, written in English and supplemented by Chinese interpretation in language to cater to the teaching requirement, focusing on the regional features and taking into account of the discipline systematicness at the same time and the difference between north and south in geography. We greatly appreciate Professor Roger Gates from South Dakota State University of United States for linguistic modification and polish in English.

This edition might appear mistakes or other problems for the limitation of authors' practical experience and academic accomplishment .Since it is the first attempt to compile a bilingual textbook, we appreciate if readers give us more advice and comments to perfect and optimize the next edition.

By Prof. Hu Tianming,
Aug 2016, In South Dakota State University of America

为主，辅以中文释义，以满足教学需要；地域上，既注重区域特色，又兼顾学科的系统性和南北差异，适用于草业科学专业教学和相关科研人员参考使用。感谢美国南达科他州立大学的 Roger Gates 教授给予全书的英文修改及润色。

由于作者生产实践经验和学术素养所限，编写双语教材亦属新的尝试，全书还可能存在这样或那样的问题，敬希读者不吝指正，以便再版时更臻完善。

呼天明
2016 年 8 月
于美国南达科他州立大学

Contents 目 录

Introduction
绪　论

Forage is defined as "edible parts of plants, other than separated grain, that provide feed for animals, or can be harvested for feeding." It includes herbage (leaves, stems, roots and seeds of nonwoody species), browse (buds, leaves, and twigs of woody species) and mast (nuts and seeds of woody species). In addition to their function as main feeds for livestock, feed forages also play important roles in maintaining and improving soil physical and chemical properties, thereby influencing soil fertility and conservation. Careful use of forage plants can provide environmental benefits of improved water quality and quantity and enhanced air quality. It is generally recognized that there are four origin centers of forages as follow: ① European (excluding Eastern Mediterranean climate zones). ② Eastern Mediterranean basin and the Near East (winter frost). ③ The African savannan (tropical steppes). ④ Tropical America.

1. Importance of forages in a changing world

The role and importance of forages continually change as the societies evolve and new technologies are developed for the plant and animal sciences. Global economic development and accompanying improvements of living standards has led to significant changes in food preferences; Demand for animal food has continued to increase day by day. "To people, food is the first; to animal, forages are the first." Forages are essential for the development of food animal production.

A western proverb says: God provided us two treasures, one is legume plants, which are rich in protein with root nodules which fix nitrogen; Another is ruminant livestock, because they can ferment plant fiber that humans cannot use directly into animal protein. These treasures provide advantages for grassland animal husbandry, and illustrate the dual forage roles of both feed and land conservation.

Forage characteristics of high light use efficiency contribute to grain and livestock production systems. Research in the loess plateau (Ren Jizhou, 2002) demonstrated that dedicating 1/4 to1/3 of the arable land area to grow forages in a farming system increased grain yield per unit area by

牧草，指具有一定饲用价值、以草本为主的野生或栽培植物，包括草本、半灌木和灌木，除了作为家畜的主要饲料外，对改良土壤理化性质、维持土壤肥力、防风固沙、保持水土、绿化环境和调节空气质量也具有重要作用。一般认为，牧草有 4 个起源中心，即欧洲（不包括地中海气候带）中心、地中海盆地和近东（冬霜）中心、非洲萨瓦纳（热带干草原）中心和热带美洲中心。

1. 牧草在世界变革中的重要地位

牧草的作用和重要性随着社会的发展和动植物生产新技术的开发不断变化。随着世界经济发展和人民生活水平的不断提高，动物性食品的需求量与日俱增。“民以食为天，畜以草为本”，牧草是畜牧业发展不可替代的物质资源。

西方有句谚语：上帝给人两件宝，一是豆科植物，它本身生产丰富的蛋白质，它的根瘤是天然氮肥发生器；二是反刍家畜，它的瘤胃则是天然的发酵罐，可将人类不能直接利用的植物纤维转化为动物蛋白质。这正是草地畜牧业的两大优势，也很好地阐释了牧草兼具饲料和养地功能的双重作用。由于很多牧草具有高光效的特点，发展牧草生产有利于建立稳产高产的农牧业生产体系。任继周（2002）在黄土高原的研究证明，把 1/4 ~ 1/3 的耕地拿来种草，进行草田轮作，粮食单产提高 50% ~ 60%，总产提高 30% ~ 40%。土壤有机质 3 年内增加 1/4，化肥施用量减少 1/3，农牧业产值均提高两倍左右。

可以说，没有优质的牧草，就不可能有现代畜牧业的健康发展，更无法确保食品安全。近年来频发的动物

50% ~ 60%, and total output increased by 30% ~ 40%. Soil organic matter increased 25% within 3 years, requiring 33% less fertilizer, while grain and livestock production doubled.

Good quality pasture is necessary for healthy modern animal husbandry production, and ensuring food safety. Several animal food safety incidents in recent years remind food producers that the fundamental tasks of modern agricultural production include not only meeting the basic human needs, but also improving the quality continuously and ensuring the safety of the supply.

In addition, forages serve multiple functions which benefit soil and water conservation and mitigate greenhouse gas accumulation effect and other ecological benefits.

Forages are the bridge and link between agriculture and animal husbandry, development of forage grass industry is the objective requirements of the development of modern agriculture.

2. World supply and demand of forages

With the further development of forage product, forage industry enters a new stage of development. Forage products on the international market, alfalfa products account for the largest proportion of world trade. Global forage exports are concentrated in the American region, including the United States, Canada, Argentina, Chile and Brazil. Europe is the world's second-largest forage product exporters. Products include alfalfa powder and granule, corn (used as feed and silage), hay and other forage products. Asian forage product exporters mainly include China, Lebanon, the United Arab Emirates, Iran, and Korea. The principle export products include alfalfa meal and pellets, other forage products. But Asian exports constitute a small world share.

Increasing domestic forage use and decreasing arable land area will tighten forage product supplies available. Meanwhile, as global forage product demand increasing, which lead to a rising trend of the international trade prices.

性食品安全事件给世界敲响了警钟，现代农业的根本任务不仅是满足人类的基本需求，而且尚需不断提高产品质量和保证产品的安全供给。此外，牧草还具有涵养水源、保护土壤、防风固沙、减少水土流失、降低温室效应等多种改善生态的作用与功能。

2. 世界牧草生产格局与供求态势

在畜牧业生产发达的国家，牧草属于作物生产的重要组成部分，在农业生产中占据重要地位。美国在 20 世纪 50 年代就将紫花苜蓿列入战略物资名录，草产业已成为美国农业中的重要支柱产业。国际市场上，苜蓿草产品在世界草产品贸易中所占的份额最大。

海关数据显示，2013 年中国进口苜蓿总计 75.56 万 t，首次突破 75 万 t 大关，同比增长 70.89%。进口燕麦草总计 4.28 万 t，与 2012 年进口 1.75 万 t 相比，2013 年同比增长 144.29%。国内苜蓿和燕麦草生产量虽然也逐年在增加，但与目前大规模建设的牧场对牧草的需求还是有很大的差距，因此进口数量增加是必然趋势。

随着对牧草产品开发的深入，牧草产业已迈入了一个新的发展阶段。在国际草产品市场上，美国是最大的草产品出口国，日本是最大的草产品进口国。世界草产品的出口贸易主要集中在美洲地区，主要出口国包括美国、加拿大、阿根廷、智利和巴西。欧洲是世界草产品出口的第二大出口地，主要出口苜蓿草粉及草颗粒、玉米（饲用 + 青贮）、干草及其他草料产品；亚洲草产品出口国主要有中国、黎巴嫩、阿联酋、伊朗、韩国等，出口的主要草产品包括苜蓿草粉及草颗粒、其他草料产品，但占世界

3. Current situation and prospects of forage production in China

Current situation of forage production in China

Forage production in China started very late as an industry due to historical and practical factors, but has experienced twists and turns in the short development process. Due to the long term influence of traditional farming culture, animal husbandry has been developing relatively slow. In the late 1990s of the 20th century, the international market demand for forage increased and strategic adjustment of agricultural industry structure in China making forage briefly flourished as a new industry. However, the strong implementation of the 2004 national policy on grain security gave a big pound response to the forage grass industry directly. 2008 "melamine" event alarmed for animal food safety, once again pulling the quantity and prices of forage products such as alfalfa. But from the perspective of supply and demand conditions, high quality forage is still a restriction factors in the development of animal husbandry in China. Alfalfa products imports rose from 19,000 tonnes in 2008 to 1.2 million tonnes in 2015, continue keep rising, imports countries are concentrated in the United States, and Canada etc. developed grassland animal husbandry countries. "It is not worth to transport forage grasses and grains with a long distance." Therefore, forage grass industry is urgent needed to speed up in China.

According to the continental monsoon climate in China, an alfalfa production and processing industries belt in Northeast, North, Northwest, and grasses production and processing priority region in Qinghai-Tibet Plateau and south of China has been formed. China's forage grass industry is booming in recent years, many forage production and processing enterprises have sprung up, and initially formed a relatively complete industrial chain that covering breeding and reproduction of forage seed, forage cultivation, processing, storage and marketing. But objectively speaking, the forage grass industry in China is still much undeveloped, small scale of production, market mechanisms are not perfect, most of the forages are poor quality, lack of competitiveness in the

份额极小。

世界草产品未来供给趋紧，一方面，可供出口的草产品数量减少，另一方面，可耕地面积减少，牧草生产资源紧张。同时，世界草产品市场需求旺盛，因而贸易价格呈上升趋势。

3. 我国牧草生产现状与前景分析

我国牧草生产现状

由于历史和现实等因素，我国牧草生产作为产业发展较晚，但却经历了波折的发展历程。由于长期受农耕文化的影响，畜牧业一直发展很慢。20 世纪 90 年代末，国际市场对牧草的需求增加和国内农业产业结构实行战略性调整，使得牧草作为一个新兴产业出现了短暂的兴盛。2004 年国家粮食安全政策的强有力实施，对牧草产业形成了不小的冲击。2008 年 "三聚氰胺事件" 为动物性食品安全敲响了警钟，再一次拉动了苜蓿等主要草产品价格的快速回升。但从供需状况看，优质饲草依旧是我国畜牧业发展的一大短板。我国苜蓿进口量从 2008 年的 1.9 万 t 增加到 2015 年的 120 万 t，一路持续走高，进口国主要集中在美国、加拿大等草地畜牧业发达国家。"百里不运草，千里不运粮"。我国牧草产业亟待崛起。

我国以大陆季风气候为主，现在已基本形成东北、华北、西北一条苜蓿草产品生产加工优势产业带和青藏高原、南方禾草两大生产加工优势区。近年来，我国草产业迅速崛起，涌现出很多大型牧草种植和加工企业，并初步形成了牧草种子繁育、牧草种植、产品加工、贮运销售等一个相对完整的产业链条。但从现阶段看，我国牧

international market. China's annual output of forage is about 60 million tons currently, only about 4 million tons enter into circulation, product types include bale, grass pellet, grass block, grass powder and wrapping silage. Most of the marketing forage products come from Inner Mongolia, Heilongjiang, Jilin, Sichuan and Gansu province, mainly are alfalfa and *Leymus chinensis*, among which alfalfa takes amount of 90% and 80% of them are under level 3 according to the quality standards of products in the international market, whereas the top level alfalfa hay products in the United States account for more than 70%.

Prospects for good quality forage production

With the rapid development of China animal husbandry, particularly dairy industry, the demand of forage grass also increased greatly. According to statistics, the annual food grain demand for people does not exceed 200 million tons in China, but the feed grain demand has reached 300 million tons, and is expected to reach 500 million tons by 2020, apparently, the real big gap exists in feed. To meet the needs of food production, only relying on traditional agriculture that target for grain production is powerless. Fully utilize the grassland resources that are 4 times of arable land and the potentiality of grass-crop rotation, increase the ratio of forages and herbivorous livestock, and might be the fundamental solution to problem of China's food security.

China is a mountainous country, lands suitable for farming is less than 12%, the other 80%~90% land resources, include grassland, forest, beaches etc., have important potential for livestock production. Nearly 5.3 million hectares of grassland and forestland can be used in China, and the area of only 6 grass agro-ecological zones in the South of China can be equivalent of two New Zealand;16 provinces of the South pasture production potential is even greater. If this potential is developed, may increase new farmland (equivalent) of 16 million hectares, 24.22 million tons of crude protein, enough to make up for the gap of protein. On the basis of ensuring grain security, implement grassland agricultural system, with "grass-ruminant" model to replace the existing "grain-pig" model, will have important significance for agriculture and animal husbandry, the good quality forage protein deficiency

草产业还非常落后，生产规模小，市场机制还不健全，所生产的大部分豆科牧草产品质量较低，缺乏在国际市场上的竞争能力。目前我国年产牧草 6000 万 t，进入流通领域的商品草仅约 400 万 t，产品类型包括草捆、草颗粒、草块、草粉及裹包青贮。商品草生产的重要省区为内蒙古、黑龙江、吉林、四川和甘肃，以苜蓿和羊草为主，其中苜蓿占 90%，且 80% 为 3 级以下。而美国一级品苜蓿干草占苜蓿干草产品的 70% 以上。

优质牧草生产前景

随着中国畜牧业尤其是奶业的快速发展，中国牧草产业的需求也逐渐增加。据统计，目前我国人民直接用的口粮不超过 2 亿 t，而饲料用粮已达 3 亿 t，预计到 2020 年将达 5 亿 t，真正缺口很大的是饲料粮。要满足粮食生产需求，仅靠以籽实生产作为全部内涵的传统农业是无能为力的。施行草地农业，充分发挥中国 4 倍于农田的草地资源和农区草田轮作的潜力，将“人畜共粮”改为“人畜分粮”，将“籽粒型农业”改为“籽粒 – 营养体型”农业，加大草食家畜比重，发挥草地农业系统节约、高效的长处，粮饲分开，应是确保中国食品安全、农民增收、根本解决中国粮食安全问题的出路。

中国是个多山国家，适宜农耕的地区不足 12%，80% 以上的土地资源，包括草地、林地、滩涂等，都有不可忽视的畜牧业生产潜力。全国可利用草地、林地近 80 亿亩（1 亩≈ 667m^2），仅将南方 6 个草地农业生态区利用起来，就可相当于两个新西兰；而南方 16 个省区的草地生产潜力就更大了，这个潜力开发出来，可新增农田（当量）2.4 亿亩，生产粗蛋白质 2422 万 t，足可弥补现在的缺口。在确保粮食安全的基础上，实

problem will be solved, and the national concern issues of "agriculture, farmer and rural area" would also be greatly eased.

The Central file No.1 in 2015 and 2016 clearly put forward that accelerating the development of forage grass based animal husbandry, supporting the silage corn, alfalfa and other forage crops planting; Carrying experiments and demonstration areas about change grain-oriented to grain-forage style, and combined animal husbandry with planting; Push the coordinated development of grain plants, cash crops and forage crops; Accelerate the implementation of returning grazing land to grassland, Pastoral areas of disaster prevention and mitigation; development and utilization of grassland in the South project, all of above statements point out the direction for further development of forage grass industry in China.

As the more and more standard milk and the animal products market, the demand for forage grass products will rapidly increasing. China is the second largest grassland country in the world, but the productivity and the utilization level are far more behind the developed countries. In view of this, either national policy guidance, or geographical advantage for export, from national policy guidance, the forage grass industry in China will face a new round of leap-forward development.

4. The characteristics, contents and learning methods

Forage cultivation is a plant science which directly supports livestock production. It is also a very practical and comprehensive subject. Two broad topics include general agronomic theories and cultivated forage plants. General agronomic theories will demonstrate forage growth and reproduction, forage growth, development and the environment, pasture establishment and management. Consideration of cultivated forage plants will introduce the Leguminosae, Poaceae, and other families, growth and development of fodder crops, forage value, biological characteristics and cultivation techniques.

Forage cultivation is an interdisciplinary subject. Background knowledge of botany, plant taxonomy, plant ecology, plant physiology, soil fertility and agro-meteorology will be highly beneficial to complete this course.

施草地农业系统，以“牧草－反刍家畜”系统取代现有的“粮－猪”系统，对农业和畜牧业都将具有重要意义，优质蛋白质饲料缺乏的难题将迎刃而解，举国忧虑的“三农”问题也将大为缓解。

2015 年和 2016 年的中央 1 号文件明确提出，加快发展草牧业，支持青贮玉米和苜蓿等饲草料种植，开展粮改饲和种养结合模式试点，促进粮食、经济作物、饲草料三元种植结构协调发展；加快实施退牧还草、牧区防灾减灾、南方草地开发利用等工程，进一步为我国牧草产业发展指明了方向。

随着奶业市场和其他畜产品市场的不断规范，我国对草产品的需求会快速增加。从面积上看，我国是世界第二草原大国，但目前我国草地资源远未达到合理、高效的利用与开发。由此看来，无论是国家政策导向，还是草产品出口地缘优势，我国牧草产业将出现新一轮的跨越式发展。

4. 学科性质、内容与学习方法

就学科性质而言，牧草栽培学是一门主要服务于畜牧业的植物学科，也是一门具有很强生产实践性的综合性应用学科。主要内容包括总论和各论两大部分，总论包括牧草的生长繁殖特性、生长发育与环境、人工草地的建植与管护等；各论部分主要介绍豆科、禾本科和其他科牧草及主要饲料作物的生长发育规律、生物学特性、饲用价值、栽培技术等。

牧草栽培学是一门多学科交叉的学科，学好并掌握植物学、植物分类学、植物生态学、植物生理学、土壤肥料学、农业气象学等专业基础课，对本课程的学习将大有益处。

Chapter 1

Classification of Forages

牧草的分类

1.1 Taxonomic classification 按植物分类系统划分

According to the binomial classification system created by Swedish botanist Lin Nai (Carl von Linne, 1707—1778), forages are divided into the following three categories.

依据瑞典植物学家林耐（Carl von Linne，1707—1778）确立的双名法植物分类系统可将牧草划分为以下三类。

1.1.1 Legumes

Legumes are important forage, because of high crude protein content and unique capacity to fix atmospheric nitrogen and resulting benefits for conservation, having been used for agricultural production since ancient times. Common legumes include alfalfa (*Medicago sativa* L.) (Fig. 1.1), sainfoin (*Onobrychis viciaefolia* Scop), erect milkvetch (*Astragalus adsurgens* Pall.), white clover (*Trifolium repens* L.), red clover (*Trifolium pretense* L.), sweet clover (*Melilotus suaveolens* Ledeb.Crownvetch (*Coronilla varia* L.), hairy vetch (*Vicia villosa* Roth), Chinese milk vetch (*Astragalus sinicus* L.), Stybsanthes guianensis (*Stylosanthes guianensias* SW) etc.

1.1.1 豆科牧草

豆科牧草是栽培牧草中最重要的一类牧草，由于其粗蛋白含量高，特有的生物固氮性能和保持水土的功效，使其在远古时期就用于农业生产。常见的豆科牧草有紫花苜蓿（图 1.1）、红豆草、沙打旺、白三叶、红三叶、草木樨、小冠花、毛苕子、紫云英、柱花草等。

Fig. 1.1 Alfalfa
图 1.1 紫花苜蓿

1.1.2 Grasses

Grasses have a relatively short cultivation history. Because of a great diversity of species, grasses constitute more than 70% of the cultivated forages and are the principle component of pastures used for mowing and grazing. Grasses are also most often used for natural grassland renovation. Widely used grasses include smooth bromegrass (*Bromus inermis* Leyss) (Fig. 1.2), elymus (*Elymus dahuricus* Turcz), perennial ryegrass (*Lolium perenne* L.) and sudangrass(*Sorghum* sudanense) etc; Grasses used both as forage and for turf include Kentucky bluegrass(*Poa pratensis* L.), red fescue(*Festuca rubra* L.), perennial ryegrass (*Lolium perenne* L.) and so on.

1.1.2 禾本科牧草

禾本科牧草栽培历史较短，但种类繁多，占栽培牧草 70% 以上，是建立放牧刈草兼用人工草地和改良天然草地的主要牧草。目前利用较多的禾本科牧草有无芒雀麦（图 1.2）、披碱草、多年生黑麦草、苏丹草等；作为草坪绿化利用的牧草有草地早熟禾、紫羊茅、多年生黑麦草等。

1.1.3 Forages of the other families

Grasses and legumes consist of the principle plants used for forages, both by number of species used and area planted. Other species remain very important alternatives, such as chicory (*Cichorium intybus* L.) (Fig. 1.3), cup plant (*Silphium perfoliatum* L.) of Compositae, sweet potato (*Dioscorea esculenta* (Lour.) Burkill) of Polemoniaceae, potato (Solanum tuberosum) of Solanaceae, carrot (*Daucus* carota) of Umbelliferae, and Rumex(*Rumexacetosa* L.) of Polygonaceae and feeding beet of *beta vulgaris* L. etc.

1.1.3 其他科牧草

指除豆科和禾本科以外的其他科牧草，在种类、数量、栽培面积方面，都不及豆科和禾本科，但仍是饲草中非常重要的辅助草种，如菊科的菊苣（图 1.3）、串叶松香草，旋花科的甘薯，茄科的马铃薯，伞形科的胡萝卜，蓼科的酸模，藜科的饲用甜菜等。

Fig. 1.2 Smooth bromegrass
图 1.2 无芒雀麦

Fig. 1.3 Chicory
图 1.3 菊苣

1.2 Classification based on growth characteristics 按生育特性划分

Based on morphology, growth habit, and utilization purpose, forages can be grouped into the following types.

1.2.1 Classification based on the development speed and lifespan

According to the development speed and lifespan of forages, it can be divided into the following three types.

1.2.1.1 Annuals

Annuals refer to the plants that can complete the whole growth and development process and die after flowering and fruiting in the seeding year. Such as annual ryegrass (*Lolium multiflorum* L.), hybrid sudangrass [*Sorghum sudanense*(Piper) Stapf.] and silage corn (*Zea mays* L.) etc.

1.2.1.2 Biennials

Biennials refer to the plants that only have the phase of vegetative growth in the seeding year, blossom and have seeds in the next year, and then die. The common biennial forages include white sweet clover and yellow sweet clover.

1.2.1.3 Perennials

Perennials refer to the forages that the lifespan can reach for more than two years. According to the length of the lifespan, they can be classified into short-lived perennial forages, medium-lived perennial forages and long-lived perennial forages.

(1) Short-lived perennial forages

The average lifespan of short-lived perennial forages can last for 3 to 4 years, their highest yield appear in the first to second year. Such as elymus, perennial ryegrass and red clover.

(2) Medium-lived perennial forages

The average lifespan of medium-lived perennial forages can last for 5 to 6 years, most of the grasses and legumes belong to this category, such as timothy, tall fescue, orchardgrass, astragalus adsurgens, white clover, birdsfoot trefoil etc., the highest yield appear in the second to third year,

根据牧草生长发育的形态、生长习性和利用目的的不同，主要划分为以下类型。

1.2.1 依据牧草发育速度和寿命长短划分

1.2.1.1 一年生牧草

指在播种当年即可完成整个发育过程，开花结实后死亡的牧草。如一年生黑麦草、苏丹草、青饲玉米等。

1.2.1.2 二年生牧草

指播种当年仅进行营养生长，第二年才开花结实，之后死亡的牧草，又称为越年生牧草，如白花草木樨、黄花草木樨。

1.2.1.3 多年生牧草

寿命两年以上，根据寿命长短又可分为短寿命牧草、中寿命牧草和长寿命牧草。

（1）短寿命牧草

短寿命牧草平均寿命 3 ~ 4 年，如披碱草、多年生黑麦草、红三叶等，其产草量第一、二年最高。

（2）中寿命牧草

中寿命牧草平均寿命 5 ~ 6 年，大部分禾本科牧草和豆科牧草都属于这一类，如猫尾草、苇状羊茅、鸭茅、沙打旺、白三叶、百脉根等，其产草量第二、三年最高，第四年产量下降。

（3）长寿命牧草

长寿命牧草平均寿命为 10 年或 10 年以上，一般利用 6 ~ 8 年，在其生长的第 3 ~ 5 年产草量最高，如无芒雀麦、草地早熟禾、羊草、冰草、苜蓿等。

the yield decrease after the fourth year.

(3) Long-lived perennial forages

The lifespan of long-lived perennial forages can last for 10 years or more, which can be using for 6 to 8 years, the growth peak appear in the 3rd to 5th year. Such as alfalfa, smooth bromegrass, Kentucky bluegrass, *leymus chinensis* and wheatgrass etc.

1.2.2 Classification based on tiller origination

1.2.2.1 Rhizomatous grasses

Besides the aerial stems, rhizomatous grasses posses many horizontal rhizomes under the soil surface at depths of 5~20 cm. New tillers sprouts develop from the internodes of rhizome, grow upward and break the soil surface, then develop into a new plant and build its new rhizomes. Each new branch also has its own rhizomes and roots. These grasses has high ability of asexual propagation, they are trampling resistant and suitable for grazing (Fig. 1.4). Examples include *Leymus chinensis*(Trin.) Tzvel. and smooth bromegrass.

1.2.2.2 Bunch grasses

Tiller nodes are located 1~5 cm below the soil surface.

1.2.2 依据牧草分蘖及成枝方式划分

1.2.2.1 根茎型禾草

除地上茎外，在地表下 5~20 cm 处具有水平横走的根茎，由此根茎的节处向上长出穿出地表的枝条，每个枝条发育成新的植株，再形成新的根茎（图 1.4）。这类牧草具有很强的无性繁殖能力，耐践踏，适合于放牧地，如羊草、无芒雀麦。

1.2.2.2 疏丛型禾草

分蘖节位于地表以下 1~5 cm 处，分蘖芽形成的侧枝与主枝以锐角方向向上生长，能产生多级分蘖。对土壤通气性要求不甚严格，但对水分要求较高，因此抗旱性稍差（图 1.5）。如披碱草、老芒麦、鸭茅、猫尾草等。

1.2.2.3 根茎－疏丛型禾草

在地表以下 2~3 cm 处的分蘖节形成短根茎，由此向上新生出枝条，每个枝条又以同样的方式进行分蘖(图 1.6)。

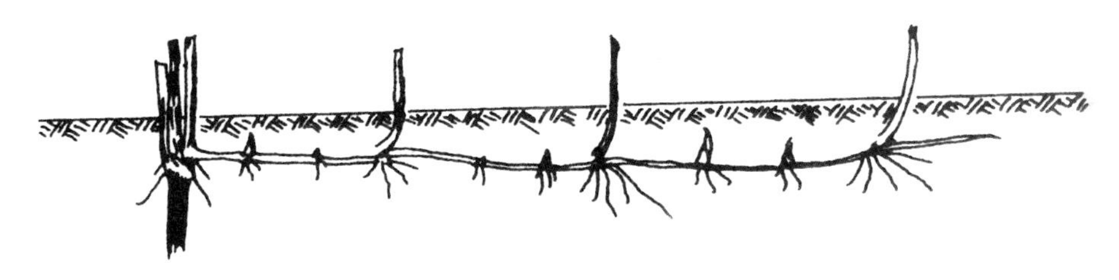

Fig. 1.4 Sketch of rhizomatous grasses
图 1.4 根茎型禾草示意图

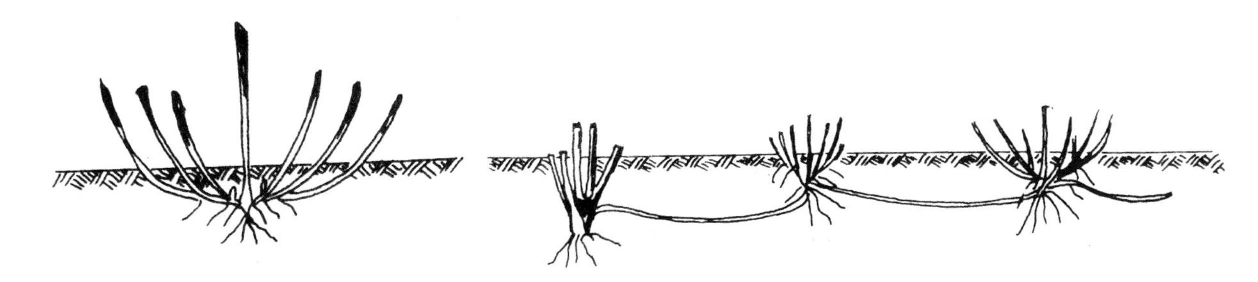

Fig. 1.5 Sketch of bunch grasses
图 1.5 疏丛型禾草示意图

Fig. 1.6 sketch of rhizome-bunch grasses
图 1.6 根茎－疏丛型禾草示意图

New shoots formed by the lateral tiller buds grow upward at a sharp angle, which can produce branched tillers. Aeration requirements less stringent, more water is needed and poor drought resistance (Fig. 1.5). Examples include Siberian wildrye elymus, orchardgrass and timothy. Some bunch grasses tolerant very moist to nearly saturated soils, but have high moisture requirements. Other bunch grasses display good drought tolerance (e.g. *stipa spp.*) maybe not cultivated. Once established, stoloniferous and rhizomatous grasses depend on vegetative reproduction and tend to be quite persistent.

1.2.2.3 Rhizome-bunch grasses

Short rhizomes form below the soil surface at depths of 2~3 cm. New shoots grow upward producing new tillers from each shoot. Such grasses can form a flat and elastic and difficult to be destroyed, which often be used for lawn establishment (Fig. 1.6). Examples include *Festuca rubra* L. and *Alopecurus aequalis Sobol*.

1.2.2.4 Sod forming grasses

Tiller section located above the soil surface, internodes are very short, lateral branches close and parallel to the main branch, grow upward, forming dense stand, these grasses are very drought tolerant and winter hardy(Fig. 1.7). Examples include *Stipa bengiana and Achnatherum splendens.*

1.2.2.5 Tap-rooted legumes

Tap-rooted legumes display vertical and stout taproots, which mainly reach a depth of 2 meters. A swelling in the lower part of the stem (1~3 cm below the soil surface) is called the root crown. New shoots originate from the tap crown growing upwards and forming sparse branches (Fig.1.8). Typical tap-rooted forage legumes include red clover and alfalfa.

1.2.2.6 Rhizomatous legumes

Rhizomatous legumes root to a depth less than 1 m. Horizontal roots occur from 5 to 30 cm below ground level. Lateral root buds produce upward growing stems which branch above ground (Fig. 1.9). Examples include *Medicago falcata*, licorice, *cephalanoplos segetum* and coronilla.

1.2.2.7 Stoloniferous forages

Horizontal tillers of stoloniferous forages originate from

这类草能形成平坦而有弹性和不易被破坏的草地，是草坪用的优良禾草。如紫羊茅、看麦娘等。

1.2.2.4 密丛型禾草

分蘖节位于地表以上，节间很短，侧枝紧贴主枝平行向上生长，形成密集的丘状株丛，具有很好的抗旱和抗寒性（图 1.7）。如针茅、芨芨草。

1.2.2.5 轴根型豆科牧草

具垂直而粗壮的主根，入土深达 2 m。在茎的下部（土表以下 1~3 cm）有一膨大部分，称为根颈，根颈上的更新芽向上生长，形成多枝的株丛（图 1.8）。属于这一类型的豆科牧草主要有红三叶、紫花苜蓿、草木樨等。

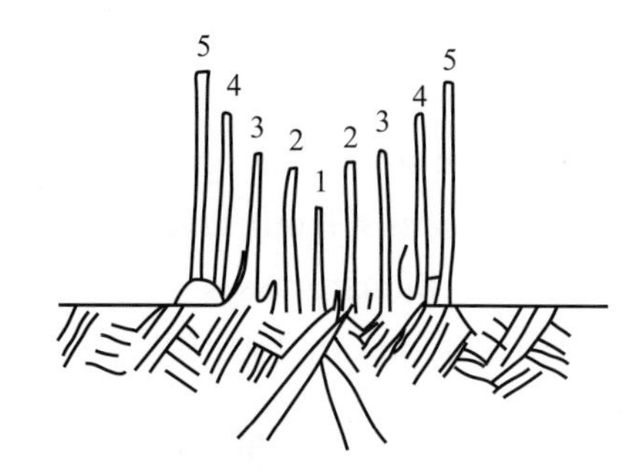

Fig. 1.7 Sketch of sod forming grasses
图 1.7 密丛型禾草示意图

Fig. 1.8 Tap-rooted legumes
图 1.8 轴根型豆科牧草

1.2.2.6 根蘖型牧草

入土深不到 1 m。具垂直根且在地面以下 5~30 cm 处生出水平根，在其上形成更新芽，向上生长形成枝条（图

Fig. 1.9 Rhizomatous legumes—A two year old kura clover plant showing rhizome production and a daughter plant which has produced another daughter plant
图 1.9 根蘖型豆科牧草——库拉三叶草的地下根茎

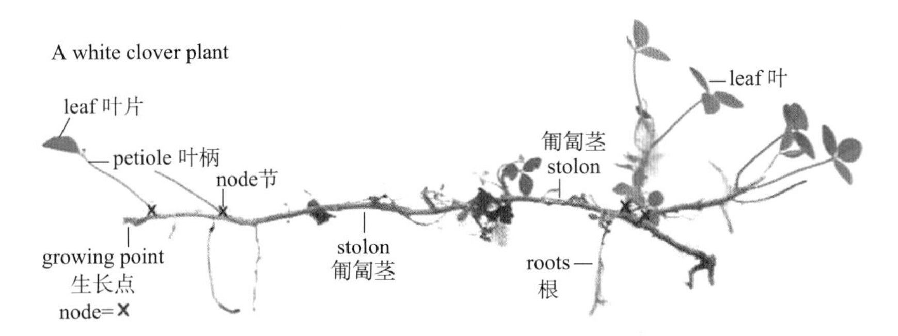

Fig. 1.10 Sketch of stoloniferous forages
图 1.10 匍匐型牧草示意图

leaf axils of the mother plant's root crown. Nodes of these prostrate stems grow downward and form adventitious roots. Axillary buds grow upward and become stems and foliage of new plants (Fig. 1.10). Examples include white clover, *cynodon dactylon*, zoysia grass.

1.2.3 Classification based on leaf distribution and plant height

1.2.3.1 Top-grass

The plant height is over 50 cm, the grass bundle is composed by the reproductive branch and long nutrition branches, more leaves and branches distribute in the upper 1/3 parts of the plant, the canopy shape looks like an inverted cone, which is often used for cutting (Fig. 1.11). Examples include *Leymus chinensis*(Trin.) Tzvel., *Elymus dahuricus*Turcz., smooth brome grass, orchard grass, *Pennisetum purpureum* Schum., *Urariacrinita* (L.) Desv. ex DC., alfalfa, sweet clover, and sainfoin.

1.9），如黄花苜蓿、甘草、刺儿菜、多变小冠花等。

1.2.2.7 匍匐型牧草

由母株根颈、分蘖节或枝条的叶腋处向周围生出平伏于地面的匍匐茎，匍匐茎的节可向下长出不定根，腋芽向上产生枝条或叶簇，从而形成新的植株(图 1.10）。如白三叶、狗牙根、结缕草、鹅绒委陵菜等。

1.2.3 依据叶片的分布和植株的高矮划分

1.2.3.1 上繁草

株高在 50~100 cm 以上，株丛多由生殖枝和长营养枝组成，叶片和枝条多分布在植株 1/3 以上的部位，株型呈倒锥形，大多数牧草属于这一类，适宜于刈割型人工草地（图 1.11）。如羊草、

Fig. 1.11 Top grass—sweet clover
图 1.11 上繁草——黄花草木樨

Fig. 1.12 Bottom grass—Kentucky bluegrass
图 1.12 下繁草——早熟禾

1.2.3.2 Bottom grass

Bottom grass plants are short, generally not exceeding 50 cm in height. Plant clumps are composed of mostly short vegetative shoots, with a large number of leaves concentrated at the lower part of the plants. After mowing, stubble accounted for 20%~60% of the total production (Fig. 1.12). This kind of forages is more commonly used for grazing purpose. Examples include Kentucky bluegrass, stipa, wheatgrass and white clover.

1.2.3.3 Rosette forming forages

Leaves of rosette forming forages grow out from the root to form circular basal clumps. No leaves grow from stems. These plants often occur on the degraded grassland that because of long time of heavy grazing (Fig. 1.13). Examples include dandelion, plantain and phoenix saussurea Stella.

1.2.4 Classification based on the plant erectness

1.2.4.1 Erect forages

The main stem of erect forage plants is perpendicular to

披碱草、无芒雀麦、鸭茅、象草、猫尾草、苜蓿、草木樨、红豆草等。

1.2.3.2 下繁草

植株矮小，高度一般不超过 50 cm，株丛多半是短营养枝，大量叶片集中于株丛基部，刈割后的留茬数量大，这类牧草适于放牧利用（图 1.12）。如草地早熟禾、针茅、冰草、白三叶、扁蓿豆等。

1.2.3.3 莲座状草

根出叶成叶簇状，没有茎生叶或茎生叶很小。由于它们比较矮小，因此在长期放牧而退化的草地上最普遍（图 1.13）。如蒲公英、车前、风毛菊等。

1.2.4 依据直立性划分

1.2.4.1 直立型牧草

主茎垂直于地面生长，整个株型呈直立状，大多牧草属于此类(图 1.11)。如苜蓿、草木樨、黑麦草等。

the ground. The plant is erect. Most forages belong to this classification. Examples include alfalfa, sweet clover (Fig. 1.11), ryegrass etc.

1.2.4.2 Plagiotropic forages

Lateral stems of lax forages grow from the root crown, and then grow parallel to the ground surface before growing upward. Examples include Bermuda grass (Fig. 1.14) and zoysia grass.

1.2.4.3 Winding or vining forages

The main stem of vining forages degenerated into tendrils or winding stems, which need to attach on the other erect plants to grow up (Fig. 1.15). Examples include crownvetch and hairy vetch.

1.2.4.2 斜生型牧草

根颈处产生的枝条穿出地面后，先贴地生长一段后再向上直立生长。如狗牙根（图 1.14）、结缕草等。

1.2.4.3 缠绕型牧草

主茎退化成卷须或缠绕状茎，须依附在其他直立物上才能向上良好地生长（图 1.15）。如小冠花、毛苕子等。

Fig. 1.13 Rosette forming forages—plantain
图 1.13 莲座状草——车前

Fig. 1.14 Plagiotropic forages—Bermuda grass
图 1.14 斜生型牧草——狗牙根

Fig. 1.15 Winding or vining forages—hairy vetch
图 1.15 缠绕型牧草——毛苕子

1.3 Classification based on the water requirement 按牧草的需水量划分

1.3.1 Drought tolerance forages

The drought tolerance forages need less water and can tolerant long term drought, these forages like wheatgrass, Russian wildrye grass, stipa, Haloxylon mmodendron, caragana and lespedeza

1.3.2 Mesophyte forages

The water requirement of mesophyte forages is between the drought tolerance forages and the water grasses, such as Siberian wildrye, *Leymus chinensis* (Trin.), smooth bromegrass, orchardgrass, perennial ryegrass, alfalfa, red clover and so on.

1.3.3 Hydrophilous forages

More water is needed during the growth period; hydrophilous forages can grow on the land with low temperature and poor drainage, such as reed canary grass, white clover and Rumex etc.

1.3.1 耐旱牧草

需水量较少，能长时间忍受干旱，如冰草、新麦草、针茅、梭梭、柠条、胡枝子等。

1.3.2 中生牧草

需水量介于耐旱牧草和喜水牧草之间，如老芒麦、羊草、无芒雀麦、鸭茅、多年生黑麦草、紫花苜蓿、红三叶等。

1.3.3 喜水牧草

需水量较多，能生长在低温而排水不良的土地上，如虉草、白三叶、杂交酸模等。

1.4 Classification based on the origin and growth habits 根据原产地及习性的不同划分

1.4.1 Tropical forages

Native to tropical or south subtropics, the tropical forages grow mainly between South and North latitude 30°, such as *Stylosanthes* spp., siratro (*Macroptilium atropurpureum* Urb.), elephant grass, broadleaf paspalum, bermudagrass (*Cynodon dactylon* L.) and *Brachiaria decumbens* Stapf etc. The optimum temperature of most tropical forages can be as high as 30 to 35℃, growth decreases below 15℃ and the aerial part of plants withered by frost damage around 0℃.

1.4.2 Temperate forages

Native to temperate zones, the optimum growth tem-

1.4.1 热带牧草

原产于热带或南亚热带，主要生长在南纬 30° 和北纬 30° 之间，如柱花草、大翼豆、象草、宽叶雀稗、狗牙根、俯仰臂形草等。大多数热带牧草最适生长温度可高达 30 ~ 35 ℃， 15 ℃以下时生长发育减弱，降至 0 ℃时地上部分受冻害枯萎。

1.4.2 温带牧草

原产于温带，最适生长温度为 20 ~ 25 ℃，以喜冷凉气候为特点。生

perature of temperate forages ranges from 20 to 25℃, characterized by cool climate. The biological zero temperature is around 5℃, very good resistance to frost, but with poor heat tolerance, the growth of temperate forages will be inhibited over 35℃. The common forage species suitable for mid temperate regions (including some cool temperate zone) include smooth bromegrass, *Leymus chinensis* (Trin.), wheatgrass, Siberian Wildrye (*Cinelymus sibiricus* L.), alfalfa, *Astragalus adsurgens* Pall., sainfoin and *Melissilus ruthenicus* L. etc. Besides smooth bromegrass, alfalfa and Siberian Wildrye (*Cinelymus sibiricus* L.), the forages that suitable for warm temperate regions include *Festuca arundinacea*, orchardgrass (*Dactylis glomerata* L.), meadow fescue (*Festuca Dratensis* Huds.), red clover, white clover, *Lespedeza bicolor* Turcz、crownetch(*Coronilla varia* L.) etc.

长起点温度为 5 ℃左右，对霜冻有较强的抵抗力，但耐热性差，超 35 ℃生长受抑制。适于中温带(包括部分寒温带)的牧草有无芒雀麦、羊草、冰草、老芒麦、紫花苜蓿、沙打旺、红豆草、扁蓿豆等。适于暖温带的牧草有无芒雀麦、苇状羊茅、鸭茅、草地羊茅、紫花苜蓿、沙打旺、红三叶、白三叶、胡枝子、多变小冠花等。

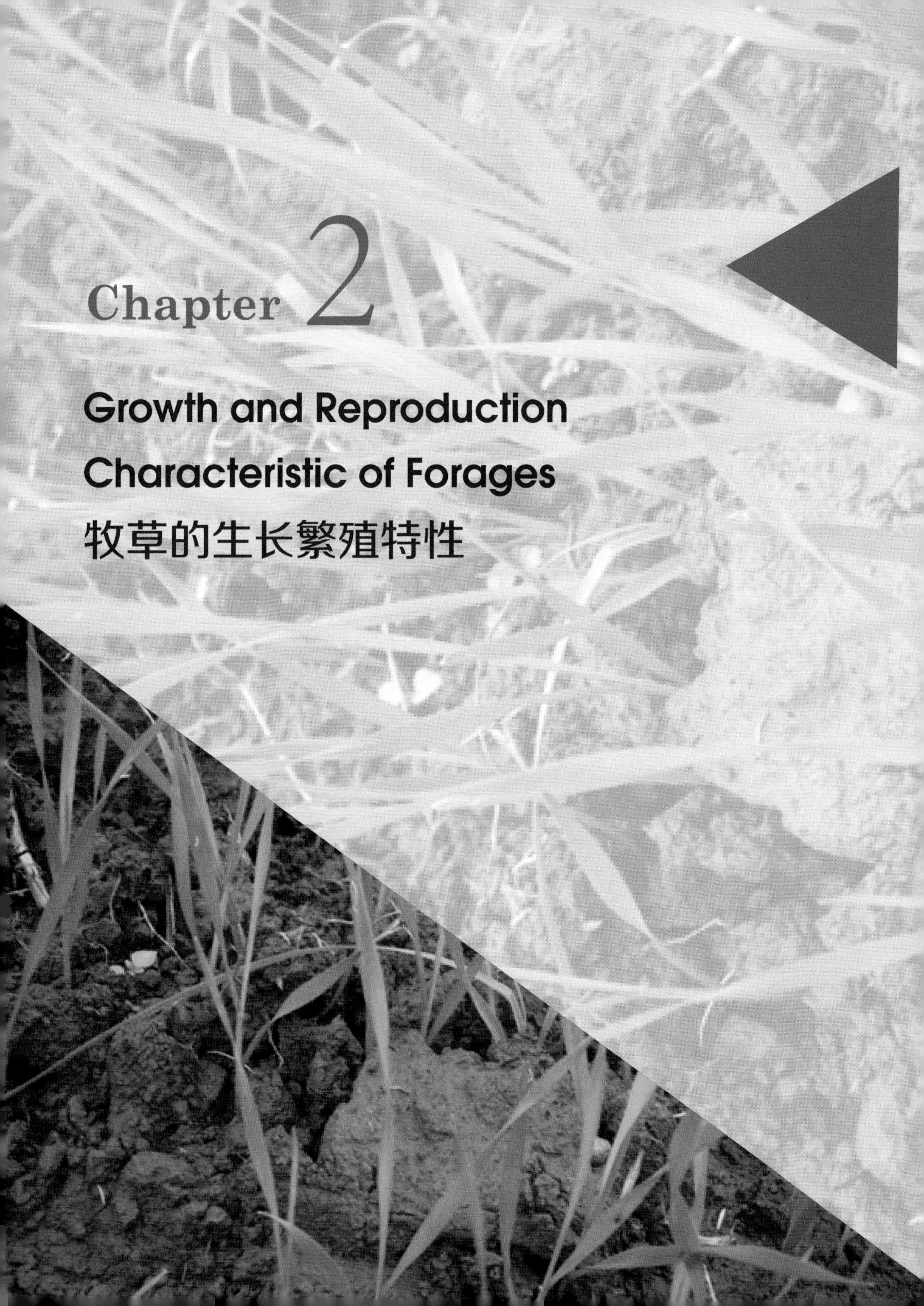

Chapter 2

Growth and Reproduction Characteristic of Forages

牧草的生长繁殖特性

2.1 Vegetative growth characteristic of forages 牧草的营养生长特性

2.1.1 Seed germination and emergence

2.1.1.1 Concept of seed germination

Seed germination refers to the stages of embryo growth and development from the relatively static state to vigorous physiological metabolism, manifested in breaking through the seed coat and outward extension in shape (Fig.2.1). It includes imbibitions, respiration, substance transformation, emergence and growth, etc.

2.1.1 种子的萌发与出苗

2.1.1.1 种子萌发的概念

种子萌发是指种胚从相对静止状态恢复到生理代谢旺盛的生长发育阶段，形态上表现为胚根、胚芽突破种皮并向外伸长，发育成新个体（图 2.1）。包括吸胀、呼吸、物质转化、出苗和生长等过程。

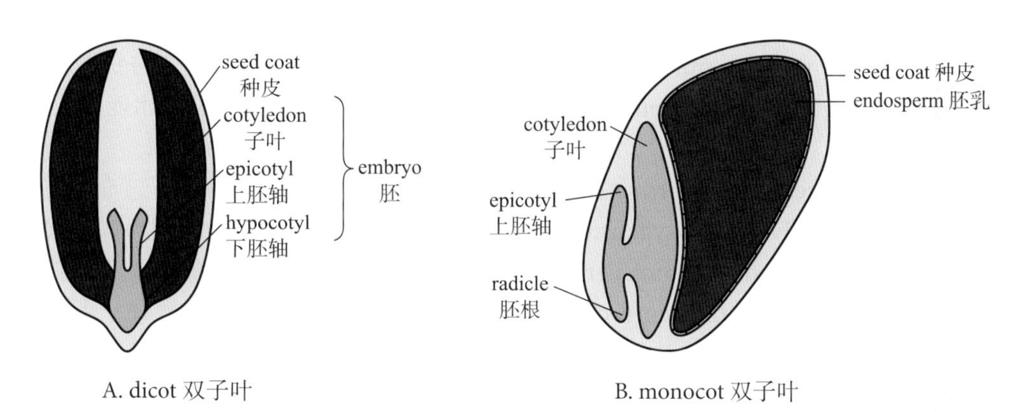

Fig. 2.1 Seed structure
图 2.1 种子的结构

2.1.1.2 Conditions necessary for seed germination

In addition to internal factors, external conditions such as water, suitable temperature, oxygen and light are also required for seed germination.

(1) Sufficient water

Before the various biochemical changes and physiological activities of seed germination can begin, it must absorb sufficient water to help the **radical** and **plumule** break through the **seed coat** easily. Stored nutrients change from insoluble state to soluble through **hydrolysis** or oxidation ways, then are transported to the **embryo** position for absorption and utilization.

Water absorption varies among different species. In gen-

2.1.1.2 种子萌发的条件

种子的萌发，除了本身必须具备生活力这一内在因素外，还要求一定的外界环境条件，主要包括水分、温度、氧气、光照等。

（1）充足的水分

种子萌发时需要吸足水，以进行各种生物化学反应和生理活动，使得胚根、胚芽容易突破种皮。通过水解或氧化，贮藏的营养物质从不溶解的状态转变为溶解状态，运输到胚的生长部位供其吸收和利用。

不同的种子萌发时吸水量不同。

eral, seed with high fat content require less water, while high protein seeds require large quantity water; seed with high starch content absorb moderate water to promote germination. Relatively, legume seeds can absorb more water than grass seeds. Water absorption differs in different portions of the seed. For instance, the embryo of grass seeds can absorb about twice water as that of the **endosperm**. Therefore, good **soil moisture** must be guaranteed when seeding, as it can promote the seed germination.

一般来说，脂肪类种子需水最少，蛋白质含量高的种子吸水较多，淀粉质种子吸水量居中。豆科牧草种子比禾本科种子吸水量大。种子的不同部位及不同成熟度，吸水量也存在差异，如禾本科种胚的吸水量约为胚乳的两倍。因此，良好的土壤墒情可促进种子萌发。

(2) Enough oxygen

All physiological activities of germination need respiration to provide energy. Respiration and oxygen of demand increase after water absorption of seeds. During respiration, seed absorb oxygen to initiate oxygenolysis of stored nutrients. After complicated process, nutrients are finally decomposed to CO_2 and H_2O and releasing energy to support physiological activities.

Sufficient soil oxygen can improve seed germination. Forages germinate normally when oxygen content around the seed is above 10%, most seeds will not germinate if oxygen content is lower than 5%. The oxygen requirement for seed germination varies from species to species, which depends on their **phylogeny**.

（2）足够的氧气

种子萌发时，一切生理活动都需要能量的供应，而能量来源于呼吸作用。种子吸水后呼吸作用增强，需氧量增大，种子在呼吸过程中，要吸入氧气，把细胞内贮藏的营养物质氧化分解，经过复杂的变化，最终变成 CO_2 和 H_2O，并释放出能量，支持各种生理活动。

一般情况下，充足的氧气可促进种子的萌发。通常草种要求其周围空气中含氧量在10%以上才能正常萌发，低于5%时多数种子不能萌发。牧草饲料作物种类不同，种子发芽时需氧情况也不尽相同，这与其系统发育有关。

(3) Appropriate temperature

Appropriate temperature is necessary for the catalysis of various **enzymes** during seed germination. Reactions rates will be slow down or halt at low temperatures, and increase with temperature rising. **Three fundamental points temperature** of plant refer to optimum temperatures, the minimum and maximum temperature it can endure. For most grass seed germination, minimum temperature ranges from 0~5 ℃, maximum temperature is 35~40 ℃, and the optimum temperature is 25~30℃. Generally, the optimum germination temperature is related to the habitat and **originates** area of the species. Minimum germination temperatures for **warm-season grasses** is 5~10 ℃, and maximum temperature is 40 ℃. The minimum temperature for cool-season grasses is above 5 ℃. The maximum temperature is 35 ℃ and optimum temperature is 15~25 ℃. Therefore, the most suitable

（3）适宜的温度

种子萌发需要适宜的温度，以利于各种酶的催化活动。温度低时反应变慢或停止，随着温度的升高，反应加快。牧草生命活动过程中所要求的最适温度及能够忍耐的最低温度和最高温度称为温度三基点。多数牧草种子萌发时所需的最低温度为0～5℃，最高温度为35～40℃，最适温度为25～30℃。一般来说，牧草种子发芽的温度与其原产地和长期所处的生态条件有关，一般暖季型牧草发芽最低温度为5～10℃，最高温度为40℃；冷季型牧草发芽的最低温度为>5℃，最高温度为35℃，最适温度为15～25℃。所以，在生产中必须根据牧草种子发芽温度范围和当地的气

seeding time is decided according to the germination temperature range and local weather conditions.

Seed germination needs not only good moisture but also suitable temperature. The external conditions vary from each other of species. Legume seeds require relatively low temperature and more water; on the contrary, grasses seeds need higher temperature and less water. In addition, practice shows that **variable temperature** can benefit the seed germination. The reasons are as follow:

① The solubility of oxygen in water will increase at low temperature.

② Swell-shrinking occurs during variable temperatures can split the seed coat and let water and oxygen enter the seed easily.

③ Variable temperatures can promote enzyme activity, which benefits to the metabolism of stored nutrients, often used for embryo growth.

④ The temperature difference inside and outside of the seeds can promote gas exchange, accelerate respiration and germination for variable temperatures seeds.

⑤ During germination process, under constant temperature situation, most stored nutrients are used for respiration and less for embryo development, while under variable temperatures, high temperatures of daytime can accelerate biochemical reaction and respiration, most stored nutrients are transformed into soluble form; low temperatures at night decreases respiration and soluble nutrients are mainly used for embryo growth. A common-used variable temperature pattern to promote forage seed germination is 15 ℃ and 30 ℃ or 20 ℃ and 30 ℃. Low temperature duration is longer than that of high temperature, generally 16~18 h at low temperature and 6~8 h at high temperature. Affects of variable temperature on forages seed germination are not always the same; some are very good while there is no effect on others.

(4) Light

Most forage seeds have no strict light requirement for germination. They can germinate in either light or dark. Many seeds are influenced by light during germination, especially most wild grass seeds and fresh seeds. According to

候条件，确定适宜的播种期。

种子的萌发需要水热条件相互配合，萌发时不同种子对外界环境条件要求也有所不同。豆科牧草种子对温度要求较低，而对水分要求较高；而禾本科牧草种子则与此相反，对温度要求较高，而对水分要求较低。

另外，大量试验和实践表明，变温有利于促进牧草种子的萌发，其可能原因如下：

① 低温时氧气在水中的溶解度增大。

② 变温时种皮因胀缩而被撕裂，有利于水分和氧气进入种子内部。

③ 变温使种子内酶的活性提高，有利于贮藏物质的转化，用于胚的生长发育。

④ 变温使种子内、外部温度不同，可促进种子内外气体交换，使种子呼吸作用旺盛，促进发芽。

⑤ 恒温发芽时，贮藏物质大部分用于呼吸作用，而少量用于胚的发育。变温条件下，高温加快生化反应，呼吸作用旺盛，贮藏物质大量转化为可利用的可溶性物质；低温则使呼吸作用减弱，可溶性物质主要用于胚的生长。

牧草种子发芽最常用的变温处理为 15℃和 30℃或 20℃和 30℃，处在每种温度下的时间长短不一，低温时间比高温时间要长些。通常每昼夜在低温下 16~18 h，在高温下 6~8 h。变温对各种牧草种子发芽的作用效果不尽相同。对某些牧草种子效果很好，但对另一些种子则无效。

（4）光照

大多数牧草、饲料作物种子发芽时对光并不存在严格要求，无论在光下或暗处均能萌发。但是，不少种子，特别是大部分的野生牧草种子和新收获种子发芽时，不同程度上受光的影响。根据

sensitivity to light, the seed can be divided into the following three categories.

① Light-dependent seed

Most grass seeds are light-dependent or light-favored seed, such as *Agropyron*, *Poa*, *Bromus* and *Agrostis*. Light-dependent seed are usually small and having little storage, which can only germinate at the soil surface and emerge rapidly to allow for photosynthesis, avoid nutrient depletion.

② Light-inhibited seed

Light-inhibited seeds can not germinate in light at the soil surface. Light inhibits seed germination, interrupting or it completely. Grain amaranth (*Amaranthus hypochondriacus*) and mount grass (*Oryzopsis munroi*) will not germinate as a result of light inhibition.

③ Light-insensitive seed

Light-insensitive seed is insensitive to light or dark and can germinate in either light or dark. Barley(*Hordeum vulgare*), oats(*Avena sativa*) and most forage seeds belong to this category.

2.1.1.3 Process of seed germination

The process of seed germination can be divided into five steps: **imbibition**, hydration and enzyme activation, **cell division** and enlargement, embryo penetrate the seed coat, seedling growth.

(1) Grass seed germination

The botanical form for a grass seed is **caryopsis**, which only inclues a single grain and the pericarp and seed coat are difficult to separate. Seed germination begins with water absorption to swell the seed and activate the protoplasm from a static state. Nutrient metabolism and respiration are initiated. The **coleorhiza** first penetrates through the seed coat, and then one or several radicles grow out with the reaching out of coleorhiza and **coleoptile** one after another. The first portion emerging from the soil is the tubular-coleoptile after germination. It is a **metamorphic leaf** which has sheath only but without **blade**, it can protect the germ. At the same time there are 3 to 5 leaf **primordium** growth points in the embryo and they also grow out from new leaf primordium in order. Next the first green leaves appear.

种子萌发时光敏感性状况，可将种子分为以下三类：

① 需光种子

又叫喜光种子。禾本科牧草种子大部分属于需光种子，如冰草属、早熟禾属、雀麦属和翦股颖属等种子在有光条件下发芽较好。需光种子一般较小，贮藏物很少，只有在土表有光处萌发才能保证幼苗快速出土进行光合作用，不致因养料耗尽而死亡。

② 忌光种子

不能在土表有光处萌发，光对种子萌发起抑制作用，使其发芽迟缓或根本不能发芽，籽粒苋和落芒草等均可受光抑制而不发芽。

③ 对光反应不敏感种子

这类种子萌发时对光或暗反应不敏感，有光无光均可发芽。大麦、燕麦和多数牧草种子均属此类。

2.1.1.3 牧草种子的萌发过程

牧草种子的萌发过程大致可分为以下5个阶段：吸胀、水合与酶的活化、细胞分裂和增大、胚突破种皮、长成幼苗。

（1）禾本科牧草种子萌发

禾本科牧草种子实际是指植物学中的颖果。颖果仅含一粒种子，果皮与种皮结合紧密不易分开。种子萌发时先吸收水分，使种子膨胀，原生质从静态进入活跃状态，促进了物质转化过程和呼吸作用。胚根鞘首先穿过种皮，随之长出一条或数条胚根，胚根鞘与胚芽鞘相继伸出。所以，发芽后的种子首先露出地面的是一个管状的胚芽鞘，它只是具有叶鞘而没有叶片的变态叶，有保护胚芽的作用。同时，在胚中就已有3 ~ 5个叶原基的生长点，也开始按顺序长出新的叶原基，之后将在地面出现第一批绿色叶片。

(2) Legume seed germination

Legume seed consists of two cotyledons and embryo wrapped with the seed coat. The cotyledons are the primary leaf primordia, which function as the nutrient storage site instead of the endosperm. The embryo includes the radicle, hypocotyl and germ. As a legume seed is emerging, the radicle initially forms the taproot. The hypocotyl elongates to push the cotyledon and embryo out of ground. Then the first true leaf which is a simple leaf appears from the germ. Subsequently, the second **trifoliolate leaf** extends to a rosette form with accumulated growth.

（2）豆科牧草种子萌发

豆科牧草的种子由种皮包裹着的两个子叶和胚所组成。子叶是原生叶原基，是代替胚乳贮藏养分的场所，胚包括胚根、胚轴和胚芽。种子萌发时，首先是胚根突破种皮后向下生长形成主根，然后胚轴伸长，将子叶和胚芽推出土面。之后从胚芽中出现第一片真叶，为单叶，随后出现第二片真叶，即为三出复叶，再经过一段时间生长，形成莲座状叶丛。

2.1.2 The development of seedlings

2.1.2.1 Seedling development of legumes

There are two types for legume seeds germination: **epigaeous** and **hypogeal** (Fig.2.2). Cotyledons of most legumes are epigaeous, such as alfalfa (*Medicago sativa.*), clover (*Trifolium repens*), sainfoin(*Onobrychis viciaefolia*).

In epigaeous germination, the hypocotyl elongates and epicotyl arches out of the soil. When the arch extends above the soil surface and is exposed to the sunshine it straightens opens the cotyledon and contributes to photosynthesis.

Germination of a few legumes is hypogeal germination such as common vetch (*Vicia gigantea*), hairy vetch (*Vicia*

2.1.2 幼苗的发育

2.1.2.1 豆科牧草的幼苗发育

豆科牧草出苗有两种不同的类型，子叶出土和子叶留土（图 2.2）。大多数豆科牧草的子叶是出土的，如苜蓿属、三叶草属、红豆草属等。如果幼苗是子叶出土类型的，下胚轴（在初生根与子叶节间的胚轴部分）会延长并形成一个弓形伸出土壤。当弓形到达土壤表面暴露于阳光下时展开，子叶张开，开始进行光合作用。

少数豆科牧草，如箭筈豌豆、毛苕子和山黧豆，萌发时是子叶留土，其下

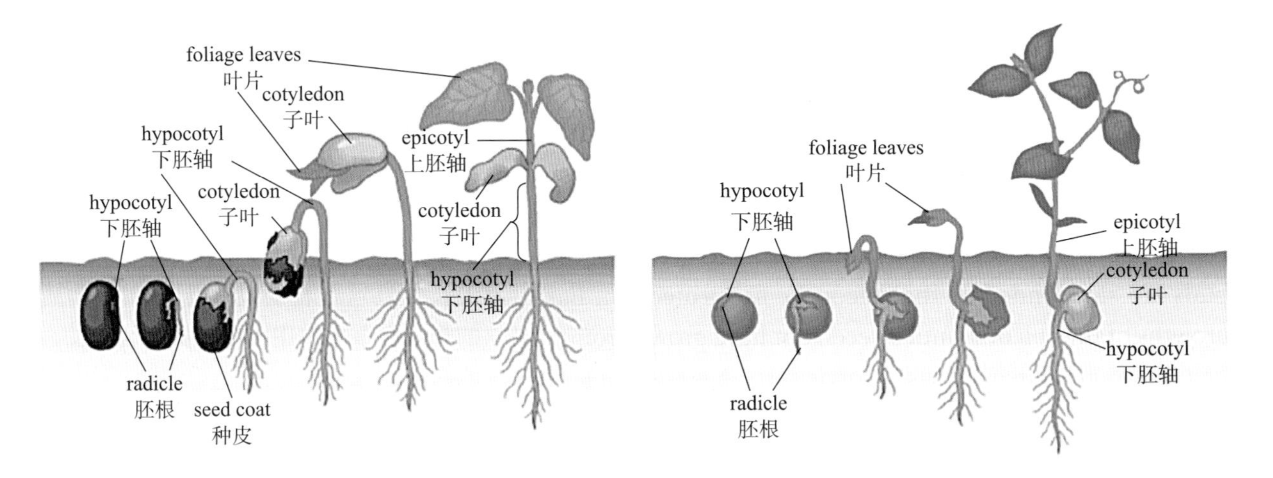

Fig. 2.2 Cotyledons position of legume seeds germination
图 2.2 豆科种子萌发的子叶类型

Fig. 2.3 The epicotyl arches out of the soil
图 2.3 上胚轴弓出地面

villosa), *Lathyrus sativus*. The hypocotyl does not elongate, while the epicotyl forms as an arch and the germ extends out of the soil (Fig.2.3). These forages have one or several underground sections with axillary buds. They also possess meristem and nutrient sources, which can be used for regeneration.

An advantage of epigaeous germination is to protect the emerging buds, especially the cotyledon have the function of photosynthesis. The legume cotyledons usually can keep their nutrition function up to 3 to 4 weeks after germination. The cotyledon and one or two developing vegetative buds finally form the root crown.

2.1.2.2 Seedling development of grasses

All the cotyledons of grasses are hypogeal type. With the imbibitions of seeds, the first sign of emergence is appearance of the radical (Fig. 2.4). The radicle then grows downward into deeper layer of the soil. Epicotyl appears afterwards and upwards grows until the coleoptile emerges. Immature buds are surrounded by the coleoptile for protection. Early growths of the epicotyl until 5 seed root have grown from the first internode above the based cotyledon. These roots play an important role in seedling's water and mineral absorption.

The primary root and seed root are temporary. Permanent roots appear from the second to sixth internodes and develop into typical fibrous roots, originating from the cotyledon at the soil surface.

胚轴并不伸长，上胚轴形成一个弓形且胚芽伸出地面（图 2.3）。子叶留土的豆科牧草有一个或几个具腋芽的地下节，具有分生组织和能量来源，可用于再生。

子叶出土的优点是子叶可保护幼嫩的芽出土，特别是子叶能进行光合作用。豆科牧草的子叶通常在出土后 3 ~ 4 周保持营养功能，子叶和一、二个低位叶叶腋中的发育营养芽最终形成根颈。

2.1.2.2 禾本科牧草的幼苗发育

所有禾本科牧草的种子萌发后子叶不出土。随着种子吸水膨胀，发芽的第一个标志是胚根的出现（图 2.4）。随后胚根向下生长。子叶上边的幼茎（上胚轴）出现，并向上生长直到胚芽鞘出土，胚芽鞘包围着幼叶的顶芽，起保护作用。早期上胚轴的生长要等到 5 条种子根从第一个节间基部的子叶上长出。这些根对幼苗水分和矿物质的吸收起重要作用，使植株茁壮成长，能自养。

初生根和种子根是临时根系，其后长出永久根。永久根从第 2 到第 6 节间出现并发育成典型的须根型。因此它们是在子叶上和土壤表层中发生的。

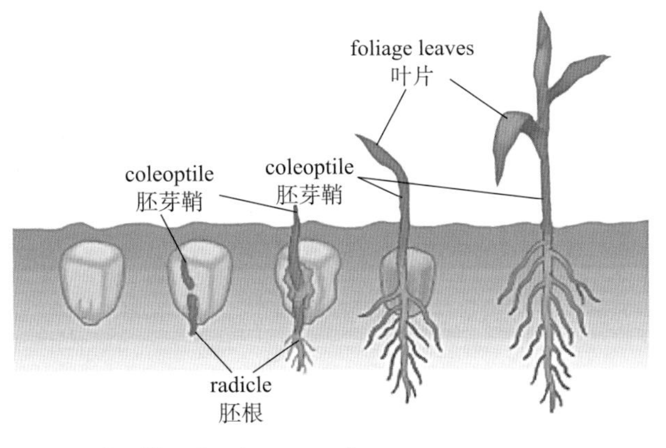

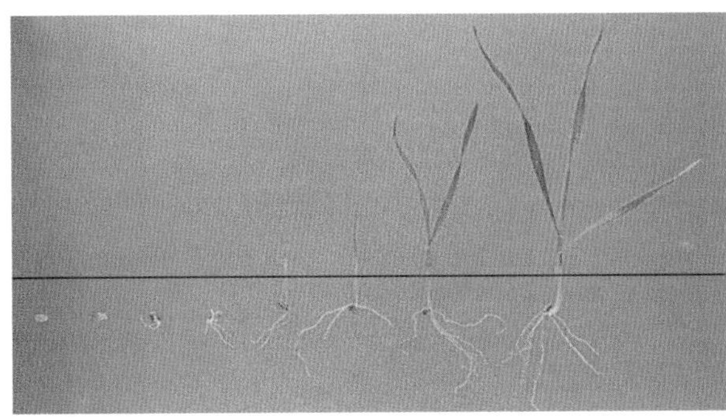

Fig. 2.4 Seedling development of grasses
图 2.4 禾本科幼苗发育

2.2 Development of vegetative organs 营养器官的建成

2.2.1 Root

2.2.1.1 Root structure and morphology

The root develops from the embryonic of the seed radicle constituting the underground part of plants. All the roots underground are called **root system**. Roots can be divided into the **taproot** and **lateral root**. Taproots develop from the radicle, and it branches into lateral roots. Lateral roots which branch from the taproot are called **secondary root**, so lateral roots which branch further are called secondary lateral roots, and lateral roots which branch from secondary lateral roots are called tertiary lateral roots, and so on.

Root systems are classified as **taproot system** and **fibrous root system** (Fig. 2.5) according to the taproot development condition. If a root system has a very developed and obvious main root, all the lateral roots branch out from the main thick roots, this kind is called taproot system. Many dicots have taproot systems including alfalfa (*Medicago sativa*), clover (*Trifolium repens*), Chinese milk vetch (*Astragalus sinicus*), soybean (*Glycine max*) and pumpkin (*Cucurbita moschata*). If a root system which is mainly composed by adventitious fibrous roots, we call this kind as a fibrous root system. Its taproot is underdeveloped. Most monocots have a fibrous root system including sudan grass (*Sorghum suda-*

2.2.1 根

2.2.1.1 根的形态结构

根是由种子幼胚的胚根发育而成的器官，构成了植物的地下部分。一株植物地下部分所有根的总体称为根系。根可分为主根和侧根。主根由胚根发育而成，主根上分出侧根。由主根分出的侧根称为一级侧根，由一级侧根分出的侧根称为二级侧根，由二级侧根分出的侧根称为三级侧根，以此类推。

根系根据其主根发达与否，可分为直根系（圆锥根系）和须根系（图 2.5）。根系中有一条较粗壮的主根，在主根上着生各级侧根的根系称为直根系。如紫花苜蓿、三叶草、紫云英、大豆、南瓜等双子叶植物的根系。根系中主根不发达（不明显），主要由不定根（须根）组成的根系称须根系。如苏丹草、黑麦草、羊茅、玉米等大部分单子叶植物的根系。从基部茎节等根以外的部位发生的根称不定根。

有些牧草的根，因行使特殊的生理功能，在形态上发生了很大的变异。如

Fig. 2.5 Taproot (left) and fibrous root (right)
图 2.5 主根系（左）和须根系（右）

nense), ryegrass (*Lolium perenne*), tall festuca (*Festuca arundinacea*) and maize (*Zea mays*) . Roots originating outside basal internodes are called **adventitious roots**.

Due to specialized physiological functions, forage plant roots display great variability in morphology. For example, the fleshy tap root of radish (*Raphanus sativus*), carrot (*Daucus carota*) and sugarbeet(*Beta vulgaris*), and the root tuber of sweet patato(*Ipomoea batatas*), cassava(*Manihot esculenta*) and sunroot(*Jerusalem artichoke*), store large quantities of nutrients. They are called storage roots. Storage roots are rich in nutrients; they are important forage crops.

2.2.1.2 Main physiological function of root

Root plays an important role in plant fixing and supporting. The root absorb large amount of water and nutrients from the soil. Root has the effect of nutrient storage. Root can synthesize amino acid, cytokinin, auxin, etc., which can be transported to the shoot and meet the need of shoot growth and development.

In the whole life cycle of forage, vegetative growth period has higher root vigor. When get into reproductive growth

萝卜、胡萝卜、甜菜等的肉质直根，以及甘薯、木薯、菊芋等的块根，适应贮藏大量的营养物质，称为贮藏根。贮藏根中含有丰富的营养物质，是饲料作物的重要收获和利用对象。

2.2.1.2 根的主要生理功能

根系深扎土壤起着固定和支撑植株的作用。根系从土壤中吸收大量水分和养分。根系具有贮藏养分的作用。根系还能合成氨基酸、细胞分裂素和植物生长素等，将之输送到地上部，满足地上部生长发育的需要。在牧草的一生中，营养生长时期根系活力较强；进入生殖生长时期后，如豆科作物现蕾开花，禾本科作物拔节抽穗后，根系便逐渐衰老，活力降低，生理功能相应衰减。

stages, such as at budding and flowering of legumes, and after the elongation heading of grass, roots will gradually aging, the vigor decreased and physiological function weakened accordingly.

2.2.1.3 Root growth

(1) Grasses

Grass seeds begin to germinate when environmental conditions are suitable. Initial growth of the radical, leads to formation of main roots. Lateral root growth followed. These initial roots are called the primary root and serve to absorb and supply moisture and nutrients.

Primary grass roots appear gradually. Under field conditions, the first primary grass root appears 8~10 d, the second 10~16 d, and the 3rd and 4th appear 12~23 d and 23~33 d after sowing.

Primary and secondary roots are distinguished mainly due to the enlargement at the base of secondary roots. Unlike the primary root, the number of secondary roots is not fixed, but entirely depends on environmental conditions and extent of cultivation. In the seeding year, secondary root growth of perennial grasses is rapid. Depending on the type of grass, roots are concentrated between 0~30 cm; they take about 65% to 95% of the total root biomass.

(2) Legumes

Legumes germinate rapidly when soil water and temperature conditions are suitable. Germination are faster than that of grasses. Legumes root development begins from the radicle. As seeds germinate, the radicle first elongates growing downward, and then as the embryo develops, the radicle differentiates developed to root. A taproot will subsequently branch to form lateral roots. When the shoot begins to grow leaves, roots elongate further and branch roots increase, begin to form **nodules**. If a rosette form, roots will have a longer development period; root length will exceed shoot length much or more than ten times. When the thickness of the taproots' upper part increases gradually and swells near the soil surface, the initial root crown appear and form a completed legume root system.

Legumes roots grow faster than shoots in seedling stage.

2.2.1.3 根系的生长

（1）禾本科牧草

禾本科牧草播种后，当环境条件适宜时开始萌发，首先是胚根的生长使主根形成。不久在其上长出侧根，这些萌发后最初的根统称为初生根，它肩负着牧草早期水分、养分的吸收与供应。

禾草初生根是在一段时间内陆续发生的，在大田条件下，禾草的第 1 条初生根在播后第 8 ～ 10 d 出现，第 2 条在播后 10 ～ 16 d 出现，第 3、4 条在播后第 12 ～ 23 d 和 23 ～ 33 d 出现。

次生根不同于初生根之处主要是由于次生根的基部膨大。此外，次生根的数目不像初生根是固定的，而且其数目的多少完全取决于环境条件和农业技术措施。播种当年，多年生禾草次生根生长速度的总趋势是快的。禾本科牧草视牧草种类不同，根系大部分集中在土表 0 ～ 30 cm 处，占总根量的 65% ～ 95%。

（2）豆科牧草

豆科牧草在土壤水、热条件适宜时很快萌发，其萌发速度要快于禾本科牧草。豆科牧草根系的发育是从胚根开始的。当种子萌发时，胚根首先伸长，并向下生长，其后随着胚芽的生长，胚根进一步发育为根，并于其上产生分枝根，形成直根系。当地上部开始形成真叶时，根系进一步伸长，分枝根增多，并开始形成根瘤。当地面形成莲座叶丛时，根系会有一个较长时期的发育过程，根系长度会超过地上部几倍甚至十几倍。随着主根上部增粗，在接近地面处膨大，形成初期的根颈，豆科牧草完整的根系也就形成了。

豆科牧草苗期根系生长的速度比地上部分快。一年生豆科牧草的根系入土深度较多年生豆科牧草的浅，其生长发

Annual legumes rooting depth is shallower than that of perennial legumes, but the growth and development patterns are similar to each other. Roots grow continuously from seed emergence until the plant matures and dies. Most legumes have a crown. New stems originate from the crown and it plays an important role in plant regeneration and nutrient storage. The crown varies size with species, stage of development, age and environmental conditions. Most legume root have nodules, most of them occurring at soil depth of 0~20 cm. Nodule shape, size and color differ with forage species and age.

To sum up, compared with the annual grass, perennial grasses have large root system and roots can extend up to 1.5~2.0 m deep. Legumes root develop even more deeply. For example, sainfoin(*Onobrychis viciaefolia*) that lives more than three years, the taproots may reach to more than 4 m deep. Rooting depth of annual grasses is generally within 1 m. Annual legume may reach 1~1.5 m deep. Most perennial grass roots continue to grow every year and the root mass generally increases with the age increasing of the stand.

(3) Forages root lifespan

Forage plant root persistence influences water and nutrients absorption from the soil. Longer root persistence ensures high and stable yields.

For annual forages plants, the root can only survive for a single growing season and dying with the shoot at the same period. Differently, perennial grasses roots will not die after the first growth cycle and will continue to grow until the end of the whole lifecycle.

(4) Affecting factors of the root growth

Forage root growth is influenced by external factors in addition to their biological characteristics. Among the external factors, the most important factors include soil moisture, soil temperature, **soil fertility**, light, and **defoliation** from grazing or clipping.

育和多年生豆科牧草有相似的规律性，即出苗后至枯死前根系始终生长。大部分豆科牧草具有根颈。根颈是形成枝条的部分，对于植株再生和贮藏营养物质等具有重要作用。各种牧草根颈大小发育不等，这与其生育年龄和生态条件有关。豆科牧草的根大多生有根瘤，根瘤大部分分布在土壤深度为 0 ~ 20 cm 的范围内，其形状、大小及色泽因牧草种类、生活年限和生育期而异。

多年生牧草与一年生牧草相比，其根系发育有以下特点：

多年生牧草根系入土较深。一般多年生牧草根系入土达 1.5 ~ 2 m，尤其是豆科牧草入土更深，如生活三年以上的红豆草，主根入土深度可达 4 m 以上。而一年生豆科牧草只有 1 ~ 1.5 m。大多数多年生牧草根系每年都继续生长，多年生牧草根系质量一般表现出随草地年龄的增加而增大。

（3）牧草根系的寿命

牧草根系寿命的长短在牧草生活中具有重要意义。如果它们寿命长，说明能长期从土壤中吸收水分和养分，从而保证牧草在一个较长的时期内能够提供高水平而稳定的产草量。

一年生牧草在地上部开花、结实完成其发育周期后，在冬季死亡，即一年生牧草的根系寿命短，它在一个生长季后与地上部同时死亡。多年生牧草根系在生活第一年的生活小周期后并不死亡。除部分侧根细枝有部分死亡外，整个根系随生活大周期的死亡而死亡。

（4）影响根系生长的因素

牧草根系的生长除其生物学特性外，还受外界条件的影响。在这些环境因素中，土壤水分、土壤温度、土壤养分、光照及刈割放牧等是其重要的影响因素。

2.2.2 Stem

The stem is the plant part which grows aboveground as the seed germinates and embryo develops. The stem forms the axis of plant. Leaves emerge from the stem at region formed node. The portion of the stem between two nodes is called **internode**. The intersection of the **petiole** with the stem is called the **leaf axil**. The stem and its branches support and adjust the distribution of leaves. Stems also serve to transport nutrients. Stems can be **aboveground** or **rhizomatous**.

2.2.2.1 Classification based on the growth habit of the aboveground stems (Fig. 2.6)

(1) Erect stem: Perpendicular to the ground, the most common type in the field.

(2) Inclined stem: Initially inclined but become upright later, such as ruthenica (*Melilotoides ruthenica*).

(3) Prostrate stem: Prostrate on the ground level, such as wolf's milk (*Euphorbia humifusa*) and thornbushes(*Tribulus terrestris*).

(4) Stolon: Grow prostrate on the ground, may develop adventitious roots and new shoots at nodes, such as white clover (*Trifolium repens*).

(5) Climbing stem: use small roots, petioles, tendrils

2.2.2 茎

茎是种子幼胚的胚芽向地上生长的部分，为植物体的中轴，通常在叶腋生有芽，芽萌发后形成分枝。茎和枝上着生叶的部位叫节，两节之间的茎叫节间，叶柄与茎相交的部位叫叶腋，茎和分枝支持和调整叶子的分布，同时也是物质输导的通道，可分为地上茎和地下茎。

2.2.2.1 根据地上茎生长习性划分（图 2.6）

（1）直立茎：指垂直于地面的茎，为最常见的茎。

（2）斜升茎：指最初偏斜，后变直立的茎，如扁蓿豆。

（3）平卧茎：指平卧于地上的茎，茎节和分枝处不再生根。如地锦草、蒺藜。

（4）匍匐茎：指茎平卧于地上，但茎节和分枝处生出不定根。如白三叶。

（5）攀缘茎：指用小根、叶柄或卷须等特有的变态器官攀缘于他物而上升的茎，如豌豆。

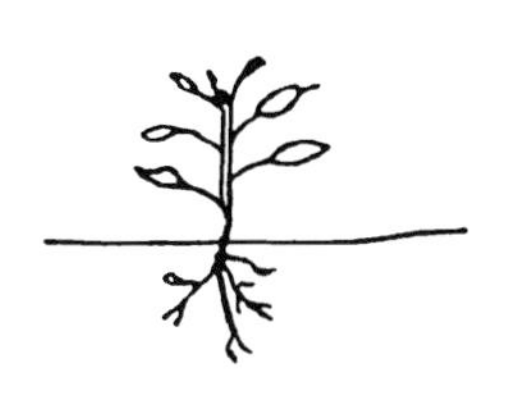

erect stem 直立茎

decumbent stem 攀缘茎

twining stem 缠绕茎

creeping stem 匍匐茎

prostrate stem 平卧茎

decumbent stem 平行茎

Fig. 2.6 Types of stems
图 2.6 茎的类型

and other unique modified organ to climb on other plants or structures, such as pea (*Pisum sativum*).

(6) Twining stem: Wrap around other plants or structures to grow upward, such as the morning glory (*Pharbitis nil*).

2.2.2.2 Classification based on the specialized rhizomatous stems

(1) Rhizomes: Which extend for straight or creeping growth in the soil, and possess obvious nodes and internodes, such as chinese wildrye(*Leymus chinensis*).

(2) Tubers: Shortened and swollen rhizomes, such as potato (*Solanum tuberosum*).

(3) Corms: Swollen and oblate rhizomes, such as taro (*Colocasia esculenta*).

(4) Bulbs: flat or disk-shaped rhizomes, such as onion (*Allium cepa*).

2.2.3 Leaves

Leaves are the primary site of photosynthesis in plants, which provides organic compounds and energy to support growth and development of forages.

2.2.3.1 Leaf structure

Fully developed leaves of **dicotyledonous** plants have three parts: **lamina**, **petioles** and **stipules**. Leaves that have all three parts are called complete leaves, such as clover or birdsfoot trefoil (*Lotus corniculatus*). Leaves that absent any part are called incomplete leaves.

Grass leaves consist of four parts: blade, sheath, ligule and auricle (Fig. 2.7). A leaf with blade and sheath is called complete leaf. Leaves with only sheath are formed incomplete one. Parallel veins are generally conspicuous on grass leaves. The largest central longitudinal vein is the midrib. Branching of midrib is named lateral veins.

Leaves contain three structural parts: **epidermis**, **mesophyll** and **veins**. **Stomas** in the epidermis serve as channels for gas exchange and water transpiration with the atmosphere. Vein structure varies with species. Forages such as corn, sorghum and other grasses are characterized by **vascular bundles** consisting of a bundle sheath surrounded by

（6）缠绕茎：指螺旋状缠绕于他物而上升的茎，如牵牛。

2.2.2.2 根据地下茎变态划分

（1）根状茎：指延长直伸或匍匐生长于土壤中的地下茎。有明显的节和节间，如羊草。

（2）块茎：指短缩肥大的地下茎，如马铃薯。

（3）球茎：指肥大而扁圆的地下茎，如芋头。

（4）鳞茎：一种扁平或圆盘状的地下茎，如洋葱。

2.2.3 叶

叶是植物进行光合作用的主要场所，牧草与饲料作物生长发育所需的有机物质和能量主要来自于叶片的光合作用。

2.2.3.1 叶的结构

双子叶植物的成熟叶在形态上具有叶片、叶柄和托叶三个部分。三部分俱全的称为完全叶，如三叶草、百脉根等的叶。缺少任何一部分的叶称为不完全叶。禾本科植物的叶由叶片、叶鞘、叶舌和叶耳组成（图 2.7）。凡具有叶片和叶鞘两部分的为完全叶；叶片退化，只具叶鞘的为不完全叶。叶片上有许多清晰可见的脉纹称为叶脉。叶片中央纵向最大的一条叶脉称为中脉。中脉的分支称侧脉。

叶片从其构造上可分为表皮、叶肉和叶脉三部分。表皮上有气孔，它是植物与外界进行气体交换和水分蒸腾的主要通道。不同植物叶脉的构造有所不同。禾本科牧草中，玉米、高粱等 C_4 作物的维管束周围仅由一层薄壁细胞组成维管束鞘，内含大量叶绿体，并与含有叶绿体的叶肉细胞紧密相邻，组成了花环形结构，光合作用效率较

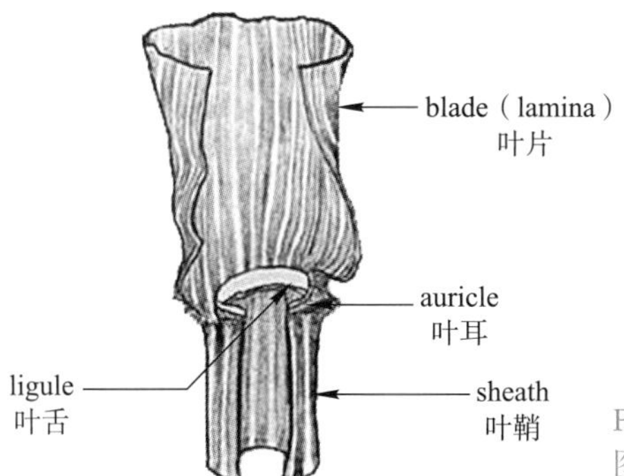

Fig. 2.7 Leaf structure of grasses
图 2.7 禾本科叶片结构示意图

a layer of **thin-walled cells** containing a large number of **chloroplasts**. Bundle sheath cells are immediately adjacent to the mesophyll cells that contain chloroplasts and compose the "garland" structure. This arrangement leads to higher photosynthetic efficiency. Barley, ryegrass and other C_3 grasses possess two layers of cells in the bundle sheath. Inner cell walls, close to the vascular bundles are thicker, almost free of chloroplasts. Cell walls of the outer layer are thin, containing small amount chloroplasts, without the "garland" structure of C_4 plants. This structural difference contributes to lower photosynthetic efficiency of C_3 forages than that of C_4 plants.

Petioles are the structure that connecting the blade to the stem, with primary functions of support and conductance. Stipules are appendages growing at basal petioles where is close to the stem.

2.2.3.2 Types of blade

(1) Simple leaf

Only a leaf blade on a petiole, this is called simple leaf, such as sweet potato, pumpkin, rape, etc. Most gramineous forages are simple leaf type, the blade is long, flat and narrow (the leaves of corn and sorghum are wider slightly and bent over, with wavy leaf margin), it is sheath rather than petioles that connect to the blade, leaf sheath is narrow and long ,which holding around the internode, support and protect the stem.

(2) Compound leaf

A leaf with more than two completely independent leaf-

高；大麦、黑麦草等 C_3 作物的维管束鞘则有两层细胞，紧靠维管束的内层细胞壁较厚，几乎不含叶绿体，外层细胞壁薄，含少量叶绿体，无 C_4 作物的“花环”结构。这一差异是造成 C_3 作物光合效率低于 C_4 作物的重要原因。

叶柄是叶片与茎相连的柄状部分，主要起输导和支持作用。托叶则是在叶柄基部紧靠茎的地方着生的附属物。

2.2.3.2 叶片的类型

（1）单叶

一个叶柄上只生一个叶片的，称为单叶，如甘薯、南瓜、油菜等。大部分禾本科牧草为单叶，叶片大多狭长扁平（玉米、高粱的叶片宽长略弯披，叶缘呈波浪状），连接叶片的是叶鞘而非叶柄，叶鞘狭长围抱着节间，具有支持和保护茎秆的作用。

（2）复叶

一个叶柄上着生两个以上完全独立小叶的，称为复叶，如大豆、苜蓿、毛苕子等。复叶与茎连接的叶柄称为总叶柄，各小叶的叶柄称为小叶柄。复叶根据小叶着生的方式，又可分为羽状复叶、掌状复叶和三出复叶。

2.2.3.3 叶的主要生理功能

叶的主要生理功能是光合作用。

lets on one petiole is called compound leaf, such as soybean, alfalfa and hairy vetch. The stem structure connecting compound leaves with the stem is a total petiole; each leaflet's petiole is called **petiolule**. Depending on the leaflet pattern, compound leaf may be classified as pinnate, palmate and trifoliolate leaves.

2.2.3.3 Physiological function of leaf

The main physiological function of leaves is photosynthesis. Using solar energy, the chloroplasts in leaf tissue absorb CO_2 and H_2O, which synthetised into glucose and other organic compounds. Light energy is converted into chemical energy and stored while O_2 is simultaneously released. Leaf photosynthesis provides organic compounds and energy to support plant growth and development.

Another important leaf function is evaporation. Most of the water absorbed by roots is released to the atmosphere in the form of water vapor. Leaf transpiration facilitates transport of water and mineral nutrients and maintains leaf temperature; it plays an important role in plant growth. However, excessive transpiration is detrimental to plant growth and development.

Leaves also serve as the site of the synthesize of amino acids and other organic.

2.2.3.4 Leaf area index

Leaf area index (LAI), also called leaf area coefficient, refers to the total plant leaf area per unit of land area (Leaf area index = the total leaf area/land area)

Leaf area index reflects the growth status of plant communities, and its size is directly related to total yield.

叶组织中叶绿体利用光能，将吸收的 CO_2 和 H_2O 合成为葡萄糖等有机物，将光能转变为化学能贮藏起来，同时释放 O_2。叶片的光合作用是植物生长发育所需有机物质和能量的主要来源。

叶的另一个重要功能是蒸腾作用，根系吸收的水分绝大部分以水汽的形式从叶面扩散到体外。叶的蒸腾作用能促进体内水分、矿质盐类的传导，平衡叶片的温度，在植物生活中有着积极的意义。但过度蒸腾对植物的生长发育不利。

此外，叶片还有利用光合产物合成氨基酸等其他有机物的功能。

2.2.3.4 叶面积指数

叶面积指数（leaf area index，LAI），又称叶面积系数，是指单位土地面积上植物叶片总面积占土地面积的倍数。即：叶面积指数 = 叶片总面积 / 土地面积。

在田间试验中，叶面积指数是反映植物群体生长状况的一个重要指标，其大小直接与最终产量高低密切相关。

2.3 The Rules of vegetative growth 牧草营养生长规律

2.3.1 Interrelation among the organs

2.3.1.1 Tillering and branching of forages

(1) Tillering of grasses

Following emergence and appearance of 3 to 4 leaves, branches may come out from the mother plant base or underground stem node, root crown. These branches are called tillers. Tillers originated within sheath are called intravaginal tiller. Branches piercing the sheath to emerge are called extravaginal tiller (Fig. 2.8). Intravaginal tiller usually grows vertically on the side of the leaf sheath along the mother shoot, which is characteristic of bunchgrasses such as orchardgrass, wheatgrass, causes the individual plants to appear as distinct clumps or bunches with open spaces between them. Extravaginal tiller grows laterally through the older leaf sheaths to emerge a short distance from the main stem. Rhizomes and stolons from similar lateral buds are extravaginal, which form loose bunches, such as smooth bromgrass, Kentucky bluegrass. Some grasses develop both type tillers

2.3.1 牧草器官生长的相关性

2.3.1.1 牧草的分蘖和分枝

（1）禾本科牧草的分蘖

禾本科牧草出苗后，当长出 3 ~ 4 片叶时,即可自母株的地表或地下茎节、根颈、根蘖上形成侧枝，这种现象称为分蘖。分蘖时枝条位于叶鞘内者称为鞘内枝，而枝条穿破叶鞘生长者称为鞘外枝（图 2.8）。前一类的枝条通常沿着母枝向上生长，形成稠密的株丛（密丛型）；后一类枝条，开始时水平方向生长，然后再向上生长（根茎型），如无芒雀麦；或同母株以锐角向上生长形成疏松的株丛（疏丛型），如鸭茅、冰草；也有禾本科牧草在同一株丛发育两种类型的枝条，如根茎 - 疏丛型的牧草。

禾草出苗长出 3 ~ 4 片叶时，主枝上第一个接近地表的节就是分蘖节，它

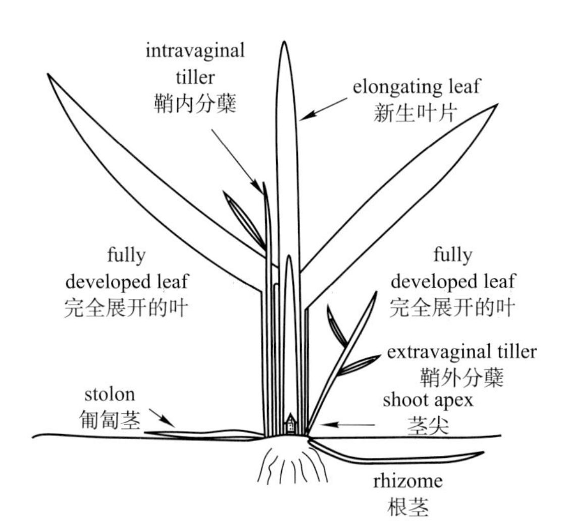

Fig. 2.8 Hypothetical grass plant showing the origin and position of tillers, stolons and rhizomes
图 2.8 分蘖、匍匐茎和根茎的产生部位

with in a clump.

When grass plant grows up to 3 to 4 leaves, the first node closest to the soil surface on main stem base is called tillering node, which is composed of many nodes with short internode. A young, white transparent, triangular sprout which is covered in the leaf sheath, then form another tiller. Grass tillers have their own roots after 2-3 leaves appear in the sheath. The main stem sheath is pushed aside or underneath new tiller. It is replaced by new tiller, sometimes quickly wither and become fibrous sheath residues at the base of tillers.

Tillers can promote the plant for better use of soil nutrient, because each tiller has its own roots. Tillering also expands the available nutrients part area and strengthens the capacity to occupy space. More tillers, more productivity, while indirectly promote vegetative propagation of individual plants.

The tiller growth is generally concentrated in spring and summer. Summer and autumn tillering starts from flowering stage, produces new vegetative shoots from buds. Tiller may continue until halted by low temperatures. Secondary tillering occurs in spring of the second year. These tillers production will last until the flowering period. No matter quantity or vigor of tillers produced during these two tillering periods vary greatly. Tillers formed during summer and fall grow vigorously because these tillers developed from the main stem that completed the process of flowering and fruiting, but the spring tillers grow from the young vegetative branch.

The stem base part near the soil surface, including many very short internodes, highly compressed together and with tillering ability, is called tiller section. Tillering ability of grasses depends on the size of the tiller section, which determines the number of tillers inside the leaf axils of main stem's lower parts.

(2) Legume branching

After legumes reach the trifoliate stage, the lower stem will enlarge to form the crown. The legume lateral buds develop from the crown which gradually extends into the soil as the plant ages. Vigorous crowns protect the plant against adverse external conditions, especially cold temperature. Le-

是由许多节间很短的节组成的，在分蘖节上形成一个幼嫩、白色透明、呈三角形的新芽，它被包裹于叶鞘内，由它形成另一个侧枝。禾草的侧枝能生根，即侧枝在叶鞘内出现 2 ~ 3 片叶后开始生根，母枝叶鞘被推向一边，或居于新枝的下面，其位置被新的侧枝所代替，有时则很快地枯萎而成为纤维状的鞘残体，残留于枝条基部。

分蘖促进植物比较完全地利用土壤营养物质，因为每个枝条都能形成自己的根系。分蘖同样有利于植物扩大营养面积，加强其占据地面的能力。通过分蘖能形成大量的干物质（饲草），同时间接地促进禾草个体的营养繁殖。

禾草的分蘖一般集中于春季和夏季两个时期。夏秋分蘖是在开花期开始的分蘖过程，从芽产生新的营养枝，这次分蘖一直延续到低温来临。第二年春季出现第二次分蘖，这次分蘖一直持续到开花期。这两个分蘖时期所形成的枝条，无论在质量上或数量上都是不同的。首先，夏秋时期形成的枝条生长旺盛而强壮，这是因为该枝条是在已达到开花结实的母枝上发育的；而春季分蘖所形成的枝条，则是在幼年的营养枝上发生的。

地表附近，茎秆基部包括若干节间极短，高度压缩在一起，并具有分蘖能力的茎节，称为分蘖带，禾本科牧草的分蘖能力取决于分蘖带的大小，而分蘖带的大小决定了母枝下部分蘖时叶腋内发育的侧枝的数量。

（2）豆科牧草的分枝

豆科牧草在三叶期后，会在茎的下部（下胚轴）与根系的连接处形成肥厚膨大的部分，称为根颈，豆科牧草的侧枝芽即发生于根颈上，根颈随着植株年龄的增长而逐渐伸入土中，

gumes branch differently from grasses. Generally legumes shoot develop from the crown, lateral braches do not form their own roots except of collar tillering and stolon types.

Collar tillering legumes produce roots from the neck part of the branches. Root may produce shoots aboveground and roots belowground. Examples include yellow clover and wild pea.

Stolons legumes form prostrate stems extending from the crown. The prostrate stem can form new leaf clusters and adventitious roots at stem nodes. White clover displays this growth pattern. Legumes may produce branches from the beginning of spring into autumn.

2.3.1.2 The regularity on concurrent emergence of leaf-tiller

Tillers originating from the main stem are called primary tiller. A tiller originating from the primary tiller is called secondary tiller, and so on. Tiller bud differentiation and development need to go through four periods: primordium differentiation, complete differentiation, intravaginal elongation and tiller. When No. *N* leaf of the mother stem grows out; the axillary buds of this leaf node begin to differentiate, No. (*n*-1) leaf node's differentiation is completed, the axillary buds of No. (*n*-2) leaf node has been elongated in the sheath and the axillary buds of No. (*n*-3) leaf node extends out of the sheath, become visible tiller. The appearance relation of grasses' tiller and main stem leaves followed (*n*-3) rule, this rule is called the regularity of concurrent emergence of leaf-tiller (Fig. 2.9). Known by the regularity of concurrent emergence of leaf-tiller, tiller usually does not occur before three leaves, but starts to tiller when 4th leaves appears. By the regularity of concurrent emergence of leaf-tiller, we can evaluate the grasses' growth status and predict tillering number.

In theory, each leaf node can produce tiller. In fact, only a few near-surface nodes do, this is because the process from primordium differentiation to development completion is not affected by external influences, but if it will develop to a tiller, this is highly dependent on environmental conditions. Forages should be managed to enhance development of tillering nodes in order to maintain dense stands and produce high yields.

以抵御不良的外界条件，特别是对越冬有益。豆科牧草的分枝与禾本科牧草分蘖形成有所不同，一是一般豆科的枝条均生自根颈处，二是所有的侧枝，除根蘖型及匍匐茎型外，都不能形成自己的根系。

根蘖型豆科牧草枝条一部分由根颈部产生，尚有大部由根蘖部长出，根蘖处向上形成地上部枝条，向下生出根系，形成疏松的草丛，如黄花苜蓿、山野豌豆等。

匍匐茎型豆科牧草在根颈处形成贴于地面的蔓生茎，茎的节上可形成新的叶簇和不定根。如白三叶、地三叶等。

豆科牧草产生分枝，由春季开始一直延续至秋季。

2.3.1.2 叶蘖同伸规律

通常把禾本科牧草主茎上发生的分蘖称作一级分蘖，一级分蘖上发生的分蘖称作二级分蘖，二级分蘖上再发生分蘖即称作三级分蘖，以此类推。分蘖芽的分化发育要经历原基分化、分化完成、鞘内伸长、分蘖出现 4 个时期，当母茎上第 *N* 叶抽出时，该叶节位腋芽开始分化，第（*n*-1）叶节位分化完成，第（*n*-2）叶节位腋芽已在鞘内伸长，第（*n*-3）叶节位的腋芽伸出鞘外，成为可见的分蘖。我们把禾本科牧草分蘖与主茎叶发生时间上的同伸关系即（*n*-3）称作叶蘖同伸规律，把这个同时出生的叶和蘖分别称作同伸叶和同伸蘖（图 2.9）。由禾本科牧草叶蘖的同伸规律可知，禾本科牧草 3 叶前一般不发生分蘖，到第 4 叶才开始分蘖。生产上可利用同伸蘖的规则诊断禾本科牧草的生长好坏和预测可能达到的分蘖数与分蘖成穗情况。

理论上讲，每个叶节位都可分蘖，但实际上，只有近地表的几个节能发生分蘖，这是由于从原基分化到分化完成

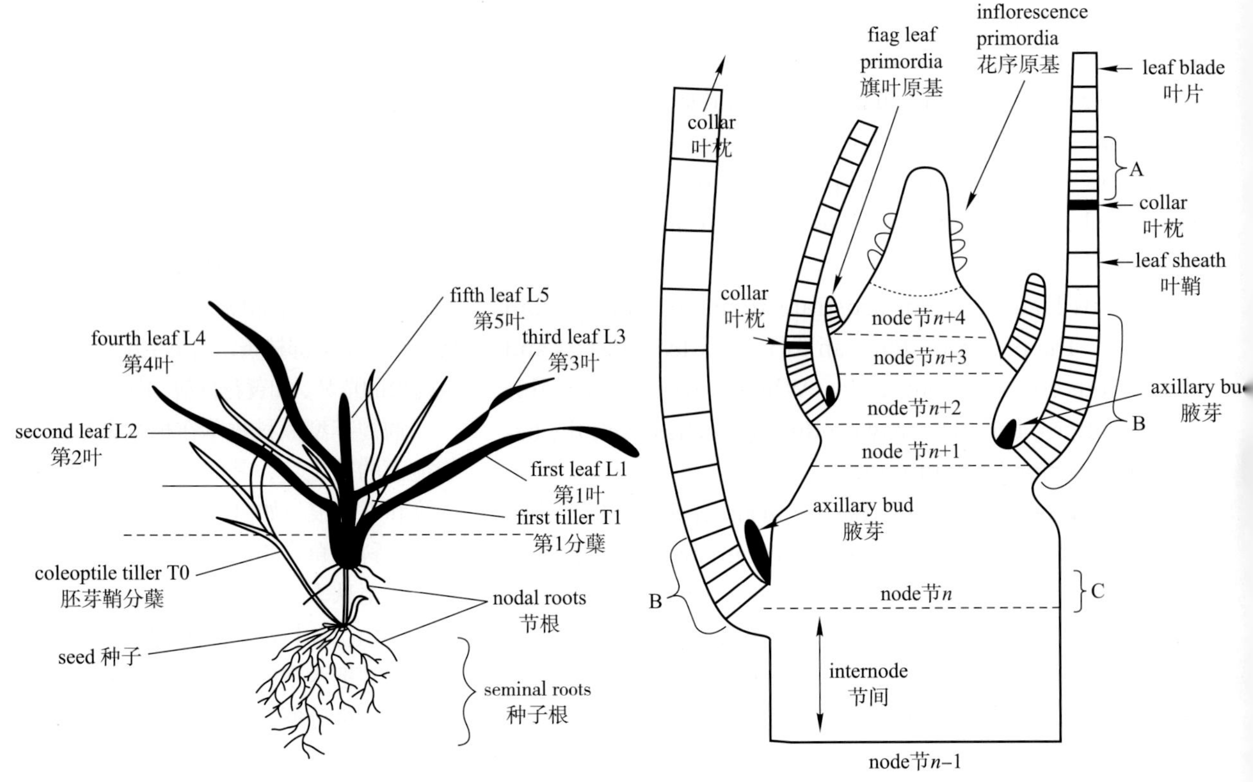

Fig. 2.9 Regularity of concurrent emergence of leaf-tiller
图 2.9 叶蘖同伸规律

2.3.2 The cycles of forage growth

Following germination and seedling emergence, forage growth transferred from heterotrophic to autotrophic growth, and goes into the vegetative growth stage. Due to the cell division and the volume increase of new born cell, the seedlings grow up quickly. At the same time, with the cell differentiation, the differentiation of other organs becomes identifiable and finally develops into a new plant.

Forage growth is generally considered an irreversible process as volume and weight increase. Growth is usually accompanied by an increase in herbage dry matter. At the time of germination, seed absorb a large amount of water. Fresh weight and volume increase significantly, but before leaves form, due to respiratory consumption of large quantities of organic matter, dry weight decreased, actually with the cell division and cell growth, it does belong to growth.

During the growth cycle of forages, no matter cells, or-

这一过程不受外界影响，要进一步发育成为分蘖则受环境条件的影响，也有可能转为休眠或死亡，从而造成该叶位的缺位现象，因此生产上应加强管理，防止分蘖节的缺位特别是低位蘖的缺位，以争取多穗和大穗，达到高产。

2.3.2 牧草生长大周期现象

随着种子的萌发与出苗，牧草从异养生长转为自养生长，进入了营养生长阶段。由于细胞分裂和新生细胞体积的加大，幼苗迅速地长大。与此同时，随着细胞的分化，牧草各器官的分化也就越来越明显，最后长成为新的植株。

通常认为，牧草的生长是一个体积和质量不可逆的增加过程。生长通常伴随着牧草干物质的增加。但要注意，在种子萌发时，由于种子大量吸收水

gans or whole plants, all the growth rates display a slow-fast-slow pattern. Initially growth is slow, growth rate then gradually increases to maximum, then decrase slowly until completely stop. We called these three stages together the big lifecycle of plant growth. If we set time as the abscissa and growth accumulation as the ordinate in graph, growth follows a sigmoid curve (Fig.2.10).

Why the plant growth shows an "S" curve? Forage plant growth is comprised of growth of individual organs, which in turn, increase based on division and growth of cells. As organ growth begins cells are mostly in the phase of cell division, cell division results from an increase in protoplast volume. Plasma synthesis process is slow, so volume increases slowly. However, when cells initiate the period of elongation, water uptake leads to rapid increase of cell volume. As cell elongation reaches the highest rate, it then slows gradually and finally stops. In addition, the plant growth speed is closely related to photosynthesis area and the vigor of plants. In initial growth stage, photosynthetic area is small. Root system development is limited and growth rate is slow. As growth continues rapid expansion of forage photosynthetic area and an enlarging root system supports markedly accelerated growth. At later stages, with plants mature, photosynthetic rates and root growth slowdown gradually and finally subsides.

分，其鲜重和体积确实也明显增加，但在绿叶形成以前，因呼吸消耗大量有机物，其干重反而减少。这时，胚内有原生质的增长和新细胞的形成，当然仍属生长现象。

在牧草生长过程中，无论是细胞、器官或整个植株的生长速率都表现出"慢—快—慢"的规律，即开始时生长缓慢，以后逐渐加快，达到最高点后又减缓以至停止。这三个阶段总合起来叫作生长大周期。如果以时间为横坐标，生长量为纵坐标绘图，则植物的生长呈"S"形曲线（图 2.10）。

器官的生长为什么能表现出生长大周期和"S"形曲线？这应从细胞的生长情况来分析。器官开始生长时，细胞大多处于细胞分裂期，由于细胞分裂是以原生质体量的增多为基础的，原生质合成过程较慢，所以体积加大较慢。但是，当细胞转入伸长生长时期，由于水分的进入，细胞的体积就会迅速增加。不过细胞伸长达到最高速率后，就又会逐渐减慢以至最后停止。此外，植物生长快慢主要与光合面积的大小及生命活

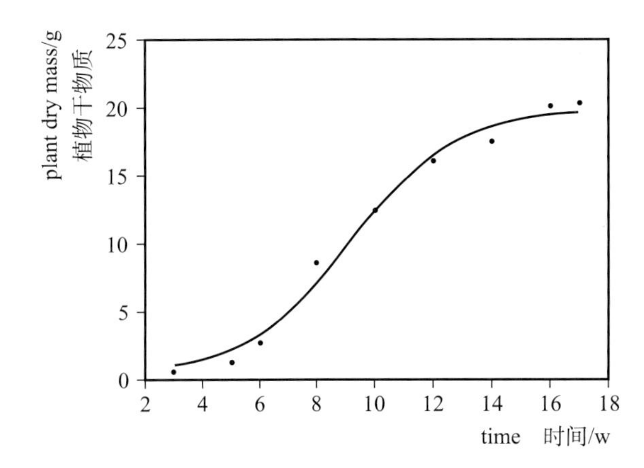

Fig. 2.10 Plant growth follows a sigmoid curve
图 2.10 植物生长"S"形曲线

2.3.3 The reproductive growth of forages

2.3.3.1 Flowering habits and pollination of legumes

(1) Characteristics of legumes flowers

Legumes are mostly entomophilous flowers. The morphology and characteristics of flower can attract insects. Big inflorescence consists of lots of florets, with bright corolla and good fragrance. Nectar is deep in the corolla of legumes. Pollen grains are large and heavy, with a sticky or rough surface, so that pollen adheres to the insect's body and to the flower stigma.

(2) Type of legume flowers

According to the structure of legume flower and whether it can go back to the original state after honeybee gather nectar, there are four types of flower.

① The keel tightly surrounds the stamens and style of a flower. Because of having great flexibility, when a bee compresses the keel, the stamen and pistil pop out of the keel. Duration of the exposure depends on the maintaining compression of the keel. After the bee leaves, the pistil returns to the original position, this type may be pollinated repeatedly. Flowers pollinated and unpollinated are not easily distinguished. Such as sweet clover and clover.

② The keel tightly surrounds the stamen and pistil. When insect compress the keel, the stamen and pistil extended from the keel and complete pollination. They do not recover their original position so that only the initial pollination of the flower is effective. Pollinated and unpollinated flowers are easily distinguished. Such as alfalfa.

③ The keel into empty tubular surrounded by female, stamens, as a speaker at the opening, when keel oppressed, pollen extend from the opening position at the top, then back to the keel cartridge, this kind of flowers need pollinators "interview" for many times to pollinate effectively, the pollination extent of flowers is difficult to distinguish. Such as birdsfoot trefoil and lupine.

④ Repeat "interview" and pollination by insects, such as wild pea.

动的强弱有关。生长初期，幼苗光合面积小，根系不发达，生长速率慢；中期，随着牧草光合面积的迅速扩大和庞大根系的建立，生长速率明显加快；到了后期，植株渐趋衰老，光合速率减慢，根系生长缓慢，生长渐慢以至停止。

2.3.3 牧草的生殖生长特性

2.3.3.1 豆科牧草的开花习性与授粉

（1）豆科牧草花的特性

豆科牧草的花大多属于虫媒花。其花的形态及开花特性具有招引昆虫的适应性。小花多而集成大形花序，具有鲜艳的花冠和浓郁香味。豆科牧草花冠深处具有花蜜，花粉粒大而重，有胶质或具不平滑的表面，这样可使蜜蜂短期内所采集的混合花粉容易黏着于昆虫的身体上及花的柱头上。

（2）豆科牧草花的类型

按照豆科牧草花的结构及蜜蜂采粉后是否恢复原来状态可以将花分为 4 种类型：

① 花的龙骨瓣紧紧包围雄蕊及花柱，具有较大的弹性，当蜜蜂采蜜压迫龙骨瓣时，雄蕊及雌蕊自龙骨瓣中弹出，其弹出时间的长短，视蜜蜂对龙骨瓣压迫时间的长短而定，当蜜蜂离开时，雄、雌蕊又恢复至龙骨瓣原来的位置，这种花能够多次授粉，而授粉与未授粉的花不易区分开来。如草木樨、三叶草等牧草的花。

② 龙骨瓣包围雄、雌蕊较紧，当昆虫采蜜压迫龙骨瓣时，雄蕊及雌蕊自龙骨瓣中伸出，授粉完毕后，它们就不能再恢复到原来的位置，所以这种花只在第一次授粉有效，授粉与未授粉的花易于区别。如苜蓿的花。

(3) The habit of legume flowering

Most legumes have indefinite inflorescences. Inflorescence form gradually from down to up of the plant. The first flower also appears from the bottom of one inflorescence. Flowers gradually bloom following a bottom-up pattern. Legumes flowering period are generally longer and extended. For example, bloom of alfalfa and sweet clover last for more than one month. The opening time of legume flower is not very concentrated within one day either. Generally legumes can bloom both in the morning and afternoon.

(4) The Pollination characteristics of Legume

① Cross-pollinated

Natural cross-pollination leads to complex in heritance patterns. For forage selection and improvement, it is difficult to obtain stable offspring. Examples include alfalfa, sickle alfalfa (*Medicago falcata*), sweet clover (*Melilotus officinalis*), sainfoin (*Onobrychis viciaefolia*), bush clovers(*Lespedeza bicolor*) , birdsfoot trefoil (*Lotus corniculatus*) and clover.

② Often cross-pollinated

Pollination methods of this type of legume is not modular. The hereditary of a plant depends on the pollination condition of the same year or earlier. Examples include white sweet clover (*Melilotus albus*), yellow lupine (*Lupinus luteus*), shamrock (*Medicago lupulina*), crimson clover (*Trifolium incarnatum*) and hairy vetch (*Vicia villosa*).

③ Self-pollinated

This type of the forage can be classified as simple genetic type. Individuals are relatively consistent with each other. Examples include *Lathyrus*, common vetch (*Vicia*) and subterranean clover (*Trifolium subterraneum*).

Most legume forages produce entomophilous flowers. Bees, bumblebees, *Nomia melanderi* and *Megachile rotundata* (leafcutting bees) are the main pollinators of legumes.

2.3.3.2 Flowering habits and pollination of grasses

(1) Flower characteristics of grasses

Most of the grass plants are wind-pollinated. Flowers are unattractive to insects, small and abundant, consisted by one to several spikelets, combined into an inflorescence. A number of dense spikelets form into a long panicles, racemes

③ 龙骨瓣成空筒状包围雌、雄蕊，如喇叭一样顶端有开口，当龙骨瓣被压迫时，花粉自顶端开口处伸出，以后又恢复至龙骨瓣筒内，这种花为了能有效授粉，需要授粉者多次“拜访”，授粉与未授粉的花至凋谢时也难于区别。如百脉根、羽扇豆等牧草的花。

④ 昆虫重复“拜访”对其授粉，如野豌豆等牧草的花。

（3）豆科牧草的开花习性

豆科牧草大部分属于无限花序，花序从植株下部向上部逐渐形成，在一个花序上也是下边的花先出现。因此花开放也是由下而上逐渐开花。豆科牧草的开花持续时间一般较长，如苜蓿、草木樨开花时间持续一个月以上。豆科牧草一日内开花动态比较分散，一般豆科牧草上、下午均能开花。

（4）豆科牧草授粉特点

① 异花授粉

这一类牧草由于天然异花授粉的结果，遗传性较为复杂，从育种工作看，通常用一次选择的方法很难获得稳定的后代。如紫花苜蓿、黄花苜蓿、草木樨、红豆草、胡枝子、百脉根及三叶草等。

② 常异花授粉

这一类牧草的授粉方式是不定型的，某一单株当代的遗传性依该单株当年及早年授粉情况而定。如白花草木樨、黄花羽扇豆、天蓝苜蓿、绛三叶及毛苕子等。

③ 自花授粉

这一类牧草的遗传类型较简单，个体之间具有相对的一致性。如山黧豆属、普通野豌豆和地三叶等。

豆科牧草大部分为虫媒花，蜜蜂、熊蜂、碱蜂和切叶蜂等是豆科牧草的主要授粉者。

or spike inflorescence. When flowering, the wind aggitates the inflorescence so that anthers protrude from the lemma exposing and drop pollen mixture to stigma, thus ensuring normal pollination of the grasses.

Different types of grasses exhibit differing flowering habits, including variation in blooming, flowering duration, open dynamic of single spikelets, flowering time within one day and blossom sequence.

(2) Flowering period and duration

Generally, the flowering period of grasses appear in mid to late June until early August in the Northern of China except the seeding year, this depending on the accumulated temperature requirement from turning green to flowering stage. It is also related to daylength during flowering period.

Flowering duration is the days from opening of the first flower in the inflorescence to the ending of flowering. Flowering duration and uniformity vary with different forage species. Some forages last for about 7 d, such as siberian wildrye(*Elymus sibiricus*), some may last up to 25 d, such as wild rye(*Hordeum brevisubulatum*). The forage depends on the biological features of flowering on one hand; On the other hand, it is related with the weather, generally, grasses do not flowering in rainy day, which will extend flowering duration.

(3) Single spike and flowering daily dynamics

Flowering duration differs from grass species. The flower time of one spikelet generally last for 6~8 d, flowering peak time generally appear at 2~5 d after flowering, but some grasses spikelets can last more than 10 days. Generally, when the temperature is above 10 ℃, grasses begin to blossom; it will stop flowering below 10 ℃.

(4) Flowering sequence

There are two types as following:

① For panicle grasses, upper spikelets open first, and then extend downward, spikelets in the base finally open. Examples include smooth bromegrass, bluegrass, and Sudan grass.

② For spike inference grasses, the spikelets of the top 1/3 inflorescence open first, then proceeds gradually downward. Examples include wheatgrass, chinese wildrye, siberian

2.3.3.2 禾本科牧草的开花习性与授粉

（1）禾本科牧草花的特性

禾本科牧草为风媒花授粉植物，花外形亦少具有吸引昆虫的特征，花小而多，由一至数朵合成小穗，再由许多小穗密集成长的花序——圆锥花序、总状花序或穗状花序等。开花时，由于风的吹动，花序摇动或彼此撞击，使得花药从稃中伸出，这样促使不同的混合花粉落于柱头上，从而保证了禾本科牧草的正常授粉过程。

不同种类的禾草开花习性不同，这表现在开花期、开花持续期、单穗开放动态、一日内开花时间及开花顺序等。

（2）开花期及开花持续期

禾本科牧草的开花期除播种当年外，北方一般多在6月中、下旬至8月上旬，这取决于牧草从返青至开花期所需积温的多少，也与开花时的日照长短有一定的关系。

开花持续期是从花序第一朵花开放至整个草丛开花结束所持续的天数。开花持续期及整齐度也因牧草种类不同而异，有些牧草花期持续7 d左右，如老芒麦；有的可达25 d，如野黑麦。这一方面取决于牧草开花的生物学特征，另一方面与开花时所处的天气状况有关，一般阴雨天牧草不开花，这会延长开花持续期。

（3）单穗及开花日动态

禾本科牧草单穗开花持续时间及大量开花持续时间因牧草种类不同而异。一个花序开花时间一般为6 ~ 8 d，大量开花时间，一般为开花后的2 ~ 5 d，但有些牧草单穗开花也可持续10 d以上。一般来说，气温10℃以上，禾草开始开花；低于10℃即停止开花。

wildrye, ryegrass, etc.

Whether panicles or spikes inflorescence, both of them follow the sequence of from down to up when spikelet flowers opening. Forages that has two kinds of inflorescence structure, its flowering order is same as that of panicles, but for bisexual flowers plants, male flowers start opening in 4~5 h, then end at the same time, such as sudan grass.

(5) The flower blooming process

When a grass flowers opens, its internal and external lemma crack gradually at first. When the opening angle is about 20°, stamens are exposed. When the angle is close to 60°, anthers protrude and droop and begin to shed pollen.

(6) Pollination features of grasses

Most grasses are cross-pollinated plants with wind-pollinated flowers. Grass pollination occurs when large number plants are flowering. If windless weather just appear at pollination stage of grasses, artificial supplementary pollination is needed and must be conducted at flowering peak. The simplest method: two people stand on both sides of the field or by machine, pull a string and skim over on the grasses from top to end of the fields. Thus on the one hand, plants would be shaken by collision and promote the spread of pollen; On the other hand, the pollen that drop on the string line can be brought to the other inflorescences, making the grass full pollination. Artificial supplementary pollination is usually conducted once or twice, the time interval is commonly 3~4 d.

2.3.3.3 The seed development of forages

(1) Fertilization

Fertilization unites male and female cells, egg and sperm cells fuse with each other forming a zygote. Fertilization is a unique metabolic process with assimilation of two sex cells. The time from pollination to fertilization varies widely in grasses and can be modified by climatic conditions. For most grasses, 12~48 h are required under normal circumstances.

(2) Seed development

Seed development process involves a series of changes, from fertilization and zygote formation until seed maturation.

（4）开花顺序

禾草开花顺序大致可以分为两种类型：

① 圆锥花序的牧草，顶端小穗首先开放，然后向下延至下部，基部小穗的小花最后开放，如无芒雀麦、早熟禾、苏丹草等。

② 穗状花序的牧草，花序上部1/3处的小穗首先开放，然后逐渐向下、向上开放，如冰草、羊草、披碱草、黑麦草等。

在一个小穗中，不论圆锥花序还是穗状花序，都是小穗下部的小花首先开放，然后顺序延至上部。具有两种花序结构的牧草，其开花顺序同圆锥花序，但两性花序开放顺序不同，经4 ~ 5 h后，雄性花才开放，然后两类花同时结束，如苏丹草。

（5）小花开放过程

禾本科牧草小花开放时，其内外稃首先逐渐开裂，当其开放的角度达到20° 左右时，雄蕊很快露出，当内外稃的角接近60° 时，花药伸出且下垂并开始散粉。

（6）禾本科牧草的授粉特点

禾本科牧草大部分属异花授粉植物，风媒花，借助于风力传播花粉。授粉阶段若遇无风天气，需要借由人工辅助授粉，人工辅助授粉必须在大量开花时进行。最简单的方法是：于牧草盛花时，由人工或机具于田地两侧，拉一绳索或线网从草层上掠过。这样一方面植株被碰撞摇动可促进花粉的传播，另一方面落于绳索或线网上的花粉在移动时可带至其他花序上，从而达到使牧草充分授粉的目的。人工辅助授粉通常进行一次或两次，两次间隔的时间一般为3 ~ 4 d。

Seed development has three aspects:

① Embryo development

The embryo is the core part of the seed, the prototype of the plant. In normal circumstances, the embryo is developed and formed by the egg in **embryo sac**.

② Endosperm development

After fertilization, polar nuclei in the embryo sac rapidly divide, forming the nucleus, and are arranged inside the embryo sac, then occur simultaneously in the diaphragm, a lot of parenchyma cells between the various nuclear are called **endosperm cells**. These cells undergo continuous division and develop into endosperm, called **karyotype**. Endosperm development is common in most of monocotyledon and dicotyledon classes of flowering plants. Another kind of endosperm development is called the cell type, namely division directly happen by the polar nuclei or secondary cell after fertilization, form the endosperm cells. This pattern is common in synpetalous flower plant of dicotyledonous plants.

Early in development, some seed endosperm is gradually absorbed by the embryo, allowing nutrient transfer to the cotyledons. The endosperm disappear as the embryo develops, forming seeds with no endosperm, such as soybeans.

③ Seed coat development

Ovules surrounding integuments, it is sometimes partly or completely absorbed by embryo during seed development, making part or all qualitatively change, after the split, form multilayer cell organization, some form **cuticle** below the **epidermis**, some cell **lignificate**, which has a strong protective effect. Such as beans, alfalfa seed coat. The micropyle in the end of the ovule forms the germination holes, the **suspensor** at the base end of ovules grow into a kind of handle. When seeds get maturation and drying, they shed from the handle, leaving a scar in the seed coat, namely **hilum**, but caryopsis of cereal is slightly different, the outside is wrapped with seed pericarp.

(3) Seed maturation

Maturity of forges seed is classified according to the change of their external morphological characteristics. The seed maturity extent is identified according to the standard of

2.3.3.3 牧草种子的发育

（1）受精作用

受精为雌、雄性细胞，即卵细胞和精细胞互相融合，形成合子的过程。受精作用是两个性细胞相互同化的过程，是新陈代谢的一种特殊方式。禾本科牧草从授粉到受精所需时间因牧草和气候条件不同而大有差异，大多数禾本科作物，正常情况下须历时 12 ~ 48 h。

（2）种子发育的一般过程

种子的形成发育过程是指从卵细胞受精成为合子开始，直到种子成熟所经历的一系列变化。

种子发育主要有以下三个方面：

① 胚的发育

胚是种子的主要部分，为植物体的雏形。在正常情况下，胚是由胚囊中的卵细胞通过性的过程发育而成。

② 胚乳的发育

胚囊中的极核或次生细胞在受精后，随即进行迅速分裂，形成大量的核，排列在胚囊内部，同时在隔膜、各个核之间形成很多薄壁细胞，称为胚乳母细胞。这些细胞不经过休眠状态连续分裂，发育成胚乳，这种胚乳发育的方式称为核型，常见于单子叶植物和双子叶植物纲中的大多数离瓣花植物。另一种胚乳发育的方式称为细胞型，即由受精后的极核或次生细胞直接分裂，形成大量的胚乳细胞，这一方式常见于双子叶植物纲的合瓣花植物。

有些种子的胚乳在发育前期，即逐渐被胚所吸收，使营养物质向子叶转移，结果胚乳消失，而胚特别发达，形成无胚乳的种子，如大豆的种子。

③ 种皮的发育

胚珠周围的珠被，在种子发育过程中有时被胚吸收掉一部分，有时全部被吸收，使部分或全部发生质变，

most seed's maturity on plant.

① The maturity stage of grass seed

Milk stage: green caryopsis, green lemmas, seed contains white milk-like inclusions, largest seed volume, high water content, embryo development is completed. A few seeds are able to germinate.

Dough stage: Glumes and lemmas begin to retreat green, inherent color of caryopsis, waxy inclusions, easily broken when pressure with nails.

Ripening period: Caryopsis is dry and tough, the seeds get smaller, and inclusions are opaque cutin, not easily broken by fingernail. A lot of grass seeds shatter naturally.

② The maturity stage of legumes seed

Green ripe stage: Pods and seeds are bright green, almost reach the final length.

Early yellow ripening: Pods turn yellow-green, green seed coat, relatively hard. Seed volume is the largest at this stage.

Late yellow ripening: Green pods color returns, natural color of seed coat and seeds get smaller, not easily pierced with nails.

Ripening period: The pod shrinks; original color and seeds get harden.

经过分裂，形成具多层细胞的组织，有的表皮下面形成角质层，有的细胞木质化，具有很强的保护作用。如豆类、苜蓿种子的种皮。原来在胚珠末端的珠孔，形成种孔，或称发芽孔。胚珠基部的胚柄发育成种柄。种子成熟干燥从种柄脱落后，在种皮上留下一个疤痕，即为种脐，但禾谷类的颖果，在种子外面还包有果皮，所以称为果脐。

（3）种子的成熟

牧草种子的成熟期是按其外部形态特征的变化划分的。鉴定种子的成熟期是否达到某阶段以植株上大部分种子的成熟度为标准。

① 禾本科牧草种子的成熟阶段

乳熟期：颖果绿色；内外稃绿色；种子内含物呈白色乳汁状；种子体积达最大，含水量较高，胚发育完成，少数种子具发芽能力。

蜡熟期：颖片和内外稃开始退绿，颖果呈固有色泽，内含物呈蜡状，用指甲压时易破碎。

完熟期：颖果干燥强韧，体积缩小，内含物呈粉质角质状，指甲不易使其破碎，很多牧草种子开始自然落粒。

② 豆科牧草种子的成熟阶段

绿熟期：荚果和种子均呈鲜绿色，种子基本足长。

黄熟前期：荚果转黄绿色，种皮呈绿色，比较硬，种子体积达最大。

黄熟后期：荚果退绿，种皮呈固有色，种子体积缩小，不易用指甲划破。

完熟期：荚壳干缩，呈固有色泽，种子变硬。

2.4 Growth period and development stages 生育期和生育时期

It is important to understand the growth and development of forages in various regions and conditions in forage production. The factors include the length of growth period, development speed, the introduction of forages, plant breeding and reproduction, cropping sequence, distribution of varieties and the suitable harvest period must be paid more attention.

2.4.1 The growth period of forages

2.4.1.1 Growth period

The growth period usually refers to the total days from seedling emergence to harvest. But for forages, it has different meaning for different forage types, different production modes and planting purposes.

(1) For seeds production, the growth period refers to the number of days from emergence of seedling or returning green to seed maturity.

(2) For annual forages production, the growth period refers to the days from seedling emergence to harvest of aboveground parts.

(3) For perennial forages, from the second year, the growth period refers to the total days from returning green to full maturity or vegetation harvest.

(4) For transplanted forage crops, such as sweet potato, growth period has two different meanings: The seedbed growth period refers to the days from emergence to transplanting. The production growth period refers to the days from transplanting to maturity.

2.4.1.2 Factors influencing on the duration of growth period

The growth period length is influenced by internal and external factors. Internal factors are mainly genetic characteristics. Different kinds of forages have different growth period lengths. In the same environment, forage growth period length of species and varieties is relatively stable. External factors include environmental conditions and cultivation

牧草生产中，了解各类牧草在各地区的生长发育状况，如生育期的长短、发育的快慢，对牧草引种、良种繁育、茬口安排及品种布局，适宜收获时期的确定等具有重要的生产指导意义。

2.4.1 牧草的生育期

2.4.1.1 生育期的概念

生育期通常指牧草从播种出苗到收获的总天数。但对于不同牧草类型、不同生产方式和种植目的，其概念又略有差异。

（1）对于牧草种子田，以籽实为播种材料又以新的籽实为收获对象的牧草，其生育期是指种子出苗（返青）到新种子成熟所经历的天数。

（2）对于以营养体为收获对象的一年生牧草，生育期指从播种材料出苗到地上部分收获所持续的天数。

（3）对于多年生牧草，从第二个生长年份，生育期指从春季萌发返青到种子完全成熟或营养体收获的总天数。

（4）对于需要育苗移栽的饲料作物，如甘薯，通常将生育期分为秧田（苗床）生育期和田间生育期。秧田（苗床）生育期指从出苗到移栽的天数，田间生育期是指从移栽到成熟的天数。

2.4.1.2 影响生育期长短的因素

影响生育期长短的因素包括内因和外因两个方面，内因主要是指牧草的遗传特性，牧草种类不同，生育期长短各异。在相同的环境条件下各牧草种和品种的生育期长短相对稳定。外部因素主要包括栽培环境条件和栽培技术措施，如气候因素、地理因素、

technique measures, such as climate, geographic factors, soil conditions and cultivation management.

(1) Environmental conditions

Forage growth period duration is not only affected by climate condition, but also relate to geographic factors such as latitude, longitude and altitude etc.

(2) Cultivation techniques

Cultivation techniques have a great influence on growth periods. Forages grown in fertile soil with adequate nitrogen, balanced carbon and nitrogen ratio and adequate water and nutrients, the plants grow rapid and delayed maturity. On the contrary, poor soil with high temperature and drought will shorten the growth period.

(3) Growth period and yield

Growth period differences between early and late maturing forage varieties primarily determined by the length of vegetative growth stage. Early maturing varieties generally form less stem and fewer leaves, leaf area index is smaller and resulting yield per plant is lower than late maturing varieties. Under favorable growing conditions, high density can increase forage output per unit area.

土壤条件及栽培管理等。

（1）环境条件

牧草生育期长短除受气候条件的影响外，还与纬度、经度和海拔高度等地形因素有关。

（2）栽培技术措施

牧草的栽培技术措施对其生育期也有很大影响，在土壤肥沃或施氮肥充足的土壤上栽种牧草时，由于土壤碳氮比、水肥适宜，茎叶常常生长过旺，成熟延迟，生育期延长。若土壤贫瘠，遇到高温干旱会引起牧草早衰，致使生育期缩短。

（3）生育期与产量

早熟品种和晚熟品种生育期长短差别，主要取决于营养生长阶段。一般地，相对于晚熟品种，早熟品种主茎形成的叶片较少，叶面积指数也小，故而单株产量低于晚熟品种。在栽培时，应相应增大其种植密度，才能确保单位面积的产量。

2.4.2 Development stages of forages

Development stages refers to the period during which forages show obvious changes in morphology, also called the phenological period. If 50% of plants in the field achieve the development stages, it is set for the standard, 20% is set as the early stage, while 80% for the late stage.

2.4.2.1 Development stages of grasses

(1) Emergence stage (returning green)

The emergence stage refer to after seed emergence, the stage of seedling grow above the ground (Fig.2.11). Returning green means the period of biennial and perennial forages germinate overwinter, green leaves start fast growing.

(2) Tillering stage

Tillering stage refers the lateral branches grow out from the base of plant (Fig.2.12). The Standard is the first tiller bud germination, and 1~2 cm out from the basal leaf axil and more than 50% of plants reach this standard.

2.4.2 牧草的生育时期

生育时期是指牧草作物一年中在外部形态特征上呈现显著变化的若干时期，也叫物候期。进行田间记载时，一般以达到某个生育时期的百分率为标准，如以 50% 为标准时期，20% 为始期，80% 为盛期。

2.4.2.1 禾本科牧草的生育时期

（1）出苗期（返青期）

种子萌发后的幼芽露出地面时为出苗期（图 2.11）。越年生、二年生和多年生禾草和饲料作物越冬后萌发，绿叶开始旺盛生长的时期称返青期。

（2）分蘖期

植株基部分蘖长出侧枝时为分蘖期。标准为第一个分蘖芽萌发，并从基部叶腋内伸出 1 ~ 2 cm（图 2.12）。

Fig. 2.11 Emergence stage
图 2.11 出苗期

Fig. 2.12 Tillering stage
图 2.12 分蘖期

全田 50% 以上植株出现分蘖的日期为分蘖期。

(3) Jointing (Elongation) stage

The first internode are starting to elongate, the nodes coming up to aboveground for 1 to 2 cm (Fig.2.13), it can be touched by hand or crack the sheath to verify.

(4) Booting stage

At the booting stage, the flag leaf sheath is completely exposed. Floral parts are wrapped in flag leaf sheath but not exposed; the upper stem looks spindly (Fig.2.14).

（3）拔节期

植株主杆的第一节开始伸长，茎节已露出地面 1 ~ 2 cm 时（图 2.13），可用手指由基部向上摸测，或剥开叶鞘证实。

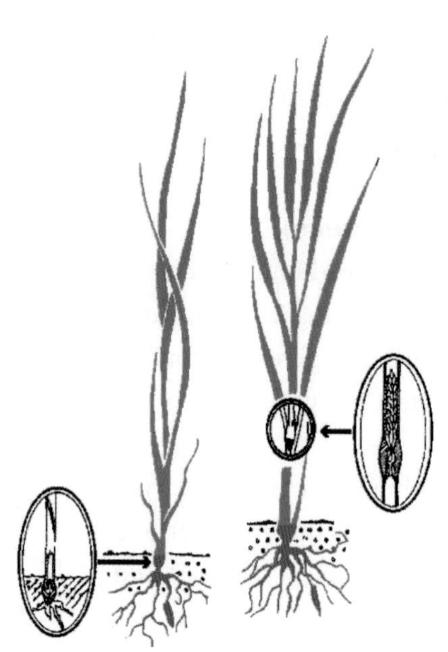

Fig. 2.13 Jointing stage
图 2.13 拔节期

(5) Heading stage

Inflorescences have emerged and are expanding (Fig.2.15).

(6) Flowering (anthesis) stage

Central spikelet petals open, filaments come out, mature anther pollen (Fig.2.16), with fertilization ability.

(7) Mature stage

At the mature stage fertilization has been completed. Embryo and endosperm have begun to develop, nutrients conversion and accumulation (Fig.2.17).

(8) Harvest stage

The actual period of forages harvest.

(9) Regrowth period

The plant regrowth after the first harvest.

2.4.2.2 Development stages of legumes

(1) Seedling (returning green)

The date that cotyledons exposed soil surface (epigaeous type) or leaf protruding soil surface (hypogeal type), shoots straight. For biennial and perennial legumes, it means from overwintering to green leaves begin to grow.

(2) Branching stage

Lateral buds begin to elongate from the base of the plant stem, with a developed leaf.

（4）孕穗期

植株的剑叶全部露出叶鞘，花被包在剑叶叶鞘中而未显露出来，茎秆中上部呈现纺锤形（图 2.14）。

（5）抽穗期

幼穗从茎秆顶部叶鞘中露出，但未授粉（图 2.15）。

（6）开花期

中部小穗花瓣张开，花丝伸出颖外，花药成熟散粉，具有受精能力(图 2.16)。

（7）成熟期

禾草受精后，胚和胚乳开始发育，进行营养物质转化、积累的过程叫成熟（图 2.17）。

（8）收获期

实际收获籽粒的时期。

（9）再生期

第一茬牧草收获后再次生长的现象。

2.4.2.2 豆科牧草的生育时期

（1）出苗期（返青期）

种子萌发，子叶露出地表（子叶出土型牧草）或真叶伸出地表（子叶留土

Fig. 2.14 Booting stage
图 2.14 孕穗期

Fig. 2.15 Heading stage
图 2.15 抽穗期

Fig. 2.16 Flowering stage
图 2.16 开花期

Fig. 2.17 Mature stage
图 2.17 成熟期

(3) Budding stage

Buds appear at the upper leaf axils.

(4) Flowering stage

The date of the flag emergence and wing petals open.

(5) Pod stage

At the pod stage, green seedpods developing.

(6) Maturity

Mostly mature brown seedpods with lower leaves dead and some leaf loss. It will crack when pressed and will rattle when shaken.

(7) Regrowth period

The plant regrowth after the harvest.

型牧草），芽叶伸直的日期为出苗期。越年生、二年生和多年生豆科牧草越冬后萌发绿叶开始生长的日期为返青期。

（2）分枝期

植株主茎基部侧芽伸长，上有一小叶展开的日期。

（3）现蕾期

植株上部叶腋开始出现花蕾的日期。

（4）开花期

植株上花朵旗瓣和翼瓣张开的日期。

（5）结荚期

植株上个别花朵萎谢后，挑开花瓣能见到绿色幼荚的日期。

（6）成熟期

大部分荚果呈褐色，植株下部叶片死亡并掉落。用手压荚有裂荚声，有些种摇动植株有响声。

（7）再生期

牧草收获利用后的再生。

Chapter 3

Environmental Aspects of Forage Growth and Development

牧草的生长发育与环境

Plant growth and development is closely related to environmental factors. Under different climatic and environmental conditions, forage growth is different; It is also various because of different forage species. Even for the same kind of forages, the growth performance is not always the same at different growth stages and conditions. Therefore, it is great significant to understand the relationship between forage growth and environment. Also, it is indispensible for forage breeding and selection, variety introduction and domestication, natural grassland improvement, artificial grassland establishment and seasonal livestock production.

Life activity of forages is the base of forage production, the whole life cycle of forages undergoing ever is called the individual development. In practice, generally a life cycle of forages refers to the process from seedling emergence, via flowering and fruiting, to seeds harvesting, but it has different meaning for those forages that take vegetative organs as the sowing material or harvest part. During the whole growth process, individual plant development always accompany with a continuous biomass accumulation and energy conversion, metabolism between plants and the environment never stop, which make the plant production. Metabolism is the engine for growth and development, and growth and development are the overall performance of metabolism.

植物的生长发育与环境因子密不可分。不同气候和环境条件下，牧草的生长发育状况不同；牧草种类不同，生长发育的状况也不同。即便同一种牧草，在不同生长阶段中，其生长发育状况也不一样。因此，了解牧草生长发育与环境之间的关系，对于牧草引种、驯化，牧草品种选育，草地改良，人工草地建设以及进行季节性畜牧业生产等都有重要意义。

栽培牧草就是利用牧草的生命活动进行生产，牧草一生所经历的生命活动周期叫做个体发育。生产实践中，人们通常把从播种出苗，经开花结实到种子成熟收获这一过程看作是牧草的一个生命周期。以营养器官为播种材料或收获对象的牧草，其生物学的生命周期有其特别的含义。牧草生长过程中，通过物质积累和能量转化，与外界环境不断地进行着新陈代谢，使得个体得到生长发育。新陈代谢是牧草生长和发育的动力，而生长和发育又是正常代谢的综合表现。

3.1 Growth and development of forages 牧草的生长和发育

3.1.1 Growth

Growth is an irreversible quantitative process result from increase in plant size, weight and quantity of individual organs, tissues and cells, it is the foundation of forage yield, so growth should be scientifically regulated to enhance yield. Growth of forage plant organs, individual plants or even plant communities may be described by indexes of size, length, thickness, weight, quantity over time. Relationships between growth and time may be quantified using linear or nonlinear

3.1.1 生长

生长是指牧草个体、器官、组织和细胞，在体积、质量和数量上不可逆的增加。生长是产量形成的基础，要想提高产量必须科学调控生长。在牧草群体、个体、器官的生长过程中，通常用大小、长短、粗细、质量、数量随时间的变化来表征，这种生长随时间而变化的关系，可分为线性关系和非线性关系。

equations.

3.1.2 Development

Development includes differentiation of cells, tissues and organs that mean substantial changes occur not only in forage morphology, but also in structure and function during development. Differentiation is a reversible qualitative process. Development includes changes at the cellular, tissues, organ and individual level during the forage plant life cycle, consider flower initiation, appearance of embryogenic cells, the process of fertilization, and development of vegetative reproductive organs. Narrow sense development is the process that meristematic tissue such as leaf primordia in forage stem apex differentiates into floral primordium. The process by which homogeneous cells or tissues produce heterogeneous cells or tissues by cell division is differentiation, also called organ development.

3.1.2 发育

发育是指牧草细胞、组织和器官的分化形成过程，也就是牧草发生了形态、结构和功能上的本质性变化。分化过程是一个可逆的质变过程。广义的发育包括个体、器官、组织和细胞水平上按生活史向前的变化，如花的发端、性细胞的出现、受精过程、胚及其他延存器官的形成等。狭义的发育特指牧草茎端的分生组织由分化叶原基转化为分化花原基的过程，如幼穗分化、花芽分化、维管束发育以及气孔分化等。分化即同质的细胞或组织通过细胞的分裂产生不同质的细胞或组织的过程，也称为器官发育。

3.1.3 The relationship between growth and development

Growth and development occur simultaneously. There are differences but also connections between them. Growth is the foundation of development, and development is the premise of growth. Quantitative indexes are usually used to describe the growth, whereas qualitative changes describe development. Growth mode depends on the qualitative changes occurring during development.

The growth and development is not necessarily complete synchronization, that is, fast growth leads to fast development, and vice versa. Sometimes due to unfavorable environmental conditions or unsuitable cultivation management technology, it is also possible that fast growth lead to slow development or slow growth but faster development.

3.1.3 生长和发育的关系

生长和发育是同时进行的，两者既有区别又有联系，生长是发育的基础，发育是生长的前提。生长一般用数量指标描述，是量的变化，而发育则只能定性描述，是质的变化。

生长和发育之间并不一定完全同步，即生长快发育也快，生长慢发育也慢。有时也会由于环境条件及栽培管理技术等的变化，出现生长快而发育慢或生长慢而发育快的情况。

3.2 The phasic development of forages 牧草的阶段发育

The theory of phasic development put forward by the Soviet Union scholar T. D. Lysenko explained that plant needs

苏联学者 T. 李森科提出的阶段发育学说（theory of phasic development）

to go through several different properties stages but related to each other step by step from seed germination to new seeds formation of the individual development process. The development stages include vegetative growth stage and reproductive stage according to the plant growth stage. It also can be divided into two development phases, namely vernalization stage and photoperiod.

认为，植物由种子萌发到新种子形成的个体发育过程中，要循序经过几个性质不同而又彼此联系的转变阶段。从植物生长的阶段上划分，生产上一般将开花前以营养器官生长为主的生长阶段称为营养生长阶段，开花后以生殖器官建成为主的阶段，称为生殖生长阶段。从外界环境条件方面划分，冬性禾本科牧草的个体发育至少有两个阶段，即春化阶段和光照阶段。

Many temperate plants must experience a period of low temperature during their seedling stage to initiate or accelerate the flowering process, this is called vernalization. The required temperature and duration for vernalization are different for different varieties, perennial grasses can be divided into winterness, springness and double nature forages.

一些温带植物在苗期必须经过一定时间的低温条件才能正常抽穗开花，该时期称为春化阶段，这种现象称为春化现象。多年生牧草由于通过春化阶段所要求的温度及持续时间的不同，可分为春性、冬性和双性牧草。

Springness forages at vernalization stage need relative higher temperature and shorter duration, then the reproductive branches can be formed in the seeding year, and the stand is dominated by reproductive shoots and short vegetative branches; Winterness forages at vernalization stage need lower temperature and longer duration, the reproductive branches formed only from the second growth year or even later, long reproductive branches account for the majority. The number of reproductive branches depends on the environmental conditions during the period of tillering, adequate water and fertilizer can produce more tillers and more reproductive branches in the next year, otherwise less. No vernalization required for double nature forages, so they can be sowed either in spring or in autumn, such as alfalfa.

春性牧草需要有较高的温度且持续时间短，播种当年即可形成生殖枝，草丛以生殖枝和短营养枝为主；冬性牧草通过春化阶段则需要较低温度和较长时间，播种第二年以后才能形成生殖枝，草丛以长生殖枝为主。生殖枝数目的多少取决于分蘖时期的环境条件，水肥充足则分蘖多，次年生殖枝也多，反之则较少。双性牧草没有春化阶段，春秋均可播种，如紫花苜蓿。

Photophase means that in some circumstances plants also need to go through a period of lighting induction before flowering after vernalization stage.

光照阶段指植物通过春化阶段后，还要通过一个感应光照的时期才能抽穗结实。

Phasic development reflects the external conditions requirement of forages during different ontogenesis stages, to target high yield and control the growth and development process scientifically, we need to formulate periodic management measures to facilitate the forage yield increasement. For example, suitable sowing temperature of winterness forages, vernalization cultivation for seed production, growth and development process regulation of biennial and perennial forages to realize the seed production

阶段发育反映了牧草在个体发育的各个阶段对外界条件的要求，从栽培技术上，围绕高产目标合理控制生长发育进程，分段制定管理措施，对牧草增产能起到促进作用。如冬性牧草适期播种的温度指标确定、种子生产中的春化栽培、调节二年生以上牧草的生长发育过程实现当年采种等，都是阶段发育理论在生产实践上的具体应用。

in the seeding year etc., are all the application examples of phasic development theory.

3.3 Environmental factors affecting forage growth and development 影响牧草生长发育的环境因素

3.3.1 Light

Light is one of the important factors that affect plant growth. It drives the process of photosynthesis which produces the carbohydrates that are needed to osmotically retain water in the cell for growth. Light influences the forage growth and development from two aspects. On the one hand, light can affect plant growth by photosynthesis. On the other hand, day length and spectrum components influence reproductive development.

Most forages are heliophile plants. Given enough light, height growth gets hindered and lateral branch development is initiated, which is tend to form dense branches. While when light is limited, leaves may not survive in the canopy. Light intensity influences forage assimilation and growth, which is reflected in yield and quality. The stronger the light is, the more dry matter plant will be produced. Under low light conditions, growth-development status is impaired and nitric acid may accumulate.

Light requirement depend on photosynthetic characteristic of plant species. C_4 plants (such as sorghum and maize) have a higher light saturation point and photosynthesis efficiency than C_3 plants (such as ryegrass, orchardgrass and red clove) (Fig. 3.1).

Light hours' rhythmicity change influence forage growth and development. Forages adapt to the changes in the relative length of light time, commonly known as photoperiodism. Some plants are adapted to flower only when the day length is shorter than a certain length, whilst others only flower then the day length is longer than a certain duration.Forages can be classified as long-day plants, short-day plants, and day-neutral plants according to the light response.

3.3.1 光照

光是影响植物生长的重要因子之一。因为光通过驱动光合作用合成糖类以供植物生长时细胞依靠渗透作用保持水分的需要。光照对牧草生长发育的影响表现为两个方面：一是通过光合作用合成有机物从量的方面影响；二是以日照长度和光谱成分角度从质的方面影响。

大多数牧草喜光，当光照充足时，主枝向上生长受阻，侧枝生长增强，牧草易形成密集短枝，株丛张开。而当光照过少时，叶片即无法生存，在冠层内形成无叶区。光照强度影响牧草同化作用和生长发育，进而影响其产量和品质。光照越强，幼小植株的干物质生产量越高。光照越少，牧草发育越差，硝酸含量增加。

对光照强度的需要还因牧草的光合特性而异，C_4 植物（如高粱、玉米、柳枝稷等）的光饱和点大于 C_3 植物（如黑麦草、鸡脚草、红三叶等），光合效率较 C_3 植物的高（图 3.1）。

光照时间对牧草生长发育的影响首先表现为日照时间长短的节律性变化对牧草生长发育的进程具有信号的作用。牧草长期适应于光照时间节律性变化，对白天和黑夜的相对长度具有相应的生理响应，这种响应称为光周期现象。根据植物对光照长度的反应，分为长日照、短日照和日中性植物。

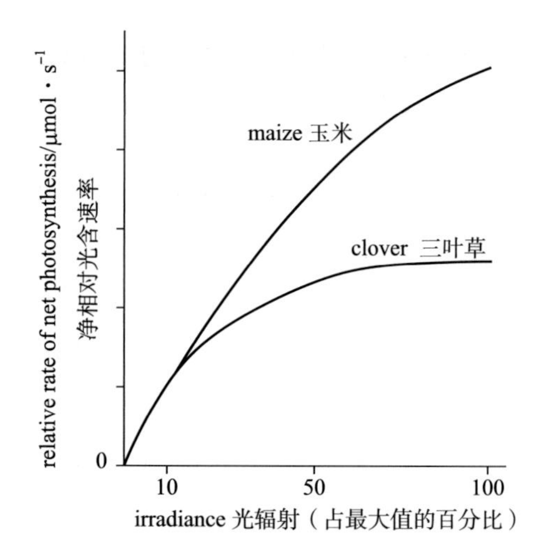

Fig. 3.1 The photosynthesis rate of clover (C_3) and maize (C_4) plants

图 3.1 C_3 和 C_4 牧草的光合速率

3.3.1.1 Long-day plants (LDP)

Long day plants are those that start blooming only within a range of relatively long photoperiods(actually short nights). These plants typically flower in the late spring or early summer as days are getting longer. When the photoperiod is less than a critical length, growth remains at vegetative stage. Artificially increasing daylength will induce flowering in advance. Barley (*Hordeum vulgare* L.), orchardgrass(*Dactylis glomerata* L.), radish(*Calathodes oxycarpa*) and perennial ryegrass(*Lolium perenne* L.) are all long day plants.

3.3.1.2 Short-day plants (SDP)

Short-day plants flowering only within a range of relatively short photoperiods (actually long nights). They cannot bloom under the long days of summer. These plants generally flowering in late summer or fall, as days are getting shorter, short-day plants will not flower if a pulse of artificial light is shined on the plant for several minutes during night/dark period; they require a consolidated period of darkness before floral development can begin(Fig. 3.2). Soybean (*Glycine max* L.), sweet potato (*Dioscorea esculenta*) and pumpkin (*Cucurbita moschata*) are short-day plants.

3.3.1.1 长日照植物

指每天的光照时数必须大于临界日长才能开花结实的植物，这类植物一般在春末夏初日照时间变长时开花，如大麦、鸭茅、多年生黑麦草等，延长光照可以促进植物提前开花结实。

3.3.1.2 短日照植物

指每天的光照时数必须短于临界日长才能开花结实的植物，这类植物一般在夏末和秋季日照时间变短时开花，如大豆、甘薯、南瓜等。这类植物要求持续的黑暗时间，如果夜间加入一束几分钟的闪光，也会阻碍植物正常开花。缩短光照可促使这类植物提前开花结实，反之则一直进行营养生长（图 3.2）。

3.3.1.3 日中性植物

对日中性植物，只要其他环境条件适宜，营养生长到一定阶段便可进入生殖生长，日照时间的长短对其生长发育无明显影响。如番茄、四季豆、蒲公英等属于日中性植物。

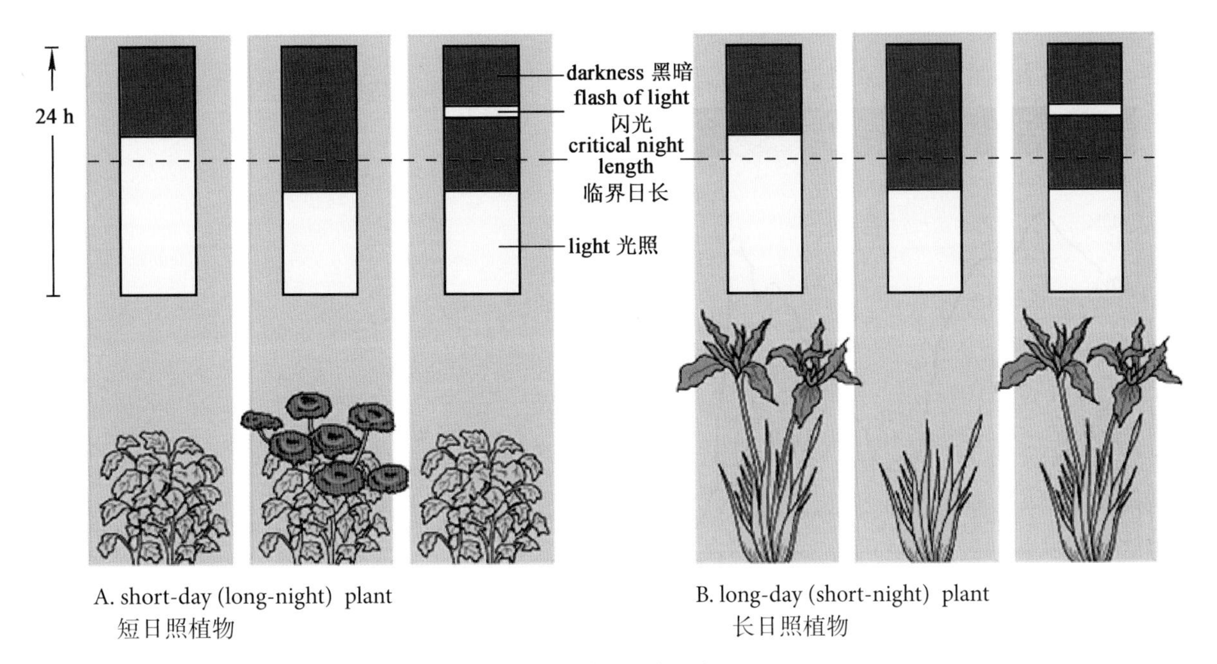

A. short-day (long-night) plant
短日照植物

B. long-day (short-night) plant
长日照植物

Fig.3.2 The photoperiodism of short-day plant and long-day plant
图 3.2 短日照植物和长日照植物的光周期现象

3.3.1.3 Day-neutral plants (NDP)

Day-neutral plants are capable of flowering under both long and short photoperiods.

Scientific experiments prove that the solar spectrum, the most important part for plant growth is visible light (400 ~ 760 nm wavelength) (Fig. 3.3).Different wavelengths of light have different effects on plant growth. For example, red light is conducive to the synthesis of carbohydrates, whereas blue light is helpful to the synthesis of proteins and organic acids. Therefore, crop quality can be improved by influencing the light quality to control the product of photosynthesis in production.

3.3.2 Temperature

Temperature is a basic environmental factor that influences forage plant metabolism. The suitable temperature is the fundamental guarantee of high yield and good quality for forages growth and development; it includes soil temperature, air temperature and body temperature of the plant. Soil temperature influences seed germination and growth, root growth and winter survival as well as growth and develop-

科学实验证明，太阳光谱中，对植物生活最重要的是可见光部分（波长在 400~760 nm）（图 3.3）。不同波长的光对植物生长有不同的影响，如红光有利于糖类的合成，蓝光有利于蛋白质和有机酸的合成。因此，生产中通过影响光质而控制光合作用的产物，可以改善作物的品质。

3.3.2 温度

温度是牧草生命活动最基本的生态因子，适宜的温度是牧草生长发育，实现高产优质的根本保证。牧草的生长包括地上部分和地下根系部分，不同生存介质中的温度对牧草生长发育的影响不同。如土壤温度、气温和植株体表本身的温度。土壤温度对牧草种子播种、根系生长以及越冬都有很大影响，从而也会影响地上部的生长发育。气温与牧草地上部生长发育有直接关系，它也间接影响土壤温度和牧草根系生

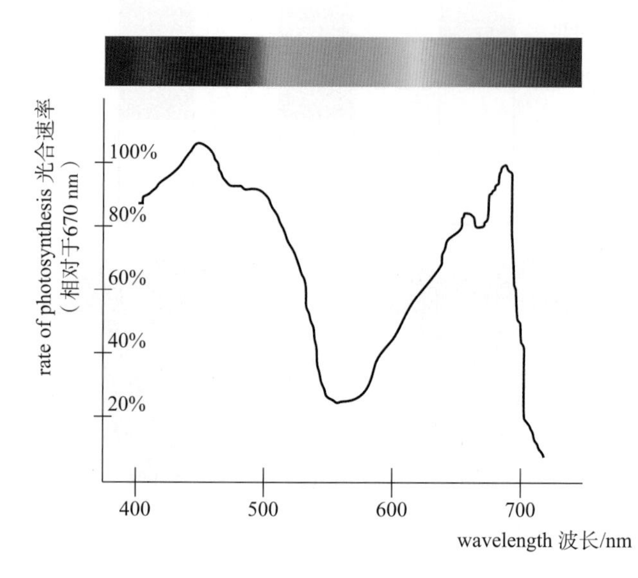

Fig.3.3 The action spectrum of photosynthesis
图 3.3 光合作用的有效光谱

ment of aboveground plant organs. Air temperature directly influences aboveground plant tissue growth and development. Root growth is influenced by air temperature through effects on aboveground growth and directly by soil temperature. Temperature directly influences physiological activities and biochemical reactions within forage plants. Soil temperature is directly related to adjacent air temperature, but differences result from the insulating mantle of vegetation and soil. The deeper of the soil layer, the greater temperature differences exist. Plant body temperature is variable. Shoots temperature is close to air temperature while roots are close to soil temperature, both of which are modified by changing environmental temperatures.

A suitable temperature range is needed to maintain forage plant survival and to ensure growth and development, which include survival temperature, growth temperature and development temperature. For most forages, the survival temperature is within -30 ~ 50℃, growth temperatures range from 5 ~ 40℃ and temperature required for development is 10 ~ 35℃. Frigid zone and temperate forages general tolerate lower temperature within above range, whereas ranges for tropical forages are higher.

长发育，是影响牧草生理活动、生化反应的基本因子。土壤热量状况和邻近气层的热状况存在着直接的依赖关系，但由于土壤、土壤覆盖层以及草地叶幕层的影响，土壤温度和气温仍有不同，而且随土层的加深两者差别加大。牧草属于变温类型，所以牧草地上部体温通常接近气温，根部温度接近地温，并随环境温度的变化而变化。

从牧草生长发育的角度划分，牧草所需温度有三种，即维持生命的温度、保证生长的温度和保证发育的温度。对大多数牧草来说，维持生命的范围一般在 -30 ~ 50℃，保证生长的温度范围为 5 ~ 40℃，而保证发育的温度在 10 ~ 35℃。一般寒带、温带牧草在此范围内偏低一些，而热带牧草则偏高一些。

一般来说，无论是牧草维持生命的温度，还是生长和发育所需温度，就其生理过程来说，都有其相应的三个基本点，即最低、最适和最高温度，简称温

Generally speaking, whether the survival temperature or the growth and development temperature, every physiological process has three basis temperature, namely, the minimum, the optimum and maximum temperature. The minimum and the maximum temperature represent the lower limit and the upper limit temperature when the plants stop growth and development. Whereas they can do best by optimum temperature. The optimum temperatures for temperate, tropical and frigid zone forages are 25～35℃, 30～35℃ and around 10℃ respectively.

Growth rates and other processes depend on the temperature pattern to which a plant is exposed, including variation not only between seasons, but also between day and night temperatures. Temperatures during the day should be near optimum for photosynthesis and growth, whereas lower temperatures at night conserve energy by reducing respiration. Temperatures variation with seasons and aboveground below ground temperatures are different. Within the normal range of growth and development conditions, day and night temperature differences may improve growth and development.

3.3.3 Water

Water is the source of all lives. Seasonal distribution patterns of precipitation, total quantity of precipitation and evaporation affect water availability and adaptation of forage species. Water is necessary for forage plant growth and development (Fig. 3.4) . Generally speaking, the importance of water for plant life activities includes two aspects of physiological function and ecological role.

3.3.3.1 The physiological functions of water

(1) Water is a major component of plasma, the plasma moisture content is about 70% ~ 90%, insufficient water will change the protoplasm from sol to gel state, cell life activities will be greatly decreased(such as seed dormancy).

(2) Water is essential for metabolism activities including photosynthesis, respiration, organic compound synthesis and utilization. Metabolism cannot happen without water. Sugars from photosynthesis often accumulate during mild to moder-

度三基点。最低温度和最高温度分别指牧草的生命、生长或发育终止时的下限和上限温度。最适温度是指牧草生命活动最适合，使其生长发育最快的温度。温度过高过低，都不利于生长。温带牧草的最适温度在 25 ~ 35℃，热带牧草在 30 ~ 35℃，寒带牧草在 10℃左右。

温度不仅随季节的变化而呈现节律性的变化，昼夜间、地上部和地下部的温度也有较大的差异。在正常的牧草生长发育温度范围以内，一般昼夜温差大，有利于牧草的生长发育和品质提高，白天高温有利于光合作用和生长，夜晚低温有助于降低呼吸消耗和保存能量。

3.3.3 水分

水乃万物生命之本。降水量的季节分布、降水总量及蒸发量影响水分的可利用性和牧草的适应性。水对牧草生长发育同样必不可少（图 3.4）。水分对植物生命活动的重要性表现在生理作用和生态作用两个方面。

3.3.3.1 水在生理方面的作用

（1）水是原生质的主要组分，原生质的含水量一般在 70% ~ 90%，如果含水量减少，原生质由溶胶变成凝胶状态，细胞生命活动大大减缓（如休眠种子）。

（2）水直接参与植物体内光合呼吸、有机物合成和分解等重要的代谢过程。在轻度和中度干旱胁迫下，由于生长减缓，光合作用合成的糖会累积。

（3）水是牧草吸收和运输无机物质和有机物质的溶剂，植物体内绝大多数生理生化过程都是在水介质中进行的。

（4）足够的土壤湿度对植物细胞保持膨压、维持固有态势非常重要。膨压与植物组织内部的水的存在状态相关，通常称为水势。

Fig. 3.4 The sprinkling irrigation system on pasture
图 3.4 人工草地喷灌系统

ate drought stress because growth is slowed.

(3) Water is a solvent that helping absorption and transport of inorganic material and organic material, most of physiological and biochemical processes are carried out in water medium.

(4) Adequate soil moisture is essential for cell turgor that drives normal growth. Turgor is related to the water status within the plant tissues, which is commonly called the water potential, the force with which water is held within a cell.

(5) Water drive the plant to absorb water and nutrients from the soil, it is also the important material maintaining CO_2 into the channel of the body of plants.

3.3.3.2 The ecological role of water

Plant temperature regulation For example, plant temperature under the scorching sun exposure can be lowered by water loss of transpiration, avoiding from high temperature burns; In addition, water can increase the air humidity, change the field microclimate, and such as irrigation in the winter can keep forage warm efficiently.

Humidity, generally expressed as air relative humidity influences forage growth and development. As relative humidity decreases, transpiration and evaporation will increase. Insufficient plant water will result in closure of stomata, reducing

（5）水是牧草从土壤中吸收水分、养分的重要动力，也是维持 CO_2 进入牧草体内的通道的重要物质。

3.3.3.2 水在生态方面的作用

水对植物体温有调节作用，如烈日暴晒下，通过蒸腾作用失水可以降低植物体温，使之免受高温灼伤；此外，水分可以增加大气湿度，改变田间小气候，如草地冬灌可使牧草保温防寒。

空气湿度，特别是空气相对湿度对牧草的生长发育有重要作用。如空气相对湿度降低使蒸腾和蒸发作用增强，甚至可引起气孔关闭，降低光合效率。若牧草根不能从土壤中吸收足够水分来补偿蒸腾损失，则会引起牧草凋零。在牧草花期，则会使柱头干燥，不利于花粉发芽，影响授粉受精。相反，如湿度过大，则不利于传粉，使花粉很快失去活力。空气相对湿度还影响牧草的呼吸作用。湿度愈大，呼吸作用愈强，对牧草正常生长发育不利。此外，空气湿度大还利于真菌、细菌的繁殖，常会引起病害的发生而间接影响牧草生长发育。

photosynthetic efficiency. If forages cannot absorb enough water from the soil to supply water needed for the transpiration, wilting occurs. During flowering, water deficiency makes stigma dry, which is harmful to pollen germination and fertilization. On the contrary, high relative humidity may also interfere with pollination, by causing the loss of pollen vigor. Air relative humidity also influences forage respiration. Higher humidity leads to greater respiration, which may be harmful to forage growth and development. What's more, high humidity promotes fungal and bacterial propagation, which may cause diseases and impact forage growth and development.

3.3.4 Soil

Soils are highly complex environments that encompass physical and chemical heterogeneity across a wide range of spatial and temporal scales. They bridge the mineral world with all the other trophic levels in the biosphere. Soil structure is central to such a fundamental linking role, as it provides the habitat for an extraordinary array of organisms and the pathway for essential resources on which they depend for their survival. Soil structure, texture, fertility and acid and alkali etc., directly affect the growth of forages.

3.3.4.1 Soil structure and texture

Soil structure is the arrangement of pore space and solids. The size, form, and strength (or persistence under degrading forces) of aggregation is used to characterize or describe soil structure. A well-structured soil will generally provide habitat for a wide range of life forms, conduit for fluids (air, water), preferential pathways for root growth. Plant growth is both strongly influenced by soil structure and is one of the factors from which soil structure arises. There are five kinds of soil structure, granular, blocky, platy, prismatic and columnar according to the size, form, and strength of aggregation. Among which granular is the best soil structure, because it provides the plant good pore space, solves the problem of coexistence of water and air and environment in soil and easy for root growth.

Soil texture describes a combination of mineral particles

3.3.4 土壤

土壤是一种广泛时空范围内包括物理和化学异质性的高度复合环境因子，桥接矿物质和生物圈中所有其他营养级。土壤结构是这个基本连接作用的核心，因为它提供了大量生物栖息地和获取植物赖以生存的关键资源的途径，土壤结构、质地、肥力和酸碱性等直接影响牧草的生长。

3.3.4.1 土壤结构与质地

土壤结构是土壤孔隙度和固体颗粒排列组合方式的特征，一般用构成土壤的聚合体的大小、形状和强度（或退化外力下的持久性）来或描述土壤结构的特征。一个结构良好的土壤通常会为各种生命形式提供栖息地、空气和水分的通道，为根系生长提供优越空间。植物生长既受土壤结构的重要影响，也是使得土壤结构形成的因素之一。根据土壤结构特点，常见的有 5 种土壤结构，分别是团粒结构、块状结构、片状结构、柱状结构和棱状结构，其中团粒结构是最理想的土壤结构，因为它为植物提供了良好的孔隙空间，解决了水和空气在土壤中的共存问题，为根系生长营造了适宜的环境。

土壤质地是指土壤中不同直径大小的矿物颗粒的组合状况，与土壤通气、保水、保肥状况及耕作的难易有密切关系。生产中常见的有黏土类、沙土类和壤土类三种，具体划分依据如图 3.5。

土壤质地是制定土壤利用、管理和改良措施的重要依据，每种质地都有自己的特性。黏土类由更多的较细黏粒构成，故土壤的黏着性好，有利于保水保肥，土壤后劲足，但透气性差，俗称“凉性土”，适宜于禾谷类作物；

of different size. It influences soil ventilation, moisture and fertilizer maintenance and the easy extent of tillage. There are three types normally, namely clay, sandy and loam and clay. The soil texture triangle is a diagram often used to figure out soil textures（Fig. 3.5）.

Soil texture provides an important basis for decision making of soil use, management and improvement measures, as each of them has their unique characteristics. Clay soils are more cohesive because they have more fine particles; they are very good at maintaining of water and fertilizer, and especially good for cereal forages. Whereas sandy soil are more loose with more coarse particles, water and fertilizer are very easy leaching out, they are more suitable for root-stock forages. Loam soil is a mixture of sand, silt and clay, and sandy clay loam is the best in landscapes, which is the optimum type for almost all the plants.

Fertile soil requires not only a good texture but also a good soil profile. The soil compaction can be described by

沙土类土壤组成颗粒大而粗糙，土质疏松，保水保肥效果差，但透气性好，更适宜于根茎类作物；壤土类是沙粒和黏粒构成比例好，适宜于各种牧草和作物生长的土壤。

肥沃的土壤不仅要求耕层的质地良好，还要求有良好的质地剖面。土壤容重（单位体积土壤的克数）是描述土壤紧实度的一个重要指标，随着土层深度增加而增加，容重越大，对植物生长越不利（图 3.6）。虽然土壤质地主要决定于成土母质类型，有相对的稳定性，但耕作层的质地仍可通过耕作、施肥等活动进行调节。

土壤质地和土壤结构直接影响土壤水分和空气量、微生物的活动等，从而影响牧草的生长发育。土壤质地对于土壤性质和肥力有极为重要的影响，而土

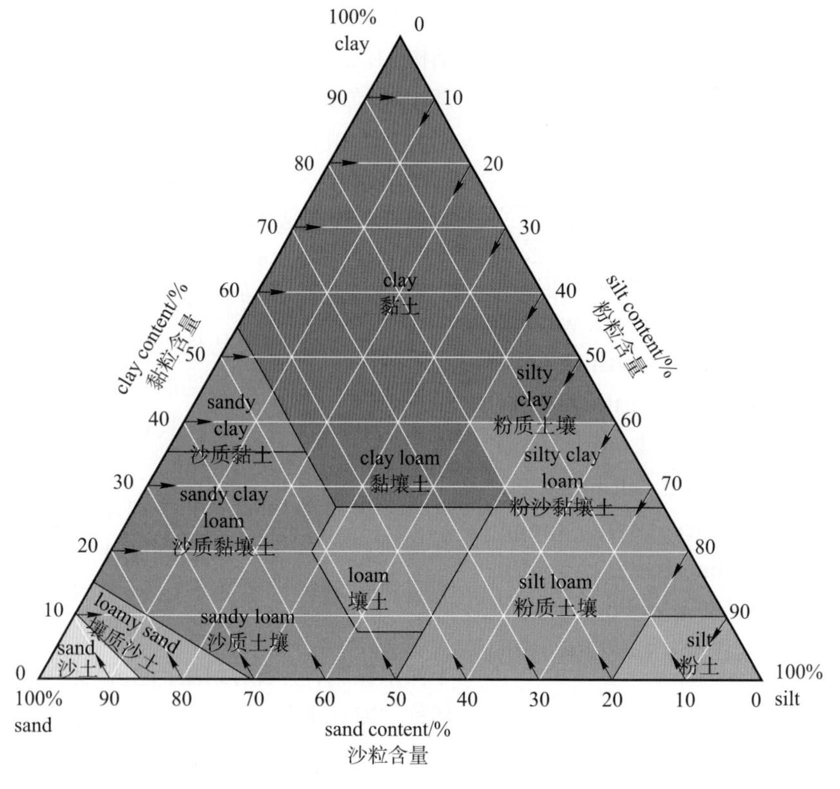

Fig. 3.5 The soil texture triangle diagram
图 3.5 土壤质地三角形示意图

Fig. 3.6 Effect of soil compaction on root growth at three different soil bulk densities
图 3.6 不同土壤容重下土壤紧实度对根系生长的影响
Low, 0.7 g/cm^3; Medium, 1.1 g/cm^3; High, 1.6 g/cm^3

soil bulk density (the soil gram per cm^3), the deeper of the soil, the more compact, which is not good for plant growth (Fig.3.6). Although soil texture is controlled by the parent material, which is stable, it can be managed with tillage, fertilization and vegetation.

Soil texture and structure influence the soil water, air and activities of microorganisms, which influence forage growth and development; but it is not the only factor that influences soil fertility. Soil fertility may be improved by increasing humus and improving the soil structure.

3.3.4.2 Soil fertility

Soil fertility is the capacity of soil to provide heat, water, nutrients, and air needed for forage growth and development. Higher soil fertility supports vigorous forage production. One task of agricultural management is to ensure soil fertility which provides adequate forage yield.

Using organic fertilizer in clay soils, for example, loosens the soil and increases permeability, resulting in higher soil O_2 and improved soil temperature, enhancing soil fertility. Another kind of fertilizer is mineral nutrients. Mineral nutrients such as C, H, O, nitrogen(N), phosphorus(P), potassium(K), calcium(Ca), magnesium(Mg), sulfur (S) etc. those are needed in large amounts are called macronutrients, among which, N, P, K are primary nutrients, Ca, Mg, S are less needed and

壤质地主要继承母质的性质，很难改变。但是，质地不是决定土壤肥力的唯一因素，因为质地不良的土壤可通过增加土壤腐殖质和改善结构性而得到改善。

3.3.4.2 土壤肥力

土壤不断提供牧草生长期间所需的热量、水分、养分、空气的综合能力称为土壤肥力。土壤肥力是土壤的基本特征，土壤肥力越高，牧草生长越茂盛。农业技术措施的中心任务之一，就是提高土壤肥力，以达到牧草生产的高产稳产。

如在黏土上多施有机肥，可疏松土壤，提高土壤通透性，增加土壤空气中的O_2，提高土壤温度，从而提高土壤肥力。另一种肥料是无机肥。根据植物对矿质营养元素需求的多少将其分为大量元素（C、H、O、N、P、K、Ca、Mg、S）和微量元素（Fe、Bo、Mn、Zn、Mo、Cu、Cl），其中N、P、K是作物需求量大而土壤又满足不了，必须依靠人工补给的元素，被称为“肥料三要素”，微量元素虽然需求量小但又必不可少（图3.7）。

be called secondary nutrients; Fe, Bo, Mn, Zn, Mo, Cu, Cl are micronutrients, as they are needed in small amounts, but essential (Fig.3.7).

Constant steeping field and salt washing, as well as planting saline tolerant plants, such as seepweed (*Suaeda glauce*) can improve soil structure, decreasing salt content, enhancing soil fertility and improving yield.

3.3.4.3 Soil pH

Soil pH directly affects the forage growth and development, every forage species grow best within a specific pH range. Normally, grasses are adapted to acid soils and legumes grow best in calcareous soils.

The microbial activities in the soil, synthesis and decomposition of organic compounds transformation and release of nutrient elements, all of them are related with soil pH, and then influence plant growth and development indirectly.

Soil pH strongly influences availability of mineral salts. Effectiveness of nutrient elements such as nitrogen(N), phosphorus(P), potassium(K), calcium(Ca), magnesium(Mg), iron(Fe), manganese(Mn), boron(B), copper(Cu), zinc(Zn) differ with various pH. The nutrient availability is various in

在盐碱地上通过不断泡田洗盐，种耐盐碱植物（如碱蓬）等可改善土壤结构，降低盐分含量，使土壤肥力不断提高。

3.3.4.3 土壤酸碱度

土壤pH直接影响牧草的生长发育，各种牧草都有其生长发育的适宜酸碱度范围。一般禾本科牧草喜中性偏酸性土壤，豆科牧草喜石灰性土壤。

土壤pH通过直接影响土壤中的微生物活动（影响养分有效性）、有机质的合成与分解、营养元素的转化与释放，从而间接影响植物的生长发育。

土壤pH对矿质盐的溶解度有重要影响。N、P、K、Ca、Mg、Fe、Mn、B、Cu、Zn等矿质元素的有效性因pH不同而不同。一般土壤呈中性或微酸性、微碱性时，养分有效性最高，对植物生长发育最有利（图3.8）。

土壤的酸碱度极易受耕作、施肥等农业措施的影响。所以，采取适当的改

Fig.3.7 Alfalfa: minus Mo (left) and plus Mo (right)
图3.7 苜蓿缺钼（左）和加钼（右）的生长对比

different pH, under neutral, slightly acidic, or slightly alkaline soil condition, the nutrients have the best effectiveness for growth and development of forage plants (Fig.3.8).

The soil pH can be easily influenced by tillage and fertilization; we often use lime in acid soils and N, P and K fertilizers in alkaline soils to regulate the soil pH.

良措施调节土壤酸碱性，是作物生产的重要技术措施。在酸性土壤上施用石灰，碱性土壤上少量多次施用 N、P、K 等便是调节土壤 pH 和考虑了不同土壤 pH 条件下养分的有效利用率而采取的措施。

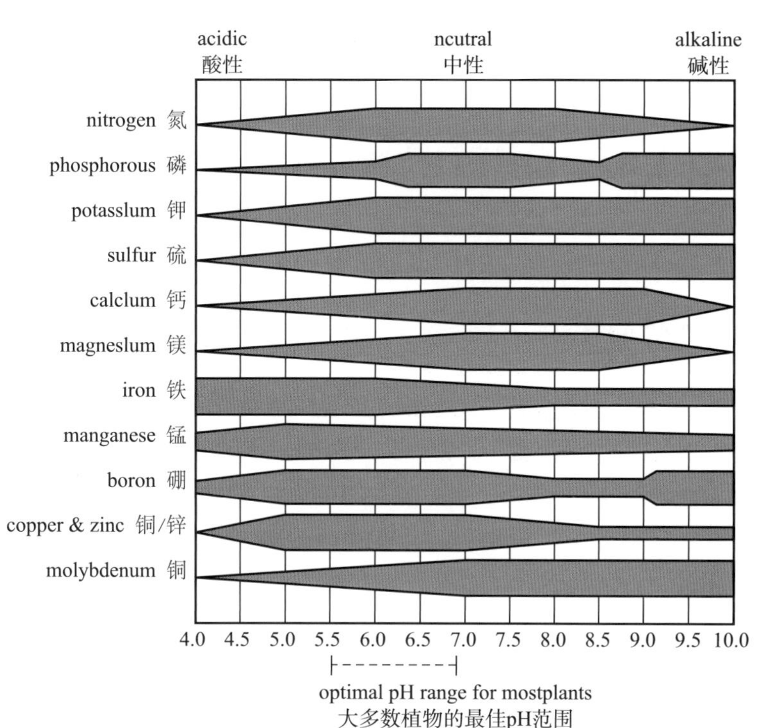

Fig.3.8 The plant nutrient availability by different soil pH
图 3.8 不同土壤酸碱度下的植物养分有效性

Artificial Grassland Establishment and Management

人工草地的建植与管护

4.1 The importance of artificial grassland 建植人工草地的意义

Artificial grassland is a kind of cultivated vegetation by applying agricultural comprehensive techniques. The Purpose is to obtain high yield and good quality forage grass, to supplement the shortage of natural grassland and meet the needs of livestock feed.

Artificial grassland is a key component of the modern livestock production systems; it plays an important role in maintaining a sustained, stable and healthy development of animal husbandry, protecting the ecologic environment and improving the level of livestock production. The proportion of artificial grassland accounting for the total grassland area is an important symbol of a country's developed degree of animal husbandry (Fig. 4.1). The area of artificial grassland accounted for 29% of total area of grassland in the United States, and accounted for 61.9% in New Zealand. According to the analysis result, every 1% increase of artificial grassland area may lead to a 4% increase of animal husbandry production in the world, while in the United States the number is more higher, nearly 10%. Artificial grassland area accounts for only about 3% of the total area of the grassland in China, and the quantity and quality are to be improved.

人工草地是采用农业技术措施栽培而成的草地，目的是获得高产优质的牧草，以补充天然草地的不足，满足家畜的饲料需要。

人工草地是现代化畜牧业生产体系中的一个关键组成部分，对维持畜牧业生产持续、稳定、健康发展，保护生态环境，提高畜牧业生产水平具有重要作用。人工草地占整个草地面积的比例大小也是衡量一个国家畜牧业发达程度的重要标志（图 4.1）。美国的人工草地（包括轮作草地）占草地总面积的 29%；新西兰的则占 61.9%。据分析，在世界范围内，人工草地占天然草地的比例每增加 1%，草地动物生产水平就增加 4%，而美国则增加近 10%。我国目前人工草地面积占草地总面积的 3% 左右，数量和质量都有待于提高。

随着人们对牧草功能认识的增强以及应用领域的扩大，人工草地概念的内涵和外延都有了较大扩展，成为集畜牧、

Fig.4.1 The ryegrass artificial grassland
图 4.1 黑麦草人工草地

With the increasingly clear awareness of forage function, and the application fields extension, the connotation and extension of the concept of artificial grassland have been greatly expanded and become a integrating of animal husbandry, ecological, economic, which integrates multi-functional, comprehensive artificial vegetation, and it shows an increasingly important role in human life.

生态、经济于一体的多功能、综合性的人工植被，并在人类生产生活中显示出越来越重要的作用。

4.2 Establishment of artificial grassland 人工草地的建植

For ease of field management, transportation and utilization, artificial grassland site generally should be located at flat, convenient transportation and close distance to residential areas and livestock shed.

为便于田间管理、运输和利用，人工草地一般应选择地势平坦、交通方便，距离居民点和牲畜棚舍较近的地段。

4.2.1 Species and cultivar selection

The intended use after establishment must be considered along with the environmental and soil factors, the biological characteristics of forages when deciding which forage species and cultivars to plant. The species or a mixture of species to plant and the cultivar(s) of each will affect production for the life of the stand.

4.2.1 种和品种的选择

在牧草种和品种的选择上，必须考虑建植人工草地的目的和与之相关的气候及土壤因素、牧草的生物学特性等，牧草的种类会影响整个草地的生产。

4.2.2 Preparing for successful establishment

Once the forage species or mixture is chosen, attention must be given to the soil environment. Seed size of most forage species is small compared with seed of traditional grain crops. This makes seedbed preparation especially critical for forage plantings. The ideal seedbed for conventional seeding should be firm, fine, moist and free from competition. Careful seedbed preparation enhances proper seed placement.

4.2.2 苗床的准备

牧草品种一旦选定，就该关注土壤环境了。与农作物种子相比，牧草种子很小，这就对苗床的要求更为严格，理想的苗床应是通过合理耕作后达到细碎平整、松紧适度、没有土块和杂草。

4.2.2.1 Tillage

Tillage practices vary but typically consist of primary tillage with a moldboard or chisel plow (Fig. 4.2), it can be done in the fall in cases where erosion is not a concern in northern area. The process of tilling a seedbed are to loosen the soil, eliminate existing vegetation, bury surface weed seeds, in corporate lime and fertilizer into the soil, and provide a

4.2.2.1 耕作

耕作方式多样，但典型的有两种，一种是以翻耕为主的表土耕作，土层上下翻动，作业幅度大（图 4.2），在北方没有水土流失的地方一般在秋季进行；翻耕可以疏松土壤，清除现有植被，翻埋杂草种子，混拌土壤与石灰、肥料，为播种和收获提供一个干净平整的地面。但对苗床进行耕翻的缺点在于土壤水分的损失很大，也增加了植被建立前土壤侵蚀的可能性。另一种是耕翻后的基本耕作，即以圆盘耙或镇压

smooth surface for seeding and harvesting operations. Disadvantages of a tilled seedbed are greater loss of soil moisture during tillage and increased potential for soil erosion until the crop is established. Another kind of tillage is secondary tillage, generally follow primary tillage, using a disc or field cultivator, usually needed to break up remaining clods and smooth the seedbed (Fig. 4.3).

机械等接近土壤表层，作业幅度小，主要用来细碎土壤，平整土壤表面（图 4.3）。

Fig. 4.2 Primary tillage
图 4.2 基本耕作

Fig. 4.3 Primary tillage—harrowing
图 4.3 基本耕作——耙地

4.2.2.2 Lime and fertilizer

Satisfactory stands and yields of forages, like crops, are obtained only when the soil is adequately limed and fertilized based on soil information. Generally, soil pH level near 7.0 can increase the availability and promotes the growth of desirable microorganisms (Fig. 3.8). Therefore, if the soil pH is lower than 5.5, lime will be needed to adjust it and also supplies calcium(Ca) and sometimes magnesium (Mg) needed for plant growth. Phosphorus (P) is particularly important for seedling growth in a new pasture. Legumes require higher levels of phosphorus than grasses. Band application improves the efficiency of fertilizer use because it is less exposed to fixation in the soil compared with broadcast P. Nitrogen is the plant nutrient required in the largest amount by grass pastures. An application of nitrogen will usually have a profound effect on pasture production. Nitrogen is generally the most limiting nutrient in forage agriculture. N inputs of soil are largely dependent on prior fixation by a legume. Compared with P and N, K is seldom a limiting factor in stand establish-

4.2.2.2 石灰处理和施肥

和农作物一样，只有在足够的土壤肥力和适宜的酸碱程度下，才能建植满意的草地和取得较高的产量。一般来说，土壤 pH 在 7.0 左右可以增加肥料的可利用率和促进理想的微生物的生长（图 3.8）。因此，如果土壤 pH 低于 5.5，需要用生石灰进行调整，同时还提供钙和植物生长所需的镁。磷对幼苗健康发育很重要，条施比撒施肥效利用率高，原因在于条施减少了化肥在空气中的暴露，增加了土壤中的固定留存。农业生产中，氮通常是最具限制性的营养成分，土壤氮的输入在很大程度上取决于前作豆类的固氮情况。与磷和氮相比，钾很少成为草地建植的限制因子，但仍然建议建植前对土壤肥力进行测定，因为饲料作物通常会从土壤中带走大量的钾。

肥料的施用方法很重要，因为施肥

ment. Still, soil testing is recommended because forage crops typically take up large amounts of soil K.Proper placement is important because direct contact of fertilizer and seed can prevent germination or kill some seedlings.

4.2.3 Seeding guidelines

Compared with the grain crops seeds, there are three special characteristics of forage seeds. Most of the forage seeds are very small and light (see 1000–grain weight in table 4.1). The appendages of gramineous seeds like glume and awn, not only affect the flow characteristics of seeding, but

不当会导致肥料直接和种子接触，阻碍种子萌发或杀死部分幼苗。

4.2.3 播种技术

与农作物相比，牧草种子具有以下三个特点：① 大多数牧草种子小而轻（表 4.1）。② 禾本科种子有的还带有稃和芒等附属物（图 4.4），不仅影响播种时的流动性，且易被风吹走，造成播种密度不匀。③ 营养物质少，大多数牧草种子贮存的营养物质只能满足种

Table 4.1 The 1000–grain weight of common forage seeds

表 4.1 主要牧草种子千粒重

forage 牧草名称	1000–grain weight/g 千粒重 /g	forage 牧草名称	1000–grain weight/g 千粒重 /g	forage 牧草名称	1000–grain weight/g 千粒重 /g
紫花苜蓿	1.4~2.3	紫云英	3.5	鸡脚草	0.97~1.34
黄花苜蓿	2.2	广布野豌豆	17.0	披碱草	4.8~5.6
山野豌豆	17.9	白花山黧豆	164.0	垂穗披碱草	2.2~2.5
普通红豆草	18~21	羊柴	9.3	老芒麦	3.5~4.9
百脉根	1~1.2	柠条	37.9	多年生黑麦草	1.5~2.0
沙打旺	1.8	达乌里胡枝子	2.2	猫尾草	0.36~0.4
小冠花	3.1~4.1	无芒雀麦	2.4~3.7	苇状羊茅	2.5
红三叶	1.5~1.8	扁穗雀麦	10~13	紫羊茅	0.7~1.0
杂三叶	0.75	高燕麦草	2.9~3.2	草芦	6.8
白三叶	0.5~0.7	草地早熟禾	0.37	偃麦草	4.1
白花草木樨	2~2.5	冰草	2.9	中间偃麦草	5.3
黄花草木樨	2.2	蒙古冰草	1.9	弯穗鹅观草	4.1
箭筈豌豆	50~60	沙生冰草	2.57	苏丹草	12.6

注：自《中国饲用植物志》1~6 卷等综合资料。

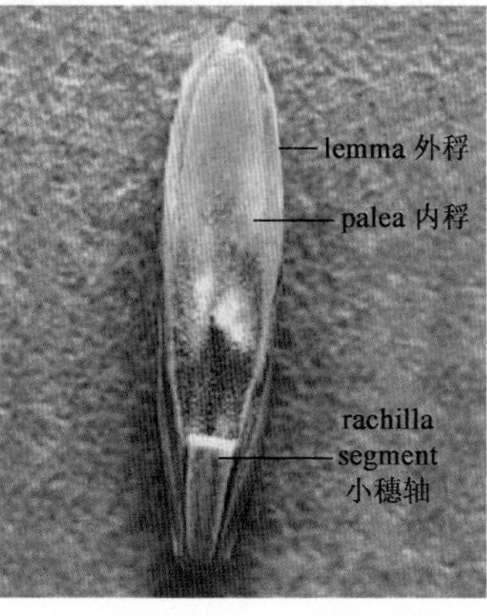

Fig.4.4 Ryegrass seed with outer appendages, enclosed by the lemma and palea

图 4.4 带附属物的黑麦草种子

also easy to be blown away and cause uneven sowing density (Fig. 4.4). The contents of nutrients less in forage seed, most of the nutrients are only utilized for seed germination. Based on the above characteristics, sowing needs more strict soil conditions (finely smooth, appropriate soil moisture) and planting techniques.

子萌发的需要。基于以上特点，播种环节对土壤条件（细碎平整，墒情适宜）及播种技术要求就更为严格。

4.2.3.1 Seed treatment

(1) Seed cleaning

To establish a good quality artificial grassland, the seeds should be pure, big granule, full developed, uniformity, strong viable, healthy, without diseases and pests. Commonly way of cleaning seeds include winnowing, water separation, etc.

(2) Removing the awn

To increase the flow characteristics of grass seeds in order to ensure the seeding quality, removing the awn first is necessary. In practices, mechanical treatment is often used. When lacking of such machines, the artificial way could be also used, for example, spreading seeds over the ground, grinding by using circular roller or stone roller, then eliminating, also can remove the awn.

(3) Dormancy seeds treatment

Many seeds cannot germinate even under ideal condition, this is called dormancy, and it is very common in forage seeds. Generally there are two types of dormancy, one is hard seeds of legumes, and another is the physiological dormancy because of the postripeness of grass seeds.

Legume hard seeds dormancy is caused by impermeability to water and gas and mechanical constraint of the seed coat. These seeds are usually called hard seeds. To ensure the sowing quality, hard seeds need to be treated before seeding. The common methods include:

① Mechanical methods like seed scarification.

② Chemical treatment, such as concentrated sulfuric acid, salt or alkali.

③ Variable temperature treatment like soaking the seeds in hot water at night and dry it in the sun during daytime.

The treatment methods for post ripeness grass seeds usually include drying in the sun for 5~6 d before seeding,

4.2.3.1 种子处理

（1）种子清选

纯净度高、粒大、饱满、整齐一致、生活力强、健康而无病虫害的种子是建植优质高产人工草地的基础。常用的种子清选方式有风选、水选等。

（2）去芒处理

播种时，为增加牧草种子的流动性，保证播种质量，必须先进行去芒处理。在生产中，可采用去芒机机械去芒，也可采用人工的办法，如将种子铺于晒场上，用环形镇压器或碾子碾压，然后筛除，也可达到去芒目的。

（3）休眠处理

牧草种子休眠是指种子在适宜的萌发条件下仍不能萌发的现象，常见的有豆科牧草种子硬实和禾本科牧草种子后熟导致的休眠两种类型。

豆科牧草种子由于种皮结构致密和具有角质层，致使种皮透水性差造成休眠，该类种子称为硬实种子。硬实种子萌发慢，有的甚至长期不萌发。为提高播种质量，播种前必须进行硬实处理。常用的处理方法有机械擦破种皮的方法，用酸、碱、盐等化学腐蚀处理的方法和温汤浸种方法。具体方法因种而异。

禾本科牧草后熟种子处理方法有：播前晒种（5~6 d）、发芽环境热温处理（30~40℃）、湿润沙藏（1~4℃冷藏或 12~14℃热藏），以上方法均可以缩短种子休眠期，促进萌发。

（4）根瘤菌接种

根瘤菌即是寄生在豆科植物根部根瘤中可以固氮的一类微生物，根瘤菌接种是指播种前将特定根瘤菌接种到与其有

heating the environment temperature(30~40℃) and burying seeds in the wet sand(cool hiding 1~4 ℃ or hot hiding 12~14℃). By these means, the dormancy period can be shorten and promoting the seed germination progress.

(4) Seed inoculation

Rhizobium is a kind of bacteria in root nodule of leguminous plants. Rhizobium inoculation refers to the introduction of the appropriate rhizobium to the legume seed prior to planting. By inoculation, the soil nitrogen and soil fertility can be increased, therefore promoting the yield and improve the quality of the legumes. Inoculation at planting time is done by adding a peat-based inoculum as a powder to moistened seed and then mixing thoroughly.

Inoculation methods: Inoculation can be done through two ways: ① Using the right commercial rhizobium bacteria in accordance with the instructions. ② Making rhizobium bacteria by hand, collecting living nodules from the root of the same (group) legumes plant or the soil, seeding after inoculation.

Note: ① The nodule bacteria is a kind of microbe, it is light avoidance and should be kept in a cool, dark and ventilated place. ② Don't place too long after inoculation. ③ The inoculation seeds can't contact with lime or high concentration fertilizer. ④ Combination between the correct rhizombium of the same group makes good inoculation effect (Table 4.2).

共生关系的豆科植物种子上的方法。通过接种可以增加土壤氮素，提高土壤肥力，从而提高豆科牧草的产量并改善其品质。

接种条件：① 适宜的土壤湿度，一般应保持在田间持水量的 60%~80%。② 通气性好，土壤空气中含氧量 15%~20%。③ 酸碱度适宜。多数根瘤菌适合中性或微碱性土壤，可适应的 pH 为 5.0~8.0。④ 适量无机氮。少量无机氮可以促进根瘤的形成，过多（1 hm^2 纯氮超过 37.5~45.0 kg）则会阻碍。⑤ 磷、钾、钙、镁、硼等微肥也有利于根瘤菌侵染。

接种方法：① 可以购买商业根瘤菌剂按说明操作。② 也可以播前从别处已种植该牧草的植株根系或土壤中获取根瘤，自制菌株接种后播种。

接种注意事项：① 根瘤菌是一种微生物，具有怕光、怕化学物等特性，接种时应在阴凉通风处进行。② 接种后不能放置时间太长。③ 已接种的种子不能与生石灰或高浓度化肥接触。④ 根瘤菌与豆科植物之间的共生关系是非常专一的，最好是用相同族类的根瘤菌接种（表 4.2）。

Table 4.2 Rhizobium groups of cross-inoculation

rhizobium bacteria group	the symbiotic legumes
Medicago	alfalfa (*Medicago sativa* L.), California burclover(*Medicago polymorpha* L.), black medic (*Medicago lupulina* L.), sweet clover(*Melilotus suaveolens* Ledeb.).
Trifolium Pisum	red clover (*Trifolium pratense* L.), white clove (*Trrifolium repens* L.), alsike clove (*Trifolium hybridum* L.)
Lupine	pea (*Pisum sativum* Linn.), chickling vetch［*Lathyrus quinquenervius* (Miq.) Litv.］, Common vetch(*Vicia sitiva* L.), hairy vetch(*Vicia villosa* Roth), broad bean(*Vicia faba* Linn.) , hyacinth bean［*Lablab purpureus* (L.) Sweet］
Lupinus	lupinus luteus(*Lupinus luteus* Linn.), white lupin(*Lupinus albus* Linn.), *lupinus angustifolius* L.

续表

rhizobium bacteria group	the symbiotic legumes
Phaseolus	kidney bean（*Phaseolus vulgaris* Linn.）, scarlet Runner（*Phaseolus coccineus* L.）
Glycine	soybean［*Glycine max* (Linn.) Merr.］, white bean［*Lablab purpureus* (L.) sweet］, black soybean（*Glycine max*）
Vigna	cowpea［*Vigna unguiculata* (L.) Walp］, sword bean［*Canavalia gladiata* (Jacq.) DC.］, *Inddigofera bungeana* Steud., mung bean［*Vigna radiata* (Linn.) Wilczek］, peanut（*Arachis hypogaea* L.）, lespedeza（*Lespedeza bicolor* Turcz）, etc.
Astragalus	milk vetch（*Astragalus sinicus* L.）

表 4.2 根瘤菌互接种族对照表

根瘤菌族	所共生的豆科植物
苜蓿族	苜蓿、金花菜、天蓝苜蓿、草木樨
三叶草族	红三叶、白三叶、杂三叶
豌豆族	豌豆、山黎豆、春苕子、冬苕子、蚕豆、扁豆
羽扇豆族	黄羽扇豆、白羽扇豆、蓝羽扇豆
菜豆族	菜豆、红花菜豆
大豆族	大豆（白豆、黑豆）
豇豆族	豇豆、刀豆、铁扫帚、绿豆、花生、胡枝子等
紫云英族	紫云英

Seed Inoculation is recommended in the following cases: poor soil condition; a particular legume species planted on a certain fields for the first time; the same kind of legume planted once again in the same field 4~5 years later. It is noteworthy that the symbiotic relationship between legume and rhizobium is unique.

在下列情况下建议最好接种根瘤菌：土壤条件不良（酸碱性太强、过于贫瘠、干旱等）；在某一块地上首次种植该豆科牧草；同一种豆科牧草相隔 4~5 年以后再次种植于同一块土地上。需注意豆科作物与根瘤菌共生关系的专一性。

4.2.3.2 Time of planting

4.2.3.2 播种期

Forage seeding are usually timed to coincide with periods of adequate rainfall and temperatures to help ensure successful stand establishment, temperature is the most important factor that decide the seeding time. In addition, the occurrence law of weeds should also be taken into account. In the North of China, these conditions typically occur in early

为确保草地成功建植，牧草播种期通常要和适宜的温度和充足的土壤墒情相吻合。此外，还要考虑到杂草的发生规律及危害程度，其中温度是第一位的。我国北方地区一般在春、夏、秋三个季节都可以播种。

spring and/or late summer to autumn.

(1) Spring planting

The advantages of spring seeding include: ① Near optimum temperatures for germination and early seedling growth. ② The ability to harvest the crop in the seeding year.

Disadvantages include: ① The increased risk of moisture stress will limit establishment. ② Heavy weed pressure compared with late summer seeding. ③ Increased risk that late frosts will damage young seedlings.

Spring plantings are typically done as early as possible, beginning in February in southern China and May in northern parts.

Frost is less of a concern with grasses than with legumes due to differences in their emergence patterns. Grasses exhibit hypogeal emergence, if exposed leaves of grass seedlings are damaged by frost, they can regrow from meristematic tissues located beneath the soil surface. Generally, the danger of frost has passed by the time the growing points emerge.

Legumes, on the other hand, usually have epigeal emergence. Late frost can kill these seedlings because their growing points emerge early in the process. Companion crops slow heat loss at night and help avoid frost injury to legume seedlings.

(2) Late summer/autumn seeding

Late summer seeding occur from August in the northern China to early November in southern areas, it varies from areas.

Disadvantages include: ① The typical loss of seeding-year forage harvest due to the short growing period before winter. ② The possibility of winter injury if plants do not become well established before a killing frost.

4.2.3.3 Seeding depth

Seeding depth is one of the key factors for the successful pasture establishment. The main factors affecting the seeding depth include forage species, seed size, soil moisture, soil type, etc. Forage seed are usually small and have limited food reserves, so they are not planted as deep as grain crop seed. Most forage seedlings are not able to reach the surface if they are planted deeper than 3 cm deep. Seedlings can emerge

（1）春播

春播的优点：① 随着气温逐渐回升，越来越接近种子萌发的最适温度，对幼苗生长有利。② 可以在播种当年收获一茬牧草或饲料作物。

春播的缺点：① 北方春季气温回升快，降雨量少，蒸发量大，土壤墒情难以保障，会给种子萌发和幼苗生长带来不利影响。② 春季杂草多。③ 晚霜有可能会对幼苗生长不利。

与豆科牧草相比，由于出苗的方式不同，禾本科牧草受到春季晚霜危害的可能性也较小。禾本科牧草属于子叶留土类型，如果地上部幼苗叶片被霜冻危害，它们仍可以通过地表以下的分生组织进行再生。一般等到生长点露出地面，晚霜的危险也已经过去了。

而对于豆科牧草，一般是子叶出土类型，之所以容易受到霜冻危害，是因为其生长点早早冒出地面。保护作物可以降低夜间的热量损失，避免豆科牧草的幼苗受到伤害。

（2）晚夏 / 秋播

晚夏和秋播一般从北方的 8 月到南方的 11 月初，因地区而异。夏秋播种的优点是：土壤墒情好，水热条件好，杂草少。但缺点是：① 由于冬前生长点较低，幼苗生长缓慢，播种当年不能或少收获牧草。② 如果播种过晚，幼苗会受到霜冻的危害。

4.2.3.3 播种深度

牧草播种深度是种植牧草成败的关键因素之一，影响播种深度的因素主要有牧草种类、种子大小、土壤墒情、土壤类型等。牧草种子通常很小，贮存的营养物质有限，以浅播为宜。但在质地粗糙及沙土、干燥的土壤或晚春播种，需要增加播种深度。

一般来说，小粒种子以 2 ~ 3 cm

from greater depths in coarser-textured, sandy soils. Greater planting depths may also be needed on dry soils or in late spring to help ensure that see have adequate moisture.

Generally, forage seed should be planted shallower. The optimum seeding depth for small grass seed should be 2~3 cm deep and big seeds by 3~ 5 cm deep. Seeding depth of legumes are usually shallower than that of grasses. In short, the smaller of the seed size, the shallower seeding depth; the

为宜，大粒种子以 3 ~ 5 cm 为宜；豆科牧草子叶顶土出苗困难，宜浅播。总之，大粒种子可深播，小粒种子宜浅播；干燥土壤稍深，潮湿土壤可浅；土壤疏松可稍深，黏重土壤则宜浅。

常见牧草的播种深度见表 4.3。

Table 4.3 Seeding depth and seeding rates of common forages

表 4.3 常见牧草的播种深度和播种量

forage 牧草名称	seeding rate 播种量 / ($kg \cdot hm^{-2}$)	seeding depth 覆土深度 /cm	forage 牧草名称	seeding rate 播种量 / ($kg \cdot hm^{-2}$)	seeding depth 覆土深度 /cm
黑麦草	15~22.5	2~3	短芒大麦草	7.5~15	2~3
多花黑麦草	15~22.5	2~3	布顿大麦草	11.25~15	3~4
鸭茅	7.5~15	2~3	高燕麦草	45~75	3~4
苇状羊茅	15~30	2~3	草地早熟禾	7.5~15	2~3
羊茅	37.5~45	2~3	扁秆早熟禾	7.5~12	2~3
紫羊茅	7.5~15	1~2	普通早熟禾	12~15	2~3
草地羊茅	15~18	1~2	加拿大早熟禾	12~15	2~3
扁穗雀麦	22.5~30	3~5	碱茅	7.5~15	1~2
无芒雀麦	22.5~30	4~5	朝鲜碱茅	30~37.5	0.5~1
草芦	22.5~30	3~5	苏丹草	22.5~30	2~3
球茎草卢	7.5~15	3~4	非洲狗尾草	7.5~15	2~3
宽叶雀稗	15~22.5	2~3	大翼豆	7.5~12	2~3
毛花雀稗	15~22.5	2~3	鸡眼草	7.5~15	2~3
巴哈雀稗	11~16	2~3	白三叶	3.75~7.5	2~3
狗牙根	6~12	2~3	红三叶	9~15	2~3
羊草	30~52.5	2~3	绛三叶	12~19.5	2~3
老芒麦	22.5~30	2~3	杂三叶	6~7.5	2~3
披碱草	15~30	2~3	野火球	15~22.5	3~4
肥披碱草	22.5~30	2~3	草莓三叶草	15~22.5	2~3
垂穗披碱草	15~22.5	2~3	地三叶	19.5~24	2~3

续表

forage 牧草名称	seeding rate 播种量 / (kg · hm^{-2})	seeding depth 覆土深度 /cm	forage 牧草名称	seeding rate 播种量 / (kg · hm^{-2})	seeding depth 覆土深度 /cm
冰草	15~22.5	3~4	花棒	9~18	3~5
蒙古冰草	22.5~30	3~4	埃及三叶草	12~15	2~3
沙生冰草	11.25~22.5	3~4	百脉根	6~12	2~3
偃麦草	22.5~30	2~3	绿叶山蚂蝗	2.25~3.75	2~3
中间偃麦草	15~22.5	2~3	银叶山蚂蝗	3.75~7.5	2~3
弯穗鹅观草	35~45	3~4	银合欢	15~22.5	3~4
纤毛鹅观草	22.5~30	3~5	葛藤	3.75~4.5	3~5
猫尾草	7.5~12	1~2	紫花苜蓿	15~22.5	2~4
大看麦娘	15~30	2~3	黄花苜蓿	15~22.5	2~4
苇状看麦娘	18.75~22.5	1~2	金花菜	15~22.5	2~4
毛苕子	45~75	4~5	箭舌豌豆	60~75	4~5
山野豌豆	45~75	4~5	柠条	7.5~10.5	2~3
紫云英	30~60	3~4	中间锦鸡儿	7.5~10.5	2~3
鹰嘴紫云英	11.25~18.75	2~3	小叶锦鸡儿	6~9	2~3
沙打旺	3.75~7.5	2~3	蓝花棘豆	15~30	2~3
草木樨状黄芪	7.5~12	2~3	二色胡枝子	7.5~15	3~4
红豆草	45~90	2~4	截叶胡枝子	9~30	3~4
外高加索红豆草	60~90	2~4	达乌里胡枝子	6~7.5	2~3
沙地红豆草	45~60	2~4	细叶胡枝子	15~22.5	3~4
扁蓿豆	30~37.5	1~2	山黧豆	60~75	3~5
黄花羽扇豆	150~200	3~5	小冠花	4.5~7.5	1~2
柱花草	1.5~3	2~3	菊苣	2.25~3	2~3
矮柱花草	1.5~3	2~3	串叶松香草	3.75~7.5	3
圭亚那柱花草	6~9	2~3	优若藜	3.75~7.5	2~3
白花草木樨	15~22.5	2~4	伏地肤	22.5~30	2~3
黄花草木樨	15~22.5	2~4	冷蒿	3~4.5	0.5
细齿草木樨	15~18	2~4	白沙蒿	3~4.5	0.5
羊柴	30~45	3~5	伊犁蒿	3~4.5	0.5

seed should be planted shallower in wet soil than in dry soil and shallower in clay soil than in loosen soil. The seeding depths of common forage species are showed in Table 4.3.

4.2.3.4 Seeding rates

Seeding rates vary widely because the size and weight of individual seeds of forage species vary. Rate recommendations are also influenced by local environmental and soil conditions. Specific seeding rates depended on many other factors, including: soil type, seedbed condition and seeding method, seed size and seed quality.

Lower rates can be used on light, sandy soils and well-prepared seedbeds. Seeding rates must be increased for seed lots with lower germination or more inert material. To take these factors into account, seeding rates should be based on pure live seed (PLS). Pure live seed and actual seeding rate can be calculated by the following formulas:

Pure live seed (PLS)(%) = seed purity (%) × germination rate (%)

The actual seeding rate = the recommended seeding rate / PLS

4.2.3.5 Seeding methods

(1) Conventional seeding

Conventional seeding use preplanting tillage to control weeds and provide a smooth, firm seedbed. Tillage loosens the soil, kills existing weeds, buries weed seeds below the surface, and allows incorporation of lime and fertilizers. The common seeding methods include:

Broadcast seeding: Broadcast seeding is a broad term that refers to many seeding techniques, all of which spread the seed evenly on the soil surface; it is commonly used in lawn establishment.

Drill seeding: Drill seeding is the most popular method for pasture establishment, the whole process include furrowing, seeding and covering can be completed by one time by mechanical way, with a certain row space. Generally the row space for herbage production range from 15~30 cm, and seed production for 45~60 cm.

Band seeding: Band seeding consists of a specially equipped drill to place a band of fertilizer 4~6 cm deep in the

4.2.3.4 播种量

由于牧草种子的大小差异很大，播种量也变化很大。播种量多少主要取决于当地的环境条件和土壤条件、苗床状况、播种方法、种子质量及利用方式等因素。

在沙地和苗床很好的情况下，可以减少播种量，但对于发芽率低或含有遗传惰性物质的种子，生产中应该加大播种量。考虑到以上因素，播种量一般应在种子用价（即种子净度和发芽率的乘积）基础上进行折算。种子用价和实际播种量可以通过以下公式计算：

种子用价（%）= 种子净度（%）× 发芽率（%）

实际播种量 = 理论播种量 / 种子用价

4.2.3.5 播种方式

（1）常规播种

常规播种是指在播前进行耕作以疏松土壤、控制杂草、混拌土壤和肥料，提供一个表面平整、松紧适度的理想苗床。常见的播种方式有：

撒播：撒播是一个广义名词，指将所有种子均匀分布在地表的播种方式，草坪建植中常采用这种方法。

条播：草地建植中普遍采用的一种方式，尤其机械播种多采用这种方法，是以一定行距一次性完成开沟、播种、覆土的播种方式。一般收获牧草行距为 15~30 cm，生产种子为 45~60 cm。

带肥播种：播种的同时，将肥料条施在种子下 4~6 cm 处的播种方式（图 4.5）。

穴播或点播：对于单株占地面积较大的大株牧草饲料作物，可以采用穴播的方式，株行距因植物种类而异。

育苗移栽：对有些不产生种子的牧草，一般采用营养繁殖，即用母株的地上匍匐茎或地下根茎、根颈和枝

soil (Fig.4.5).

Sprigging: Some forages do not produce viable seed, so this species is usually established by planting vegetative material, called sprigging. Sprig diggers are used to harvest the mixture of rhizomes, stolons, plant crowns, and shoot material that serve as the source of sprigs.

条等进行繁殖。

（2）保护播种

保护播种是指多年生牧草在一年生作物的保护下进行播种的方式（图 4.6）。一般情况下，多年生牧草苗期生长缓慢，快速生长的一年生作物可以迅速覆盖地面，减少杂草竞争，降低风蚀和水蚀。同时，保护作物也可以收获利用其籽粒，或做青贮。

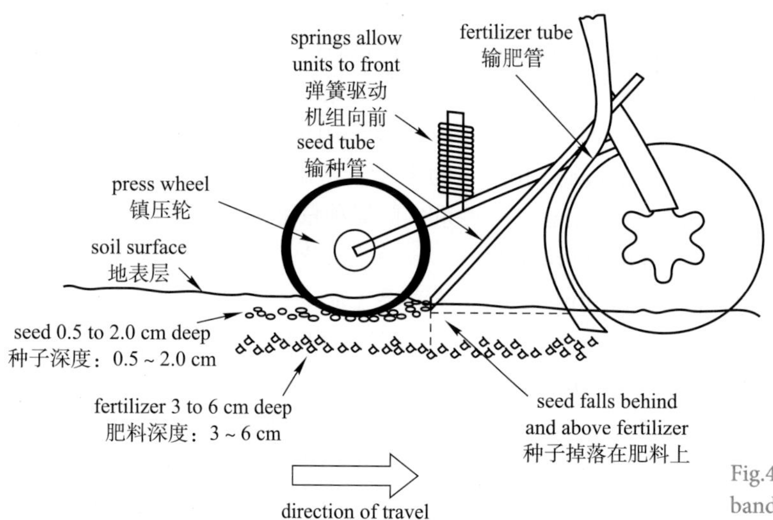

Fig.4.5 Schematic for converting a grain drill to a band seeder for forages
图 4.5 带肥播种原理示意图

(2) Companion seeding

A companion crop (also called cover crop or nurse crop) of small grain or a small grain-pea mixture is often used with spring seedlings of legumes and grasses in northern region of China to provide quick cover than the perennial forage seedlings alone, help reduce wind and water erosion, and deter weed invasion during forage establishment(Fig.4.6). It also provides a usable crop for grain, bedding, silage, or pasture.

The companion crop should have these three features: (a) less tillers in order to prevent shading of pasture; (b) mature early in order to shorten the symbiotic growth period with forages; (c) faster development in early stage to reduce weed competition. The commonly used companion crops include wheat, barley, oats, peas, etc. Theoretically, any spring-seeded small grain can be used as a companion crop, but spring oat is the most ideal one.

保护作物应具备的特点：一是分蘖要少，以防止对牧草的遮阴；二是成熟要早，以缩短与牧草的共生期；三是初期发育要快，以减少杂草的竞争。常用的保护作物有小麦、大麦、燕麦、豌豆等。理论上讲，任何春播的谷类作物都可以作为保护作物，但其中燕麦是最理想的。

（3）少耕和免耕播种

少耕和免耕比铧式犁翻耕法减少了对土壤苗床的干扰。少耕技术通过适度耕翻可以使种子和土壤良好接触，而留在土壤表面的残茬可以减少径流

Fig.4.6 Companion seeding of alfalfa and spring oat
图 4.6 春燕麦和苜蓿的保护播种

(3) Reduced tillage and no-till seedings

Reduced tillage and No-till creates less soil disturbance than moldboard plowing and seedbed preparation. Tillage sufficient to provide needed seed-soil contact is done, yet plant residuals are left on the surface to impede runoff and control erosion. These methods are recommended on sloping soils where erosion is a concern.

Equipment now available allows forage stands be established without any primary tillage. No-till drills use coulters preceding the seed openers to cut through residues or sod. This method usually depend on herbicides to limit growth of existing vegetation during establishment. Incorporation is not possible, so lime and P should be top dressed 1 year or more prior to no-till seeding.

(4) Frost seeding

Frost seeding is an method that works well with some species. Frost seeding is done by distributing seed directly onto the frozen soil surface in late winter or very early spring. Expansion and contraction associated with freeze/thaw cycles incorporates the seed into the surface layer.

Species vary widely in their ability to establish by frost seeding, the most successful were red clover, orchardgrass and perennial ryegrass, and the least successful were reed

和土壤侵蚀，尤其适用于容易发生侵蚀的坡耕地上。

现在已有可以不通过任何基本耕作直接进行草地建植的机械设备，免耕播种机是在种子掉落口之前由机器所带犁刀划开地表留茬和草皮播种的方法，这种方法通常有赖于用除草剂抑制现有植被的生长，但缺点是不能起到混拌土壤和肥料的功效，所以石灰和磷肥应该在免耕播种前一年或更长时间施入土壤。

（4）顶凌播种

顶凌播种是一种适用于北方春季部分牧草的抢墒播种方法，即于晚冬或早春将牧草种子直接播于冻结土壤的表面，借助昼夜温差的冻融交替循环使得土壤膨胀收缩，从而将种子播入土壤。

canarygrass and smooth bromegrass. Birdsfoot trefoil and timothy were intermediate.

4.2.3.6 Mixture seeding of forages

Compare to monoculture planting (only one species on one paddock), mixture seeding refers to that more than two kinds of forages are seeded on the same field at the same time, this is the most common way in pasture establishment and legume-grasses mixture is the best combination.

(1) The advantages of forages mix seeding

① Legume-grass mixtures can promote the yield by nearly 14% than that of pure stands, and improve the forage quality because of its high protein concentrations and digestibility.

② Less nitrogen needed. Legumes fix atmospheric nitrogen (N_2), reducing the need for N fertilizer. The experience showed that if the legume-grass mixtures with more than 30% legume should not receive nitrogen fertilizer.

③ Reduce the risk of forage production. Leguminous forage can benefit the soil obviously, and can effectively reduce the risk of diseases and pests, establishing faster, maintaining the stability and extending the life of the mixture.

④ Good for harvesting and processing. By mix seeding of upright and prostrate or twining type of legume-grasses, the yield loss when harvesting can be reduced and more convenient to cut.

⑤ Continues to provide high quality forage grass. The growth rate varies by different species, a good mixture can provide high quality forage output and satisfy the animals at different periods all the year.

4.2.3.6 牧草的混播

混播是指两种或两种以上的牧草在同一块地上同时播种的方式。除了种子生产采用单播外，以收草或放牧为主的人工草地一般采用混播，生产中常见的是豆科和禾本科牧草的搭配。

（1）混播的优点

① 提高产量和品质。混播组合比单播禾本科牧草有更高的粗蛋白质含量。

② 减少施氮量，培肥地力。豆科牧草固氮可以供禾本科牧草及整片草地的利用，经验证明，如果混播群落中豆科牧草的比例超过 30%，就无需再施用氮肥。

③ 降低生产风险。豆科牧草有培肥地力的明显作用，同时可有效降低单一作物及草地病虫害风险，能更快地形成草地，并能保持草地的稳定性，延长草地寿命。

④ 有利于收获和加工。直立型与匍匐型或缠绕型的豆禾混播，减少损失，方便刈割。

⑤ 持续提供高产优质饲草。不同的牧草生长速度各异，科学混播可以满足不同时间段都有优质牧草产出。

（2）牧草混播组分及配比

① 牧草混播组合选择

如何选择适宜的混播组合，是一

Table 4.4 The proportion of forage mixture group member for different utilization period

表 4.4 不同草地利用年限的牧草混播组分比例

pasture utilization period 草地利用年限 / 年	legumes 豆科 /%	grasses 禾本科%	among grasses 在禾本科作物中	
			rhizome and rhizome-bunch grass 根茎与根茎－疏丛型禾草 /%	bunch grass 疏丛型禾草 /%
short term 短期草地 (2~3)	65~75	35~25	0	100
midterm period 中期草地 (4~7)	25~20	75~80	10~25	90~75
long term 长期草地 (over 8~10)	8~10	92~90	50~75	50~25

Table 4.5 The proportion of forage mixture group member for different utilization ways

表 4.5 不同利用方式的牧草混播组分比例

utilization ways 利用方式	top grass type 上繁草 /%	bottom grass type 下繁草 /%
cutting 刈割	90~100	10~0
grazing 放牧	25~30	75~70
cutting and grazing 刈牧兼用	50~70	50~30

(2) Components and proportion of mixed seeding

① Species combination

It is a complex problem on how to choose the appropriate mixed seeding combination. Generally speaking, besides the rule of Legume-grass mixtures, the dependable factors should be considered include local climate and soil condition, good adaptability, the duration (Table 4.4) and purpose (Table 4.5) of the pasture, and the compatibility between different species.

② Principle of mixed seeding

Mixtures should follow these principles: Morphology complementation (such as top grass and bottom grass, broad leaves type and narrow type, compact and flat type, deep root and shallow root). Characteristics of growth (life span, growth rate, etc.) and contents of nutrient (legume and grass) also should be complementary.

③ Proportion of mixed seeding

Generally, the seeding rate of mixed seeding is larger than that of single-sowing. The seeding rate of each group member should be accounted for 70% ~ 80% compare with monoculture for two kinds of mixed seeding, if the mixture has three members, the seeding rate of coordinal two forages should be 35% ~ 40% respectively, and another forage should be 70% ~ 80%. Generally, the proportion of leguminous forage should be less for long term pasture.

Generally, 2 ~ 3 forage species for short term pasture; 3 ~ 5 for moderate term and no more than 5 for long-term pasture.

(3) Mix seeding

Appropriate seeding time is depended on the biological characteristics of forage, soil and climatic conditions. Because le-

个比较复杂的问题。一般根据当地的气候和土壤等生态条件选择适应性好的牧草品种，同时还要考虑草地利用年限（表 4.4）、利用方式（表 4.5）和品种间的相容性，特别应做到豆科牧草和禾本科牧草的混播。

② 混播组合的选配原则

在符合播种材料选择的基础上，还应遵循如下原则：形态学互补（上繁草与下繁草、宽叶型与窄叶型、紧凑型与平展型、深根系与浅根系）、生长习性互补（生长年限、生长速度等）和营养互补（豆科与禾本科）。

③ 掌握好混播牧草的组合比例

混播牧草的播种量比单播要大一些，如两种牧草混播，则每种草的种子用量应占到其单播量的 70% ~ 80%，三种牧草混播则同科的两种应分别占 35% ~ 40%，另外一种要用其单播量的 70% ~ 80%。利用年限长的混播草地，豆科牧草的比例应少一些，以保证有效的地面覆盖。

通常利用 2 ~ 3 年的草地，混播草种 2 ~ 3 种为宜；利用 4 ~ 6 年的草地，3 ~ 5 种为宜；长期利用的则不超过 5 种。

（3）混播技术

根据牧草的生物学特性及土壤、气候条件决定适宜的播种期。如禾本科与豆科牧草同为冬性或春性牧草，可以在秋季或春季同时播种，否则应分别秋播和春播。由于禾本科牧草苗期生长较弱，

gumes is more competitive to grasses in seedling stage, so grasses are often seeded in autumn and legumes in the next spring.

The mixture can be drill seeded in the same rows simultaneously, or interplanting successively, which is depend on the species and the establishment purpose.

(4) Mixture management

Good management practices are just as important as proper establishment techniques. Pasture establishment involves a considerable investment and returns depend on how efficiently the pasture can be managed and converted into milk or meat.

① Weed control

Weed control is critical to the development of long-lived and productive forage stands. Small seeded species generally are not strong competitors with weeds, and weeds can reduce the productivity of the sown pastures particularly during the establishment year. Therefore: Control weeds during the first year by either hand weeding or by use of herbicide. In subsequent years, keep fields clean by slashing, hand pulling or mowing of weeds.

② Cutting and grazing management

About 3 ~ 4 months after planting during the seeding year, when mixtures reach early flowring stage, it is advisable to make hay rather than graze the pastures to avoid the risk of the animals pulling out the young shoots. If grazing must be done during the establishment year, it should be light enough to enable the plant to establish firmly in the soil. For maximum benefits, utilization of the pasture no later than the early flowering stage. Graze or cut at interval of 4 to 6 weeks leaving stubble height of 5 cm.

New seedlings are also subject to winter damage. Pasturing and clipping new seedlings should end 4~6 weeks before a killing frost to allow plants to harden for winter. Less winter damage will occur with moderate grazing just before or after frost, after plants are hardy, than by continuous grazing to freeze-up.

③ Irrigation and fertility management

To attach maximum production the grass requires additional nutrients in the form of inorganic fertilizer or farmyard

易受豆科牧草抑制，可以秋播禾本科牧草而在第二年春播豆科牧草。如无芒雀麦与苜蓿混播，适宜秋播无芒雀麦，第二年早春播种苜蓿。

混播方法既可以采用同行条播，也可采用间行条播，或分先后分期播种，这要取决于所选牧草的种类。

（4）混播草地的管理

好的草地管理技术与其建植技术同等重要。混播草地管理的核心是保持混播群落的产量和持久稳定性。

① 杂草控制

对长期、高产的草地，牧草幼苗无法与杂草竞争，尤其播种当年会降低草地的生产力。所以，播种当年应通过人工或除草剂的方式控制杂草；接下来的年份应通过打碎清理、手工拔除或刈割的方式清除杂草。

② 收获和放牧管理

播种当年种植 3 ~ 4 个月后，牧草达到初花期时，建议制作干草而不是放牧，这样可以避免动物将幼小的地上植株连根拔除。如果一定要在播种当年放牧，则应该尽可能轻度放牧，确保幼苗能稳固扎根于土壤，为保证饲草质量，利用时间不宜晚于初花期。放牧或刈割间隔时间为 4 ~ 6 周，留茬 5 cm 左右。

幼苗也容易受到冻害，霜冻前 4 ~ 6 周停止对草地的利用，以确保幼苗正常越冬。霜冻前后的适度放牧比频繁放牧更能降低牧草受到冻害的危害。

③ 灌溉和施肥

为达到最高产量，草地需要无机肥和农家肥等形式的更多营养，建植当年氮肥应足够牧草的利用，但接下来的生长过程中，结合雨季施入适量的无机肥和农家肥也是非常必要的。

每次刈割或放牧后，也应结合灌

manure. During the establishment year soil nitrogen is adequate for grass productivity, but in the subsequent seasons, moderate quantity of inorganic fertilizer or farmyard manure should be topdressed during the rainy season.

After each cutting or grazing, use fertilizer combined with irrigation is necessary to ensure the regrowth of grass. Leguminous forage, in general, demand more phosphorus and potassium while gramineous grasses need more nitrogen fertilizer. In order to maintain the stability of the mixed community, it is essential to adjust the fertilization according to their growth performance in mixture community.

溉进行合理施肥，确保牧草的再生。一般来讲，豆科牧草对磷钾肥需求量更多，而禾本科牧草则需要更多的氮肥，生产中应根据它们在混播群落中的长势进行合理施肥，以维持混播群落的稳定性。

4.3 Forage-crop rotation 草田轮作

Forage-crop rotation is to arrange the forages and crops in a paddock during a certain period by a fixed order according to the combination rule of utilization and fertilization farming system. It is of great significance for using land resources reasonably, improving soil fertility and realizing the sustainable production of agriculture. The earliest prototype of forage-crop rotation is the rotation of crops and green manure crop, which has a history of thousands years in our country. This rotation system is the essence of the traditional farming system in China, leading to a transformation from traditional "grain crop-economy crop" model to "grain crop-economy crop-forages" model.

In a rotation cycle, the plants that need more nitrogen are generally arranged after legume forages and the intertilled crop followed perennial forages, thus can avoid the weeds and insects invasion. In recent years, with the development of animal husbandry, forage and combination mode in our country develop rapidly both in the North and the South of China. Because of the huge area and rich natural resources and relatively complex social economic condition, the forage adaptability varies because of the soil, climate and pasture condition. The cropping system in different climate area also

草田轮作是指在一定耕地面积和年限内，按照规定好的顺序进行轮换种植牧草和农作物的一种合理利用土地的耕作制度。实施草田轮作，对合理地利用土地资源，改善土壤肥力，实现农牧业的可持续生产具有重要意义。草田轮作的最早雏形是作物和绿肥的轮作，在我国已有几千年的历史。用地与养地结合的草田轮作制度是我国传统耕作制度的精华，也是实现由传统的"粮–经"二元结构，向"粮–经–饲"三元结构转变的重要措施。

轮作中需氮素多的作物安排在豆科牧草种植之后，多年生牧草安排在中耕作物之后，有利于防止杂草和虫害的侵袭。近年来，随着畜牧业发展，粮草结合耕作模式在我国南北方发展迅速。由于我国幅员辽阔，自然资源与社会经济条件十分复杂，土壤、气候和牧草的适应性差异较大，不同气候类型区饲草参与的种植制度也各有特点。南方逐步形成以柱花草和一年生黑麦草为主的种植

has their own characteristics. *Stybsanthes guianensis*(Aubl) and annual ryegrass gradually dominate in the South of China, whereas alfalfa, ryegrass and *Astragalus adsurgens* dominated in the north.

模式，北方则呈现以紫花苜蓿、黑麦草和沙打旺为主的种植模式。

4.3.1 Forage-crop rotation mode in southwest of China

"Italian ryegrass– rice" mode is widely used in southwest of China (Fig. 4.7); planting ryegrass can save labor and can offer high quality green fodder for livestock. The exact way is furrowing the paddies and drain first after rice harvesting in early September, then no-till or ploughing, sowing lolium multiflorum ryegrass in late September to early October, compound fertilizer 75 kg/hm^2, mowing for feeding until it grows up to 40 ~ 60 cm, topdressing urea 75 kg/hm^2 after each mowing, generally it can be mowed 4 ~ 5 times .This mode is suitable for the flat, hilly lower mountain region and the winter fallow fields in farming and pastoral areas of 300 ~ 1000 m altitude southwest of China.

4.3.1 西南区草田轮作模式

"多花黑麦草－水稻"是一种在我国南方采用较为广泛的草田轮作模式（图 4.7），在水稻田种植多花黑麦草，既可以节省劳力，又可以为家畜提供优质的青饲料。具体做法是：在 9 月初水稻收获后，稻田开沟，沥干水分，免耕或翻耕，9 月下旬至 10 月上旬撒播或条播黑麦草，施复合肥 75 kg/hm^2，水稻长到 40~60 cm 时即可刈割利用，每次刈割后及时追施尿素 75 kg/hm^2，一般可刈割 4~5 次。该模式适宜于西南地区海拔 300~1000 m 的平坝、丘陵区、低山区、半农半牧低山区的冬闲田。

Fig.4.7 Rotation of Italian ryegrass after rice harvesting in southwest of China
图 4.7 西南地区水稻和多花黑麦草的轮作模式

4.3.2 Forage-crop rotation mode in northwest of China

The common mode in Shaanxi province: "alfalfa (3-5 years) - winter wheat". As a farmer's saying: "one Chinese mu alfalfa productivity equals to three mu field, and the soil fertility does not decreased too much even after three years",

4.3.2 西北区草田轮作模式

陕西常见种植模式为："紫花苜蓿（3~5 年）– 冬小麦"。有农谚曰："一亩苜蓿三亩田，连做三年劲不散"，充分说明苜蓿对培肥地力的明显效果；"小麦 – 青贮玉米"（陕西关中）、"油

it fully explain that alfalfa have an obvious effect on soil fertility; "Wheat - silage corn" and "rape –hairy vetch" (south of Shaanxi) can improve land utilization and enriching the soil fertility.

In Xinjiang, takeing advantage of strong salt tolerance characteristics of Sudan grass, it is planted in mid to late April spring in newly reclaimed uncultivated land, and then plant alfalfa on Sudan grass residual after harvesting, till off alfalfa to plant other crops 2 ~ 3 years later, thus can better improve the soil (Fig.4.8).

菜–毛苕子"（陕西南部）有利于提高土地利用率和培肥地力。

在新疆，新开垦的生荒地上，利用苏丹草抗盐碱性强的特点，春季 4 月中下旬种植苏丹草，在苏丹草收割后的茬地上利用松土补播机播种苜蓿，2~3 年后翻掉苜蓿再种植其他农作物，可以很好地改良土壤（图 4.8）。

4.3.3 东北区草田轮作模式

东北平原区位于中国农牧交错带的

Fig.4.8 The newly reclaimed field of northern Xinjiang
图 4.8 北疆新垦土地种植模式
Rotation model: Sudan grass+alfalfa (1st yr)→alfalfa (2nd~4th yr) → silage corn (5th yr) →winter wheat (6th yr) →winter wheat+alfalfa (7th yr)

4.3.3 Forage-crop rotation mode in northeast of China

The northeast plain is located in the eastern of transition zone between cropping area and nomadic area, land use contradiction is outstanding, introducing forage grass into the crop system and developing grass industry has important significance in solving the contradictions. Based on the realities of soil and water resources loss in the maize production, decreased land natural fertility, dust of winter and spring bare farmland and years of continuous cropping, unreasonable production and business operation activities, as the main crops in farming system, alfalfa-corn strip intercropping planting patterns are developed (Fig. 4.9A, Fig. 4.9B).

东段，农作物用地与饲草料用地矛盾比较突出，引草入田，发展农区草业，对缓解该地区的人地矛盾、草畜矛盾有着重要意义。针对东北地区玉米生产中存在的黑土资源水土流失日益加剧、土地自然肥力不断下降、冬春裸露的农田极易扬尘以及常年连作等不合理的生产经营活动这些现实，把紫花苜蓿作为主要作物纳入耕作制，构建紫花苜蓿、玉米条带间作种植模式（图 4.9A，图 4.9B）。

紫花苜蓿与玉米搭配，地上与地下部分互利互补，边际效益突出，紫花苜蓿富含蛋白质，玉米籽粒是重要的能量

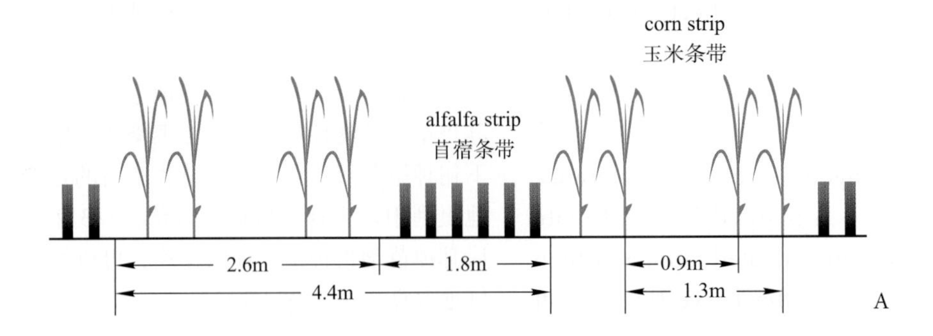

Fig.4.9 Interplant of alfalfa and maize in northeast of China

图 4.9 紫花苜蓿与玉米间作模式

Alfalfa and corn can benefit each other and provide an outstanding marginal benefit, alfalfa is rich of crude protein and corn grain is important energy feed, corn stover also has great nutritional value among the major crops straw, which can meet the nutrient requirements of different kinds of livestock. The disadvantage is that the establishment is relatively complex, cultivating, harvesting are not as convenient as monoculture.

饲料，且玉米秸秆的营养价值极高，可满足不同种类家畜的营养需求。缺点是种植时较为复杂，机械化中耕、收获不如单作方便。

4.4 Artificial grassland management 人工草地的管护

4.4.1 New established artificial grassland management

Young growing plants need time to develop a significant root system in order to compete against weeds. It is important to defer grazing for at least two growing seasons or until plants are well established.

In general, for optimum production and species persistence, most species should not be grazed until they reach a height of 20 to 30 cm and should cease before stubble height is lower than 10 to 15 cm. Grazing below the minimum stubble height will remove the plant growing point and severely limit re-growth and persistence. Considering of the yield and quality of forages, grazing or harvesting should occur at the crosspoint of the two curves (Fig.4.10).For grasses, it is the booting to heading stage, whereas for legumes, it should be at the bud to early flowering stage.

Because production on nonirrigated pastures depends on the amount and timing of annual precipitation, grazing pressure and grazing timing are critical. Compared to irrigat-

4.4.1 新建人工草地的管护

幼苗为了与杂草竞争，发达根系的建成需要一定时间，延迟至少两个生长季再进行放牧或直到草地已经良好建成是很重要的。

一般来讲，为了达到最佳产量和提高牧草的持久能力，大多数牧草在草层高度达到 20~30 cm 以前不宜放牧，同时留茬高度不能低于 10~15 cm。低于最低留茬高度坚持放牧会破坏植物生长点，严重影响牧草再生和持久性。收获或放牧时间应同时考虑产量和品质两个因素，即位于产量上升和品质下降两条曲线的交叉点（图 4.10）。对于禾本科牧草而言，是孕穗至抽穗期，豆科牧草则在孕蕾至初花期。

由于在无灌溉草地上的生产主要取决于降水量的数量和时间，所以放牧压力和放牧时间至关重要。干旱时

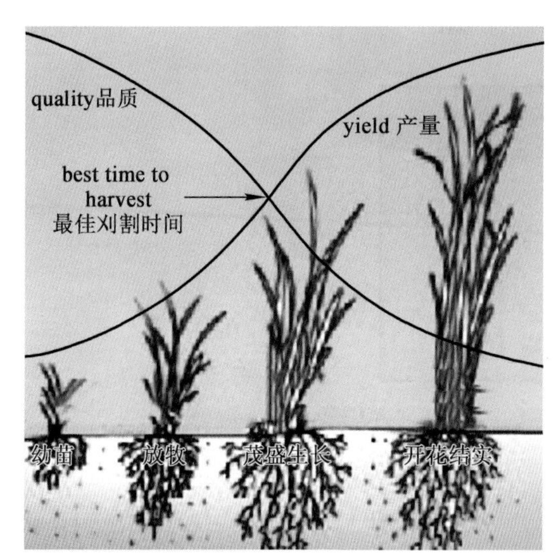

Fig. 4.10 The proper graze/harvest time
图 4.10 牧草的最佳放牧 / 刈割时期

ed pastures, nonirrigated pastures have significantly less production and recovery potential. During periods of drought, it may be necessary to stop grazing altogether. Correct grazing allows plants enough reserves to regenerate themselves.

Insects, and/or diseases should be controlled through mowing, burning, flash grazing, or pesticides as needed to maintain a healthy stand. Where stands are damaged by drought, insects, or other uncontrollable events, the stand should be replanted, overseeded, or spot planted. Thin stands may only need grazing deferment during the growing season to recover rather than replanting.

4.4.2 Mature artificial grassland management

Plants get the energy needed for growth through photosynthesis. The root system is in the dark and totally reliant on the leaves to supply the carbohydrates required for maintenance and growth. When forage plants are continuously grazed to a very short stubble, the root growth will be inhibited and even stopped if over grazing (Fig.4.11).

Over grazing not only reduces the health and vigor of the plants, causing a decrease in the regrowth rate, but it predisposes the pasture to weed invasion.

建议完全停止放牧。

病虫害可以通过刈割、焚烧、间断式放牧或农药进行控制，以维持草地的健康生长。如果草地遭到了干旱、虫害或其他不可控制的意外因素影响，需要重新建植、加大播量或斑块修复。长势差的草地只需要在生长季延迟放牧而不是一定要重新建植。

4.4.2 成熟人工草地的管护

植物通过光合作用合成所需的能量，根系位于地下黑暗的环境中，完全依靠地上的叶片供应糖类以维持生长，当牧草被持续放牧到留茬很低时，根系的生长会受到抑制甚至停止生长（图4.11）。

过度放牧不仅会影响植物的健康生长和生活力，同时也会降低再生速度，使杂草易于入侵。非灌溉草地比灌溉草地的恢复力差，恢复不仅慢，还经常需要等待降雨帮助其恢复生长。因此，当草地干旱缺水时，尽可能的

Fig. 4.11 Over-grazing stops root growth and reduces grass production. It occurs when more than 50% of the leaf mass is removed

图 4.11 过度放牧（50% 以上的地上部分被采食）使牧草根系停止生长，降低草地产量

Non-irrigated pastures are less resilient to grazing than irrigated pastures. They are slower to recover and often must wait for precipitation to be revitalized. Therefore, irrigate the pasture as possible as you can if soil moisture is limited.

Rotational grazing

Rotational grazing involves confining animals in one section of pasture or paddock while the remainder of the pasture "rests" and regrows. These paddocks are small enough that all the forage is grazed to a uniform height in a relatively short period of time. The timing of rotations must be adjusted for the growth rate of the forage.

灌溉很有必要。

轮牧

轮牧是为有计划地让其余地块“休息”和牧草的再生，而将动物限制在一个草场片区放牧的方式。放牧区应尽量小一些，以确保所有牧草可以在短时间内被采食到一个比较一致的高度，轮牧的时间需根据牧草生长速度进行调整。

Chapter 5

Legumes
豆科牧草

Legumes is a large family of seed plants, containing about 600 genera and 12 000 species. About 185 genera and 1 380 species grow in China. Legumes distribution is second only to grasses and legume forages are apt to grow all over the world. Legumes are usually scattered or grow individually in rangeland, do not generally constitute the majority of the herbaceous layer, but legumes have superior forage quality, most fix atmospheric N_2, and they are often seeded into and managed as important components of pastures and hayfields.

豆科是种子植物中的一个大科，共600余属，12 000余种，我国目前大约有185属，1 380种。在自然界中，豆科牧草遍布各地，仅次于禾本科，在天然草地通常是分散或单株成长，不是草层主要部分，但在栽培草地中却占有重要地位。豆科牧草富含蛋白质及钙，对土壤结构改良及提高肥力具有重要作用。

5.1 *Medicago* L.— alfalfa (lucerne) 苜蓿属—紫花苜蓿

The genus *Medicago* L. comprises more than 60 species of which about two-thirds are annuals and one-third are perennials. They are distributed in America, Europe, Asia and Africa. There are 12 species, 3 varieties and 6 variants in China. The predominant forages of the genus *Medicago* cultivated in China are alfalfa (*Medicago sativa* L.) and sickle alfalfa (*Medicgo falcate* L.).

Alfalfa (lucerne)

Scientific name: *Medicago sativa* L.

Alfalfa is native to Asia Minor, Iran, Transcaucasia and Turkmenistan. The center of origin is Iran. Because of its wide adaptability and high nutritional value, alfalfa is a kind of perennial leguminous forages cultivated worldwide. It was introduced to China by Zhang Qian in Han dynasty in 119 B.C. when he went to the Western Region. It was first planted in Chang'an of Shaanxi province and then spread to the Yellow River basin and the northwest regions. Alfalfa is the most cultivated forage legume in the world. In 2009, alfalfa was grown on approximately 33 million hectares worldwide. The US is the largest alfalfa producer with cultivation of 10 million hectare. There are about 3.6 million hectares in China (sixth in the world). Alfalfa is mainly distributed in the North of China. It is also planted in some southern provinces of Jiangsu, Hunan, Hubei, and Yunnan.

苜蓿属（*Medicago* L.）植物共60余种，为一年生或多年生草本，分布在美洲、欧洲、亚洲、非洲等地。我国有12个种，3个变种，6个变型，目前我国栽培最多的为紫花苜蓿和黄花苜蓿。

紫花苜蓿

学名： *Medicago sativa* L.

英文名： alfalfa或lucerne

紫花苜蓿原产于小亚细亚、伊朗、外高加索和土库曼斯坦一带，中心产地为伊朗，其适应性广，营养价值高，是一种世界性广泛栽培的多年生豆科牧草。紫花苜蓿公元前119年由汉使张骞出使西域带回中国，先种植于陕西长安，后扩散到黄河流域及西北地区。目前全世界的栽培面积约3300万 hm^2，其中美国栽培面积最大，约1000万 hm^2。我国约360万 hm^2，居世界第6位，主要分布在西北、东北、华北地区，江苏、湖南、湖北、云南等地也有栽培。

5.1.1 Botanical characteristics

Alfalfa is a perennial leguminous plant, with a prominent taproot that typically penetrates to more than 1 m underground. About 60%~70% of the root system is concentrated in the upper 0~30 cm soil layer(Fig.5.1A). Symbiotic rhizobia bacteria infect root hairs and cause development of nodules on roots. The crown is located above the root, about 1.2~1.9 cm below the soil surface (Fig.5.1B). It will grow into the soil more deeply along with the growth of plants year after year. Winter hardiness and grazing tolerance of these legumes tends to be directly related to their crown depth. Axillary buds from the crown can develop into about 25~40 branches depending on fertility level. Most alfalfa grows upright, 60~150 cm in height(Fig.5.2). Leaves are normally pinnately trifoliolate with the central leaflet having an extended stalk. Leaflets are typically smooth on the surface and are oblong with a serrated margin close to the tip. Stipules are large and narrow lanceolate. Flowers are borne in a short raceme and are most frequently purple. Pods are spiral with 8 seeds per pod (Fig.5.3). Kidney-shaped seeds are yellow or yellow-green, thousand seed weight is 1.44~2.30 g (Fig.5.4).

5.1.1 植物学特征

紫花苜蓿为多年生豆科苜蓿属草本植物。主根深度可达 1 m 以上，60%~70% 的根系分布于土层 0~30 cm 处（图 5.1A）；有共生根瘤；根颈为根最上部，位于地表以下 1.2~1.9 cm，随生长年限延长不断下延，入土深度与耐寒、耐牧能力有关（图 5.1B）；分枝由此产生，一般有 25~40 个，因肥力不同而异。茎直立或斜生，绝对高度达 60~150 cm（图 5.2），草层高度达 60~80 cm，较柔嫩；茎粗 2~4 mm，花期生长最迅速。三出羽状复叶，中间叶柄较长，叶缘上端近 1/3 处有锯齿；小叶长圆状倒卵形或倒披针形，顶端凹形有尖刺，托叶狭披针形。短总状花序，紫色，蝶形。荚果螺旋状（图 5.3）；种子肾形，黄色或黄绿色；千粒重 1.44~2.30 g（图 5.4）。

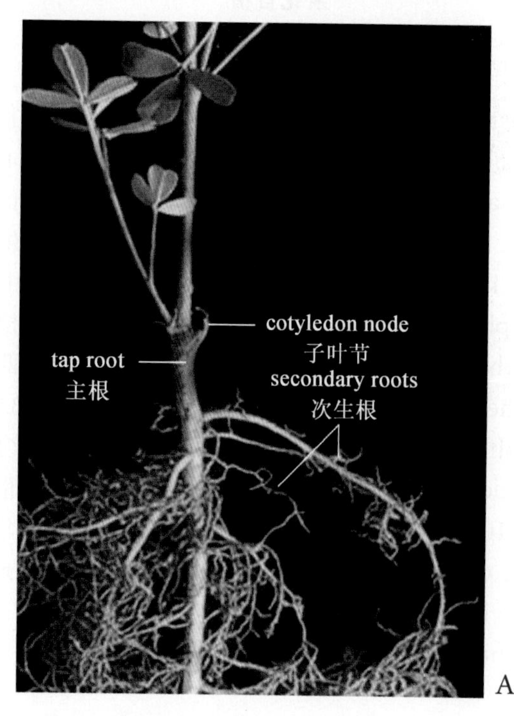

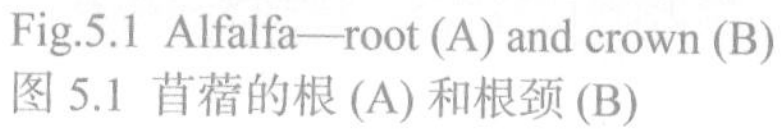
Fig.5.1 Alfalfa—root (A) and crown (B)
图 5.1 苜蓿的根 (A) 和根颈 (B)

A

B

C

D

Fig.5.2 Alfalfa—stem (A), leaf (B) and flower (C, D)
图 5.2 苜蓿的茎 (A)、叶 (B) 和花 (C, D)

Fig.5.3 Alfalfa—pod
图 5.3 苜蓿荚果

Fig.5.4 Alfalfa—seed
图 5.4 苜蓿种子

5.1.2 Biological characteristics

Alfalfa prefers warm climate and cultivar adaptation to cold ranges from nondormant to very winter hardy (-20~-30℃). The optimal temperature is 25~30℃ during growth.

For better performance, 400~800 mm yearly rainfall is required. Alfalfa is not well adapted to areas where annual rainfall is above 1000 mm. The evaporation coefficient is 800, so sufficient water is required for growth. Alfalfa tolerates drought but not prolonged flooding. It grows best in warm and dry areas with irrigation system. Alfalfa prefers full sunlight and is not tolerant to shade.

Alfalfa does best on soils with fine to medium textures that are moderately to well drained and neutral or higher (pH 7~9). Alfalfa was moderate saline tolerance and can grow on soils with 0.3% salinity. Alfalfa planted on saline soil is able to decrease the salt content.

Plant heaving is a serious problem in regions where temperatures fluctuate around 0℃ and forages are grown on imperfectly drained, fine-textured soils. Volume of liquid water expands about 10% when freezing to ice. Alternate freezing and thawing of surface soil causes vertical expansion and contraction. Plants are gripped by the freezing surface soil and lifted upward. Plants with taproots may not subside when soil thaws. The next freezing event grips the plant lower and moves the plants further upward. With repeated cycles, the taproot breaks. Plants die because crown meristems that are normally protected by the soil are elevated above the ground and exposed to freezing temperatures and desiccation (Fig.5.5).

5.1.2 生物学特性

紫花苜蓿适应性广。

温度：喜温耐寒（-20~-30℃），适宜生长温度 25~30℃。

水分：在年降水量为 400~800 mm 的地方生长良好，超过 1000 mm 则生长不良。需水较多，蒸腾系数 800，但抗旱力强，在温暖干燥有灌溉条件地方生长最适宜。

光照：苜蓿喜光不耐阴。

土壤：对土壤要求不严，适宜 pH 7~9，耐盐 0.3%。除重黏土、极瘠薄的沙土、过酸过碱的土壤及低洼内涝地外，其他土壤均能种植，最为适宜的是沙壤土或壤土。在盐碱地上种植，有降低土壤盐分的功能。种植苜蓿 5 年的盐碱土，0~30 cm 的全盐量可下降 40.62%，30~60 cm 下降 76.3%，pH 也有所下降。另据测定，苜蓿幼苗能在含盐量为 0.3% 的土壤上生长，成年植株可在含盐量 0.4%~0.5% 的土壤上生长。

冻拔现象　指在高纬度的寒冷地区，当土壤含水量过高时，由于土壤结冻膨胀而升起，连带植物也会被抬起。至春季解冻时，土壤下沉而植物留在原位，造成植物根部裸露死亡的现象（图 5.5）。

Fig.5.5 Alfalfa—heaving problem
图 5.5 苜蓿的冻拔现象

Fall dormancy is a growth character of alfalfa directly associated with cold tolerance and production. It refers to responses in alfalfa morphology and production capability to lower light and temperatures in fall. This response occurs only during following fall harvest.

Fall dormancy is directly related to the regrowth ability, cold tolerance and production. Cultivars with low fall dormancy scores have lower productivity due to later spring growth initiation and lower regrowth rate. However, high fall dormancy scores, cultivars have higher production ability. American scientists classify fall dormancy of alfalfa cultivars as a certain scores range from 1 to 11 according to regrowth ability and their morphology. Cultivars with low scores have slow growth during autumn and become the most winter hardy. Cultivars with high scores are low fall dormancy and even non-dormant.

5.1.3 Cultivation technology

5.1.3.1 Tillage and fertilization

Well-cultivated, uniform and firm seed bed is beneficial for alfalfa establishment. Organic fertilizer: 22.5~37.5 t/hm^2, superphosphate: 300~450 kg /hm^2.

5.1.3.2 Seed treatment

(1) Alfalfa seeds require exposure to sunlight for 2~3 d to break seed dormancy before planting. Alfalfa generally has a high level of seed viability and germination rate. However, **hard seed** is a common form of dormancy in forage legume seeds. Hard seeds fail to germinate because they are unable to imbibe water. Dormancy caused by hard seed may be overcome by **scarification**. Physical or chemical treatment may be used to promote moisture uptake and/or radicle emergence.

(2) Seed inoculation: All legume seeds should be inoculated with the proper strain of N-fixing bacteria before seeding. Heat, direct sunlight, and drying are detrimental to survival of rhizobia inoculants.

5.1.3.3 Seeding methods

Drill seeding or broadcast seeding is feasible. Row spacing should be 30 cm for forage production and 50 cm for seed production. Seeding rate of forage production is usually

秋眠性 是苜蓿的一种生长特性，指秋季由于光照减少和气温下降，导致苜蓿形态类型和生产能力发生变化的现象。这种现象只能在苜蓿秋季刈割后的再生过程中才能观察到，而在春季和初夏收割后却观察不到。

秋眠性和苜蓿再生、耐寒性、生产力密切相关。秋眠级低的品种，因其春季返青晚，刈割后的再生速度慢，生产能力也低。秋眠级高的品种，因春季返青早，刈割后的再生速度快，生产能力明显高于秋眠级低的品种。美国科学家将苜蓿的秋眠级分为 11 个等级，等级越低，秋眠性越强，等级越高，秋眠性越弱，甚至无秋眠现象。

5.1.3 栽培技术

5.1.3.1 土壤耕作与施肥

深翻，精细整地，平整地面。每公顷施有机肥 22.5~37.5 t，过磷酸钙 300~450 kg 为底肥。

5.1.3.2 种子处理

（1）播前晒种 2 ~ 3 d 打破休眠；种子生活力强，发芽率高，但有硬实，不易吸水，须经处理后才能发芽。

（2）根瘤菌接种（参见第四章播前种子处理）。

5.1.3.3 播种

条播或撒播，收草田和种子田条播行距分别在 30 cm 和 50 cm 左右。收草田播种量为 15 ~ 20 kg/hm^2，播种深度 1.0~2.5 cm，春、秋季均可播种。

5.1.3.4 田间管理

苜蓿苗期地上部生长很慢，而地下根系生长较快。在陕西关中地区 3 月上旬返青，5 月中旬现蕾，6 月上旬开花，6 月下旬种子成熟，生育期 110 d 左右。寿命可达 20~30 年，但一般生长到第 4~5 年即翻耕种植其他作物。虫媒异花

15~20 kg/hm^2. Seeding depth is generally 1.0~2.5 cm, depending on soil moisture and structure. Spring and autumn are generally suitable for planting.

5.1.3.4 Field Management

Alfalfa shoots grow slowly and the roots rapidly during the seedling stage, it recovers growth in early March in Guanzhong of Shaanxi province, reaches early bud stage in mid-May, blooms in early June and seeds ripen in late June. The entire developmental period is about 110 d. Typical stand lifespan is 4~5 years, but plants can persist for 20~30 years. Seed production requires cross-pollination by leafcutter bees.

(1) Loosen hard soil before sprouting. Control weeds once or twice during the seedling stage.

(2) In the seeding year, alfalfa should be harvested before plants stop growth. If plants do not reach a harvestable stage, seedlings should be protected from grazing or cutting and allowed to over winter.

(3) Before subsequent spring emergency, residue removal is recommended and the soil surface should be tilled to retain soil moisture. After cutting or harvesting seeds, soil should be tilled and fertilized. This management practice is important following the last cutting or seed harvest in fall. Irrigation should be combined with fertilization.

(4) Before winter, sufficient water is necessary. Irrigation time depends on the temperature. Generally, proper irrigation coincides with average daily temperatures as low as 7~8℃ . However, when the temperature declines to 5℃ , irrigation is no longer appropriate. Excessive winter irrigation is beneficial. Optimally, soil is thoroughly moist and moisture permeates deeply into soil on the day of irrigation.

(5) Pay attention to potential insect and disease problems, particularly, root rot and thrips.

5.1.4 Utilization

Alfalfa is a high yield legume which provides high nutrient value and very palatable. Alfalfa hay productivity in Guanzhong of Shaanxi province is about 22~30 t/hm^2 if irrigation is available.

(1) It is proper to harvest when plants are 30~40 cm high. Harvest alfalfa for hay in the early bloom stage with a

授粉。

（1）出苗前及时破除板结，以利于出苗。苗期中耕除草 1~2 次。

（2）播种当年，在生长季结束前，刈割利用一次，当植株高度达不到利用程度时，要留苗过冬，冬季严禁放牧。

（3）翌年春季萌生前，清理田间留茬并进行耕地保墒。每次刈割和收种后也要耙地追肥，灌溉区要结合灌水追肥。施用磷、钾肥可显著增加苜蓿的产量，并可提高粗蛋白质含量。钾肥对苜蓿的效果仅在施用的当年较好，第二年效果不明显；磷肥施用量过高，还有降低牧草产量的趋势。

（4）入冬时要灌足冬水，冬灌时间应以气温而定，一般从日平均气温下降到 7~8℃时开始，到 5℃左右时结束。冬灌的用水量并不是越大越好，一般以灌透且当天能渗完为宜。

（5）注意根腐病及蓟马虫害的防治。

5.1.4 利用方式

苜蓿具有产量高、适口性好、营养价值高的特性，在陕西关中地区苜蓿干草产量在灌溉条件下可达 22~30 t/hm^2。

（1）可进行青刈，株高 30~40 cm 时开始为宜；调制干草以初花期收割为好；刈割时留茬高度 3~5 cm，干燥寒冷地区最后一茬不低于 7 cm。

（2）苜蓿鲜喂牛羊会引起臌胀病。臌胀病是由于苜蓿鲜草中含有大量的可溶性蛋白质和皂素而引起的。苜蓿鲜草含有雌激素，在交配前饲喂牛、羊会降低受孕率。一般鸡的日粮中苜蓿粉可占 4%~8%；猪日粮以 4%~5% 为宜；牛日粮中可占 25%~45% 或更多；羊可达 50% 以上；在肉兔的日粮中

stubble height of 3~5 cm is appropriate. For the last fall cutting, stubble height should be more than 7 cm in cold areas.

(2) Fresh forage fed to ruminants can induce **bloat**. Bloat results from high levels of soluble protein and saponin content in fresh alfalfa forage. Fresh alfalfa also contains estrogens which can reduce conception rates in cattle and sheep when alfalfa is fed or grazed prior to breading. Alfalfa powder may contribute 4%~8% of the daily ration for chickens, 4%~5% for pigs, 25%~45% or more for cattle, more than 50% for sheep, and 30% for rabbits.

(3) To avoid bloat and **diarrhea**, livestock should not graze alfalfa when they are very hungry. Feeding grass hay prior to grazing is advisable. Grasses plants must exceed 50% in mixed stands to reduce bloat hazard.

(4) Seed should be harvested when 1/2~2/3 of the pods have turned from green to tawny. For seed production in forage fields, seeds should be harvested every 1 to 2 years. Compare with that of crops, alfalfa seed yield is quite low, about 600 kg/hm^2.

(5) Choose a sunny day for hay harvest and make sure the dries sufficiently. During storage, hay should be checked frequently in case mold develops.

(6) Alfalfa also can be processed into silage. Wilted silage or haylage is usually produced.

(7) Longevity of alfalfa stands depends on potential yield of seed and forage. Typical stand life is 4~5 years.

(8) In addition to forage production, alfalfa can also be used to extract leaf protein, make medicine and cosmetics and favored for honey production.

30%左右最佳。

（3）放牧利用时为防止反刍动物患臌胀病，应避免家畜在饥饿状态时采食苜蓿，放牧前要先喂禾本科干草，既可防治臌胀病，还能防止家畜腹泻。混播草地禾本科牧草要占50%以上的比例。

（4）收种适宜在1/2~2/3的荚果由绿色变成黄褐色时进行；收草田不能连续收种，种子田也应每隔1~2年收草一次。种子产量低，每公顷约600 kg。

（5）调制干草时，要选择晴朗天气。干草必须保持绿色状态。存放过程中应勤检查，以防霉变损失。可用裹夹碾压法（也叫染青法）调制。

（6）苜蓿亦可进行青贮利用，一般可采用半干青贮（含水量在50%左右）。

（7）苜蓿田利用年限应视种子和产草量最高年限而定，一般利用4~5年。

（8）除作为饲草外，苜蓿还可提取叶蛋白、药用、制作化妆品，还能作为蜜源植物。

5.2 *Trifolium* L.
三叶草属

The genus *Trifolium* includes about 250 species of clovers. The average longevity of *Trifolium* which has been cultivated for centuries throughout the world is about 8~10 years. Clovers originated in a broad region from southern Europe to Asia and are distributed in the temperate zone. 8 species

三叶草属，又叫车轴草属，全世界有250余种，多年生草本，平均寿命8~10年，为世界上栽培历史较悠久的牧草之一。原产于亚洲南部和欧洲东南部，广泛分布于温带。我国有8个栽培

are cultivated in China and are adapted in South or North China. Dominated cultivated species are red clover, white clover, alsike clover, crimson clover, subterranean clover and strawberry clover.

5.2.1 White clover

Scientific name: *Trifoliun repens* L.

5.2.1.1 Botanical characteristics

White clover is a perennial legume with short taproot and well-developed lateral roots. Its shallow root system is concentrated in the 0~10 cm soil layer, and contains many nodules (Fig.5.6). Stems are solid, smooth, thin and long reaching a height of 20~40 cm, and under some conditions can reach 80 cm (Fig.5.7). Its prostrate stems have secondary and tertiary branches and can develop roots as the stem nodes grow downward into the soil. Leaves from rhizome or stolon nodes are palmately, alternate trifoliolate and borne on long, upright petioles. Leaflets are obovate and serrated around the margin, and have a smooth surface that is usually variegated with a white "V" shape "watermark". The raceme has 20~40 white or pink flowers on long peduncles (Fig.5.8). Its florescence is long. Pods are small, long and slender and contain 3~4 seeds. Seeds are heart-shaped, yellow or brown and seldom purple. Thousand seed weight is 0.5~0.7 g (Fig.5.9).

种，是华南、华北的优良草种。主要栽培的有白三叶和红三叶，此外还有杂三叶、绛三叶、地三叶、草莓三叶等。

5.2.1 白三叶

学名： *Trifolium Repens* L.

英文名： white clover

5.2.1.1 植物学特征

白三叶为多年生豆科三叶草属草本植物。主根短，侧根发达，根系浅，根集中在10 cm表土层，着生许多根瘤(图5.6)。茎实心，光滑，细长，20~40 cm，最长可达80 cm(图5.7)，有二、三级分枝，匍匐生长，茎节处着地生根，侵占性强。掌状三出复叶，由根茎或匍匐茎节上生出，互生，叶柄细长直立。小叶倒卵形，叶面光滑，有"V"形白斑或无，叶缘有细锯齿。头形总状花序，自叶腋生出，花梗多长于叶柄，每花序有小花20~40朵(图5.8)，白色或略粉红，花期长。荚果很小，细长，每荚种子3~4粒。种子心形，黄色或棕黄色，间有紫色、紫红色，千粒重0.5~0.7 g(图5.9)。

Fig.5.6 White clover—root and nodule
图 5.6 白三叶的根和根瘤

Fig.5.7 White clover—stolon
图 5.7 白三叶的匍匐茎

Fig.5.8 White clover—leaf and inflorescence
图 5.8 白三叶的叶和花序

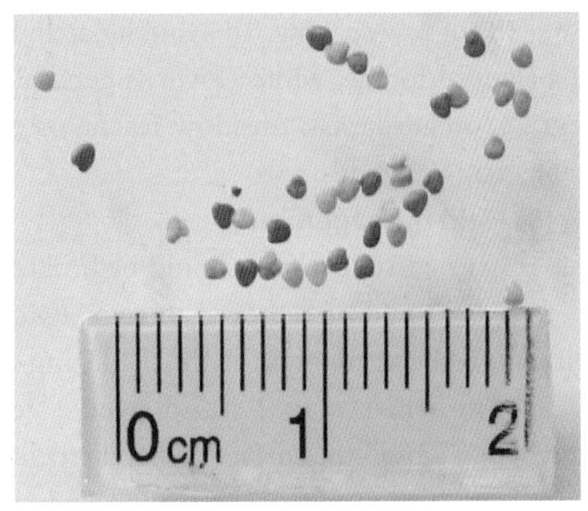

Fig.5.9 White clover—seed
图 5.9 白三叶的种子

5.2.1.2 Biological characteristics

This perennial legume is well adapted to humid, temperate climates. The optimal temperature is 19~24℃ . White clover is moderately winter hardy and heat tolerant. It prefers growing where rainfall is more than 600~800 mm. White clover is adapted to soils with fine to medium texture and poorly to well-drained. It can tolerate acid to moderately alkaline conditions. The proper pH for growth is 5.5~7.0. It is also quite shade tolerant. White clover is cross-pollinated by bee and butterfly and has high seed yield. It reproduces by seed or vegetatively, with stolons.

5.2.1.3 Cultivation technology

(1) Tillage and fertilization

Well-prepared seedbed is necessary before sowing. Deep ploughing is generally combined application of 22.5~30t / hm^2 base organic fertilizers mixed with 225~300 kg/hm^2 calcium superphosphate.

(2) Sowing

Before planting, legume seeds need to be inoculated with the appropriate rhizobium bacteria. Hard seeds should be treated through physical or chemical methods to overcome dormancy. The seeding rate is 3~3.75 kg/hm^2 for seed production, and 6~7.5 kg/hm^2 for forage production. Sowing depth is 1.0~1.5 cm. It is advised to broadcast or seed in line

5.2.1.2 生物学特性

白三叶喜温暖湿润气候，生长适宜温度为 19 ~ 24℃，耐热抗寒中等；适于在年降水量 600 ~ 800 mm 以上区域生长，耐湿，稍耐积水；耐酸性，适宜的 pH 为 5.5 ~ 7.0；较耐阴。异花授粉，蜂蝶为媒，结籽多，种子繁殖或无性繁殖。

5.2.1.3 栽培技术

（1）整地

播种前要精细整地，结合深耕施足底肥，每公顷施有机肥料 22.5~30 t，混入过磷酸钙 225~300 kg。

（2）播种

播前种子进行硬实处理及根瘤菌接种，播种量对于种子田为 3~3.75 kg/hm^2，人工草地为 6~7.5 kg/hm^2；播深为 1.0~1.5 cm；条播行距为 30 cm，或撒播。宜与黑麦草、鸡脚草、牛尾草等禾本科牧草混播建立多年生人工草地，以防止臌胀病。

with a 30 cm row spacing. To avoid bloat and produce nutrient-balanced forage, white clover is generally planted with ryegrass, orchardgrass, meadow fescue or other grasses to establish mixed stands.

(3) Field management

① Crusted soil surface should be broken before sprouting. Weeds should be removed once or twice during establishment. Pest control must be timely during seedling stage.

② The white clover stand over 2 years should be ploughed to loosen the surface soil and fertilized before turning green in spring and regrowth after the last grazing or cutting in fall.

③ Irrigation and fertilization is needed in drought, infertile condition when water is available.

④ In mixed stands, grass growth should be managed to protect white clover from inhibited growth or disappearance by cutting, grazing or enough P and K fertilizer.

（3）田间管理

① 出苗前，及时破除板结。苗期生长慢，不耐杂草，应中耕松土除草 1 ~ 2 次；及时防治害虫。

② 生长二年以上的白三叶草地，土层紧实，透气性差，在春、秋两季返青前和放牧刈割后的再生前，要进行耙地松土，并结合追肥。

③ 有灌溉条件的，在土壤干旱时，结合追肥进行灌溉。

④ 混播草地生长不协调时，应通过偏施磷钾肥，借刈割或放牧来调整生长，控制禾本科牧草生长，避免白三叶受抑制或从混播草地中消失。

5.2.1.4 Harvesting and utilization

(1) Harvesting for forage and seed

White clover should be cut in early bloom for forage production. Fresh forage yield is 37~45 t/hm^2. In some conditions, yield can reach more than 75 t/hm^2.White clover may flower for as long as 2 months. Seeds in different maturity stages are present on the same plant. Therefore, seeds should be harvested repeatedly many times or directly harvested once when 60%~70% flowers turn dark brown. Seed yield is 300~750 kg/hm^2. While clover stands are able to persist as long as 3~7 years with proper management.

(2) Forage value and utilization

White clover is palatable for all kinds of livestock because of its high nutrient value, leaf-stem ratio and soft, tender texture. It can be used directly as fresh forage, made into hay, or processed into forage powder and compound fodder.

① Used as hay : after exposure to sunlight, hay should be stored by stacking and avoid rain.

② Grazing: grazing should be scheduled at branching to bud stage or when canopy height is above 20 cm. Grazing should be stopped when the canopy layer is 5~8 cm. White

5.2.1.4 收获利用

（1）收获与采种

若用于饲草可于初花期开始刈割，一般年产鲜草 37~45 t/hm^2，高者可达 75 t/hm^2 以上。白三叶花期长达 2 月之久，种子成熟很不一致，应分期多次采种，或在 60% ~70% 的花序变为深褐色时收割。种子产量每公顷高达 300~750 kg，利于扩大种植。若田间管理及利用得当，一般能利用 3~7 年不衰。

（2）饲用价值及利用

茎叶细软柔嫩，叶多茎少，营养丰富，各类家畜均喜食。可刈割青饲或晒制青干草，或加工为草粉和配合饲料。

① 晒制青干草：干燥后及时堆垛贮存，避免雨淋。

② 放牧：分枝盛期至孕蕾期，或草层高度达 20 cm 时开始放牧，高度在 5~8 cm 时结束放牧，放牧不宜过重；每次放牧后，应停牧 2~3 周，以利再生；

clover is not tolerant to heavy grazing. Plants should be allowed to recovery for 2~3 weeks after grazing. Cattle and sheep are not supposed to graze after rain or when there is dew on the plants because of bloat hazard. White clover contains estrogen-coumadin estadiol, which is harmful to animal reproduction. Therefore, when grazed or fed to breading animals, white clover should be limited to a defined quantity.

③ When fed to ruminants, white clover should be mixed with grasses, the proportion of grasses is 50%~60%.

④ Forest/grass intercropping: white clover is adapted to planting under fruit trees and conserves soil moisture and provides nitrogen for trees.

⑤ Fertilizer and other uses: white clover can be used as green manure. It is especially useful in medium to long term field or forest rotations. White clover benefits maintenance of soil and water. It can be used in establishing lawns and gardens. It is also regarded as a landscaping plant and **nectariferous plant**. In recent years, its medicinal value has been developed and **isoflavones** can be extracted from white clover.

5.2.2 Red clover

Scientific name: *Trifoliun pratense* L.

Red clover is native to Asia Minor and southeastern Europe. It is a major cultivated forage in the regions of maritime climate such as Europe, Canada, America, New Zealand and Australia. Wild species can be found in Yunnan, Guizhou, Hubei and Xinjing provinces of China. Red clover has been planted for about 100 years in western Hubei province with about 6 600 hm^2. It is also planted in large areas of the Yungui Plateau. When red clover was introduced to Gansu province from America in 1940, it proved to be adapted to the local climate and soil condition and performed very well.

5.2.2.1 Botanical characteristics

Red clover is a perennial legume and can survive averagely 2~4 years. The taproot of red clover penetrates into soil 60~90 cm deep and well-developed lateral roots are concentrated in the 0~30 cm soil layer containing abundant nodules. Stems are round, hollow and upright or oblique. Branches

放牧牛、羊时不要在雨后和有露水时进行，以免发生膨胀病。白三叶含有雌激素香豆雌醇，使牛羊生殖困难，应控制采食。

③ 青饲反刍畜家畜时，应与禾本科牧草搭配，搭配比例禾本科草占50% ~60%。

④ 林草间作：可在果树下种植，降低果园水分损失并提供氮素。

⑤ 肥用及其他：可用作绿肥及中长期草田轮作的常用草种；还是良好的水土保持植物；还可作为城市草坪建植和庭园绿化、美化植物以及蜜源植物。近年其药用价值也被开发出来，可进行天然产物异黄酮的提取。

5.2.2 红三叶

学名： *Trifolium Pratense* L.

英文名： red clover

原产于小亚细亚及东南欧，是欧洲、加拿大、美国、新西兰和澳大利亚等海洋性气候地区的主要栽培牧草。我国云南、贵州、湖北、新疆都有野生种。湖北西部山区已有近 100 年的栽培历史，经普查种植面积约 6 600 hm^2，云贵高原也有较大面积的栽培。甘肃省于 1940 年从美国引进栽培。因红三叶生长良好，适应当地的气候和土壤条件受到群众的欢迎。

5.2.2.1 植物学特征

红三叶为短期多年生豆科草本植物，平均寿命 2~4 年。直根系，主根入土 60~90 cm，侧根发达，根系多分布在 0~30 cm 土层，多着生根瘤。茎圆，中空，直立或斜生，自根颈抽出 10~20 个，呈丛状，高 30~90 cm（图 5.10）。茎叶有茸毛，三出掌状复叶，卵形或长卵圆形，近全缘，有“V”形斑（图 5.11），叶柄长，托叶阔大，

usually number 10~20 originating from the crown and are fasciculate with a height of 30~90 cm(Fig.5.10). Stems and leaves are covered with fuzzy hairs. Leaves are palmately trifoliolate. Leaflets are oval or long oval, with mostly entire margins, and usually variegated with a white "V" shape "watermark" and long petioles(Fig.5.11). Stipules are large and membranous (Fig.5.12). The raceme from the tip of branches or axillae has 35~150 red or pale violet red flowers. Red clover is cross-pollinated. Pods are small and transverse cracked with only one seed. Seeds are kidney-shaped, oval or subtriangular, tawny or purple. Thousand seed weight is 1.5~2.0 g (Fig.5.13).

膜质（图 5.12）。头形总状花序，聚生枝梢或腋生小花梗上，小花众多，常有 35 ~ 150 朵，红色或淡紫红色，异花授粉。荚小，横裂，每荚含种子一粒，种子肾形或椭圆形或近似三角形，黄褐色或紫色，千粒重 1.5 ~ 2 g（图 5.13）。

Fig.5.10 Red clover
图 5.10 红三叶

Fig.5.11 Red clover—leaf and flower
图 5.11 红三叶的叶与花

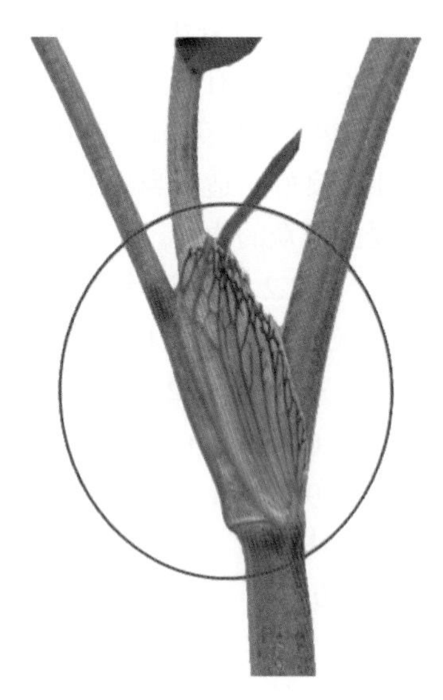

Fig.5.12 Red clover—stipule
图 5.12 红三叶的托叶

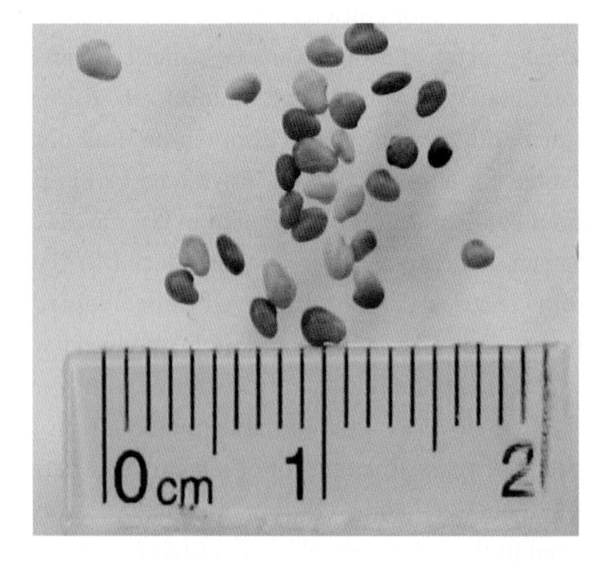

Fig.5.13 Red clover—seed
图 5.13 红三叶的种子

5.2.2.2 Biological characteristics

Red clover prefers a warm and humid climate. It is adapted to grow where it is neither too cold in winter nor too hot in summer. The optimum temperature for growth is 15~25℃. It tolerates cold temperatures as low as -10℃, but tolerance of hot temperatures is poor. It has low drought tolerance, but can grow very well in humid soil. For best growth, 600~1200 mm annual rainfall is required. It grows well in moderately acid and neutral soil and the proper soil pH for red clover is 6~7. Red clover is cross-pollinated. Bumblebees are an important pollinator. Flowering extends for 30~50 d.

5.2.2.3 Cultivation technology

(1) Tillage and sowing

well-prepared seedbed is necessary before sowing. Deep ploughing is generally combined with base fertilizer application. The seeding rate is 6~7.5 kg/hm^2 for seed production, and 9~11.5 kg/hm^2 for forage production. When establishing mixed stand, red clover should be planted with timothy in cold, humid regions, perennial ryegrass in warm, humid regions, and orchard grass in warm, dry regions. The ratio of red clover to grass should be 1:1. Red clover stands should be rotated with other crops.

(2) Field management

① In the seedling stage, red clover requires loose the soil and prompt weed remove. A nurse crop should be harvested in time to limit inhibition of red clover growth and ensure red clover to over wintering by trapping snow to provide insulation.

② Tillage is needed to loosen the surface soil before green-up in early spring and after cutting or grazing in order to improve the soil permeability.

③ Red clover needs supplemental phosphorus, potassium, calcium and other elements for healthy growth. Therefore, combined with tillage, top dressing with fertilizer is needed: calcium superphosphate 300 kg/hm^2, potash 225 kg / hm^2 or plant ash 450 kg/hm^2.

④ Irrigation is beneficial after harvest and grazing. Irrigation frequency is about 2~4 times per year.

5.2.2.2 生物学特性

红三叶喜温暖湿润气候，适宜于冬不过冷、夏不过热的地区，生长适温15~25℃，耐-10℃低温，不耐热；不耐干旱，耐湿性较好，年降水量600~1 200 mm时，生长较好；喜中性和微酸性土壤，适宜的pH为6~7；喜长日照。严格的异花授粉，虫媒花，以大黄蜂为主，每一花序每天开花2~3层，花期长达30~50 d。

5.2.2.3 栽培技术

（1）整地与播种

精细整地，结合深耕施足底肥，种子田播种量6~7.5 kg/hm^2；草地建植播种量为9~11.5 kg/hm^2。与禾本科牧草混播时，在较高寒而湿润的地区，与猫尾草配合效果较好；在温暖湿润地区，宜与多年生黑麦草混播；在温暖而稍干旱地区，则与鸡脚草混播较好。混播比例一般为1∶1。红三叶忌连作（间隔4~6年），不耐水淹。

（2）田间管理

① 苗期及时松土除草以利出苗，及时收割保护作物（留茬高度15 cm以上），减少生长抑制，以利冬季积雪，保护越冬。

② 早春返青前和每次刈割或放牧后要耙地松土，改善土壤通透性。

③ 生长过程中，所需磷、钾、钙等元素较多，结合耙地每公顷要追施过磷酸钙300 kg、钾肥225 kg或草木灰450 kg。

④ 灌区要在每次刈割、放牧利用后灌溉，2~4次/年。

⑤ 红三叶病虫害少，常见病害有菌核病，早春雨后易发生，主要侵染根颈及根系。施用石灰或喷洒多菌灵可以防治。

⑤ Red clover is seldom injured by pests or disease. It is susceptible to sclerotium disease in early spring after rain. Sclerotium mainly infects crown and root system. Liming and carbendazim are usually used to address such disease problems.

5.2.2.4 Forage value and utilization

(1) Fresh forage or grazing: red clover is palatable for all livestock for it is juicy and tender in texture and high in nutrient value. It seldom induces bloat in ruminants.

(2) Green hay: Red clover can produce hay with high quality for leaves do not shatter easily.

(3) Pulped red clover can be fed pigs to save concentrated fodder.

(4) Other uses: Red clover is regarded as a nectariferous plant due to extended flowering. It also can be used as a nurse crop due to its short longevity and early and rapid growth. It is appropriate to plant in rotations medium and long term with crops. Its extensive root system can increase soil organic matter and enhance soil fertility. Red clover can enhance soil and water conservation. It is adapted to plant under trees for it is rather shade tolerant. Red clover can be used to extract flavones, producing higher extant yields than white clover.

5.2.2.4 饲用价值与利用

（1）用于青饲或放牧：红三叶草质柔嫩多汁，营养丰富，适口性好，多种家畜都喜食，反刍家畜极少发生膨胀病。

（2）调制青干草：叶片不易脱落，可制成优质干草。

（3）打浆喂猪，可节省精料。

（4）其他用途：花期长，可作为良好的蜜源植物；寿命较短，生长发育早而快，可作保护草种；根系发达，能给土壤遗留大量的有机质，增强地力，宜于中长期草田轮作；可用于水土保持、缀花草坪；营养生长期较耐荫蔽，宜在林地树间种植；亦可用作异黄酮提取，提取产量高于白三叶。

5.3 *Astragalus* L. 黄芪属

5.3.1 Erect milkvetch

Scientific name: *Astragalus adsurgens*

Erect milkvetch is endemic in China. It has been cultivated for several decades or even a hundred of years. It originated in the Yellow River basin of Hebei, Henan, Shandong, and Jiangsu provinces and gradually spread to northern provinces through successful aerial seeding in the 1970s. Erect milkvetch is the main specie to improve barren hills and hillsides in the north area of China.

5.3.1.1 Botanical characteristics

Erect milkvetch is a perennial legume plant with a thick and deep taproot and large lateral roots. The root system

5.3.1 沙打旺

学名： *Astragalus adsurgens*

英文名： erect milkvetch

沙打旺为我国特有牧草。栽培历史达数十年乃至上百年，原在河北、河南、山东、江苏等省的黄河故道地区广泛栽培。20 世纪 70 年代随着大面积飞播成功，沙打旺逐步向北方各省区推广，是北方改造荒山荒坡进行飞播的主要草种。

5.3.1.1 植物学特征

沙打旺为豆科黄芪属多年生草本植

is distributed horizontally in the 0~30 cm soil layer. The main stem is not distinct, erect or oblique with about 10~30 branches. Clustered stems are round, hard and hollow with a diameter of 0.6~1.2 cm and height of 1.3~2.3 m(Fig.5.14). Stems are tender in the seedling stage and lignify gradually with maturity. Leaves are odd-pinnate with 7~25 oblong leaflets arranged alternatively on each leaf. Stems and leaves are covered by T-shaped white hairs. Triangular stipules are membranous. The long raceme extending from the axillae has 17~59 purple or blue-purple flowers(Fig.5.15). The flower has a tubular bell calyx, papilionaceae crown and 5~10 cm peduncle. Pods are rectangular or oval and with a downturned beak on top and a two-compartment ovary. Each pod has more than 10 seeds. Seeds are round or kidney-shaped. Thousand seed weight is 1.5~2.4 g (Fig.5.16).

物。主根粗长，入土深2 m，侧根发达，水平分布于0~30 cm土层内，根毛不多，着生大量根瘤。茎丛生，主茎不明显或斜生，全株有“丁”字形绒毛，圆硬中空，直径0.6~1.2 cm，高1.3~2.3 m（图5.14），幼时脆嫩，老时木质化，分枝多达10~30个。奇数羽状复叶，互生，每叶具小叶7~25枚，长椭圆形，茎叶上有“丁”字形白色茸毛；托叶三角形，膜质。长总状花序，腋生，花萼筒状钟形，蝶形花冠，总花梗很长，5~10 cm，每个花序有小花17~79朵，紫色或蓝紫色（图5.15）。荚果矩形或椭圆形，顶端有向下弯的喙，子房二室，内含种子10余粒；种子圆形或肾形，千粒重1.5~2.4 g（图5.16）。

Fig.5.14 Erect milkvetch
图5.14 沙打旺

Fig.5.15 Erect milkvetch—inflorescence
图5.15 沙打旺的花序

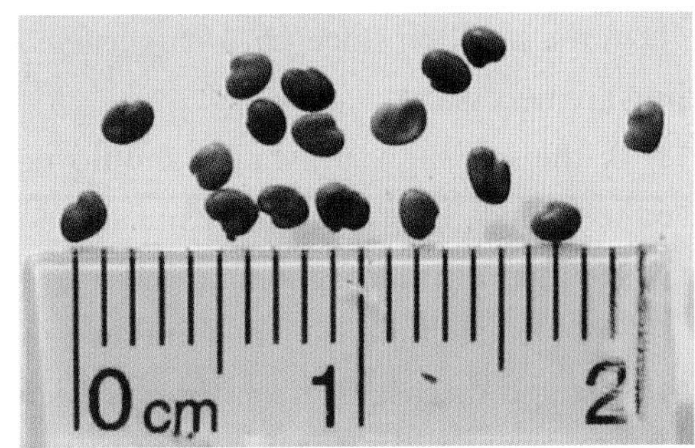

Fig.5.16 Erect milkvetch—seed
图5.16 沙打旺种子

5.3.1.2 Biological characteristics

Erect milkvetch has good tolerance of adverse environmental conditions. Therefore, it is well adapted to grow over a wide range of locations. It can be grown on poor, saline and alkaline soils (pH 9.5~10.0, salt content 0.3%~0.4%) and withstands drought, cold and sandstorms.

It prefers a warm climate and the optimal temperature for growth is 8~15℃ . However, it can grow in cold temperatures as cold as -30 ℃ .Tolerance to drought stress of erect milkvetch is better than alfalfa. It is not adapted to humid climates. Irrigation is helpful when available. It can grow in a wide range of soils, but prefers loam and sandy soils with rich calcium and good permeability.It tolerates shade conditions. Seeds can shatter easily, but tolerate wind deposition and sand burial.

5.3.1.3 Cultivation technology

Erect milkvetch is widely adapted. It can be planted in abandoned farmland and degraded, desertified and saline grasslands.

(1) Sowing

Erect milkvetch does not need fertilization for it is very tolerant of infertile soil. Seeds should be exposed to sunlight to break dormancy 1~2 d before sowing. The seeding rate depends on intended use: 1.5~3 kg/hm^2 for seed production when planted in abandoned farmland and normal farmland; 3.75~7.5 kg/hm^2 for forage production. Sowing depth is 1~2 cm with a 30 cm row spacing.

(2) Field management

① In the seedling stage (2~3 true leaves), intertilage and weeding are necessary.

② In the seeding year, stands should not be grazed but allowed to reseed. The replanting time should not be too late, otherwise it may affect the plants to over wintering.

③ Stubble height of 5~6 cm should be maintained.

④ Pest and disease control: Pesticide should be used, as needed, to prevent and control common pests, such as aphids and beetles. Diseases common to erect milkvetch are powdery mildew, stem rot and root rot. Carbendazim or tuzet are generally used to prevent and cure powdery mildew. High

5.3.1.2 生物学特性

沙打旺抗逆性强，适应性广，具有抗旱、抗寒、耐瘠薄、耐盐碱（pH 9.5~10.0，全盐含量0.3%~0.4%）、抗风沙等特点。

温度：喜温暖气候，适宜于年均温8~15℃的地区，但能抗寒-30℃。

水分：耐旱（耐旱能力比紫花苜蓿强）怕湿，需要灌溉。

土壤：要求不严，适宜于钙质丰富、通透良好的壤土和沙土。

光照：耐荫蔽。

种子易落粒，抗风蚀和沙埋。

5.3.1.3 栽培技术

沙打旺适应性广，在耕地、弃耕地和退化、沙化、盐碱化草地的各类土壤均能种植。

（1）播种

耐瘠薄，一般播种不进行施肥。播种前要晒种1~2 d，播种量视利用目的而定，种子田、缓坡、弃耕地或耕地每公顷播种量1.5~3 kg；收草地，每公顷播种量3.75~7.5 kg，播深1~2 cm，行距30 cm左右。

（2）田间管理

① 苗期：在2~3片真叶时中耕除草。

② 播种当年严禁牲畜放牧，及时补播，补播不能过晚，以免影响越冬。

③ 刈割留茬高度不宜过高，一般5~6 cm。

④ 防治病虫害：虫害有蚜虫、金龟子等，要及时进行药物防治；常见病害有白粉病、茎腐病、根腐病等。对白粉病可用多菌灵或退菌特防治；长时间积水或土壤水分过多时，易发生茎腐病和根腐病，应及时做好排水防涝工作；菟丝子是危害沙打旺最严重的寄生植物，一旦发现，要连同被害

soil water content may induce stem rot and root rot of erect milkvetch. It is necessary to drain water and prevent waterlogging in time. Dodder is a parasitic plant harmful to erect milkvetch. Once found, it should be removed, destroyed or deep-buried together with the affected plant. Spraying fungicide is useful to control damage from dodder.

5.3.1.4 Harvesting and utilizations

(1) Harvest in time: Harvest should coincide with the late stage of vegetative growth or when plant height is 80~100 cm. Because of poor regrowth, it can be cut for forage only once or twice every year and the yield is 30~120 t/hm^2. Seeds in different maturity stages are present on the same plant. Therefore, seed should be harvested repeatedly. Seed yield is 225~450 kg/hm^2.

(2) Erect milkvetch is rich in nutrients. It can be harvested for fresh forage, silage, hay and processed into powder and compound fodder. Palatability is poor because it contains 3-nitro-1-propanol and β- glycoside which make erect milkvetch slightly toxic and bitter. It is usually combined with other forages. In the right proportion it is not harmful to ruminants.

(3) Erect milkvetch is suitable for use as fresh forage and hay, but is not appropriate to graze. Feeding chickens and rabbits with excessive amounts of erect milkvetch could cause toxicity. Erect milkvetch should be mixed in reasonable proportion with other forages which have good palatability to feed cattle and sheep.

(4) Silage: Erect milkvetch is suggested as silage mixed with other forages. Fresh, crushed and fermented erect milkvetch can be used to feed pigs. Crushed and fermented erect milkvetch can be stored in a cellar to be used in the winter and spring when fresh forages are deficient.

(5) Stems can be used as fuel and fertilizer.

(6) It is useful as green manure and can be used in crop rotations. Conditions of high temperature and humidity are best to make green manure. Stubble of erect milkvetch can improve soil fertility by enriching the soil with quantities of organic matter and nitrogen.

(7) Other uses: It is a nectariferous plant with an ex-

植株全部拔除销毁或深埋，也可用菌剂喷施防治。

5.3.1.4 收获利用

（1）适宜刈割时期为营养生长后期，或当株高达 80~100 cm 时进行。再生能力较弱，一年内可刈割 1~2 次，鲜草产量达 30~120 t/hm^2。种子成熟很不一致，要分期分批采摘，种子产量为 225~450 kg/hm^2。

（2）营养成分含量丰富，可青饲、青贮、调制干草、加工草粉和配合饲料等，但因其植株含 3– 硝基 –1– 丙醇、β – 葡糖苷，有微毒，带苦味，适口性差，当与其他牧草适量配合利用时对反刍动物较为安全。

（3）适宜做干草或青饲，不宜放牧。单独饲喂沙打旺过量，易造成鸡、兔中毒，饲喂牛、羊，要与适口性较好的牧草适量配合利用。

（4）青贮：不宜单独青贮；青绿沙打旺喂猪，可打浆饲喂或打浆发酵饲喂，还可以打浆窖贮，以扩大冬春季节青绿饲料来源。

（5）茎秆是上好的燃料和肥料。堆肥时，可在短期内达到腐熟的目的。

（6）是优良的绿肥和草田轮作牧草。用于绿肥时，可进行异地翻压，在高温高湿条件下效果最好。其肥效高，茬地能给土壤遗留大量有机质和氮素，提高地力。

（7）其他利用方式：沙打旺花期长达 45~60 d，是很好的蜜源植物；茎秆外皮柔软耐拉，可削皮拧绳、织袋；种子能榨油，也可饲用；用于防风固沙，要用沙蒿、沙拐枣等作保护行。

tended flowering period. The stem cortex can be peeled for making rope and bags because it is soft and pull-resistant. Seeds can be used to extract oil and used as fodder. It also can be used as a windbreak and sand-stablization. Artemisia desterorum spreng and calligonum are often planted on the boundaries as nurse plants.

5.3.2 Chinese milkvetch

Scientific name: *Astragalus sinicus* L.

5.3.2.1 Botanical characteristics

Chinese milkvetch is an annual or biennial herbaceous plant with a thick taproot and well-developed lateral roots. The root system is concentrated in the 15~30 cm soil layer. Purple-red or brown nodules are scattered on the lateral roots. Stems are prismatic, hollow, erect or prostrate with a height of 30~100 cm (Fig.5.17). Leaves are odd-pinnate and leaflets are long oblong, entire margin with slightly concave or emarginate top. Stipules are oblong with a slightly pointed tip. The umbels grow from axillae. The pedicel is slender. Flowers are light red or purple red. The elongated pod with beak top has 5~10 seeds (Fig.5.18). Seed is kidney-shaped, yellow or black. Thousand seed weight is 3.0~3.5 g (Fig.5.19).

5.3.2 紫云英

学名： *Astragalus sinicus* L.

英文名： astragalus, Chinese milkvetch

5.3.2.1 植物学特征

豆科黄芪属一年生或越年生草本植物。主根肥大，侧根发达，密集于 15~30 cm 土层，侧根密生紫红色或褐色根瘤。茎棱形、中空、直立或匍匐，高 30~100 cm（图 5.17）。奇数羽状复叶，小叶长椭圆形，全缘，顶端微凹或微缺，托叶卵形，顶端稍尖。伞形花序腋生，花柄细长，花冠淡红或紫红，荚果细长，顶端喙状，每荚种子 5~10 粒（图 5.18）。种子肾形，黄绿色至黑色，千粒重 3.0~3.5 g（图 5.19）。

Fig.5.17 Chinese milkvetch
图 5.17 紫云英

Fig.5.18 Chinese milkvetch—pod
图 5.18 紫云英的荚果

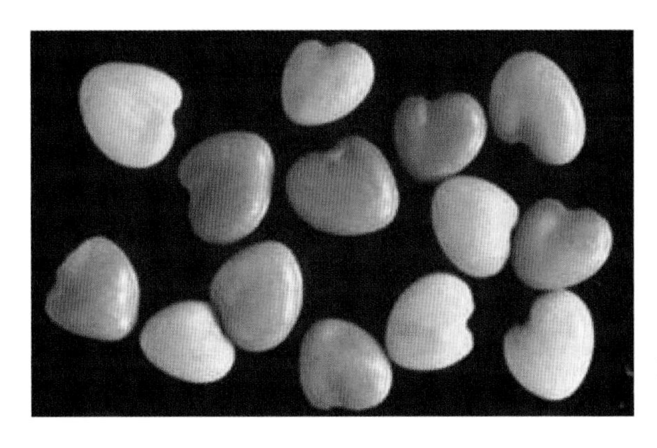

Fig.5.19 Chinese milkvetch—seed
图 5.19 紫云英种子

5.3.2.2 Biological characteristics

Chinese milkvetch prefers a warm and humid climate. It has poor tolerance of cold. The optimal temperature for growth is 15~20℃. It has poor drought and waterlogging tolerance, but it is quite tolerant of high humidity. It grows well in clay and clay loam soil with high fertility. Tolerance to acid soil is good, but not to alkaline soil.

5.3.2.3 Cultivation technology

Chinese milkvetch can be sown in spring and autumn. Before sowing, seeds should be treated to overcome dormancy and inoculated with rhizobia. The seeding rate is 30~60 kg/hm^2. When planted in mixture with grass, like annual ryegrass, the seeding rate of Chinese milkvetch is 15 kg/hm^2. In southern areas, seeds can be directly broadcast on the farmland soil surface after rice harvest. After preparation of the seedbed, sowing in rows or dibble seeding is feasible. It may promote seed germination and plant growth to fertilize with plant ash mixed with phosphate rock at the time of sowing. Phosphate fertilizer may improve N-fixation capability and enhance disease resistance. In case of disease or pest epidemic, all injured plants should be removed promptly or disease and pests should be controlled using pesticides.

5.3.2.4 Utilization

Chinese milkvetch has high moisture and good palatability with high crude protein content (20%~30%). It can be processed into silage and hay. The optimal time for cutting is the early bloom stage. Yield of fresh forage is 22~38 t/hm^2,

5.3.2.2 生物学特性

温度：喜温暖潮湿气候，不耐寒，生长最适温度为 15~20℃。

水分：耐旱性较差，较耐湿但不耐积水。

土壤：喜黏土或黏壤土，不耐瘠薄，耐酸性较强但不耐碱。

5.3.2.3 栽培技术

春播、秋播均可，播前硬实处理，接种根瘤菌。播种量为每公顷 30~60 kg，与禾本科牧草如多花黑麦草混播时，每公顷播种量 15 kg 即可。南方多采用收获后稻田直接撒播或耕翻土壤撒播，也可整地后条播或点播。在播种的同时施以草木灰拌磷矿粉（用人粪尿拌和）有利于萌芽和生长。多施磷肥能提高固氮能力和抗病力，使植株旺盛生长。如遇病虫害发生，应及时拔除病株或施用农药防除。

5.3.2.4 利用方式

紫云英鲜草鲜嫩多汁，适口性好，粗蛋白质含量很高（20%~30%），可青饲、青贮或调制干草，最好在初花期利用。鲜草产量 22~38 t/hm^2，最高达 60 t/hm^2。年刈割 2~3 次。

and could reach to 60 t/hm^2. It can be harvested 2~3 times each year.

5.3.3 Cicer milkvetch

Scientific name: *Astragalus cicer*

Cicer milkvetch is native to Europe. It was introduced to China from Canada and prefers a cold and humid climate. Cicer milkvetch produces rhizomes, or extended horizontal stems propagated underground, growing continuously as new branches (Fig.5.20). Stems are prostrate or erect with a height of 50~100 cm and branches can reach lengths of 1~2 m (Fig.5.21). Leaves are pinnate and alternate with about 16~20 leaflets. The raceme has 5~40 white or yellow flowers. Flowering occurs from June until frost. Pods are dark, bladder-shaped with 3~11 seeds (Fig.5.22). Seeds are yellow and lustrous. Thousand seed weight is about 7.7 g (Fig.5.23).

5.3.3 鹰嘴紫云英

学名： *Astragalus cicer*

英文名： cicer milkvetch

原产于欧洲，我国自加拿大引入，喜寒冷湿润气候。根在表层土中向四周匍匐，根茎芽出土后，可形成新的分枝（图 5.20）。茎匍匐或半直立，草层高度 50~100 cm，分枝长度可达 1~2 m（图 5.21）。羽状复叶，16~20 片。总状花序有花 5~40 朵，花冠白色或淡黄色，花期从 6 月到降霜。荚果黑色，膀胱状（图 5.22），含种子 3~11 粒，黄色有光泽，千粒重 7.7 g 左右（图 5.23）。

Fig.5.20 Cicer milkvetch
图 5.20 鹰嘴紫云英

Fig.5.21 Cicer milkvetch—grassland
图 5.21 鹰嘴紫云英草地

Fig.5.22 Cicer milkvetch—pod
图 5.22 鹰嘴紫云英的荚果

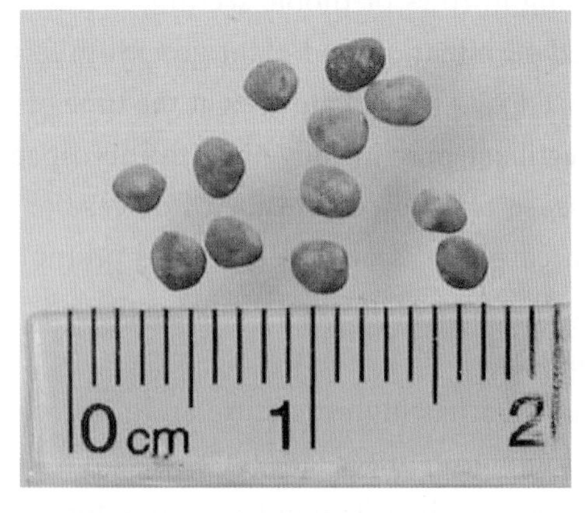

Fig.5.23 Cicer milkvetch—seed
图 5.23 鹰嘴紫云英的种子

Cicer milkvetch can reproduce by seeds and stems, and has a strong capacity for asexual reproduction. The seeding rate is 7.5~15 kg/hm^2 with a row spacing of 30~60 cm and sowing depth of 2~3 cm. In the seedling stage, prompt weed control is necessary.

Yield of fresh forage is 15~45 t/hm^2. It can be processed into hay or directly grazed by livestock. Cicer milkvetch has good palatability and does not induce bloat. It also can be used to conserve water and soil.

5.3.4 Sweetclover-like milkvetch

Scientific name: *Astragalus melilotoides* Pall

Sweetclover-like milkvetch is a perennial herbaceous plant with a long, thick taproot. Erect or oblique stems are 30~100 cm high (Fig.5.24). Leaves are odd pinnate with 3~7 leaflets. The long raceme has light purple or white flowers. Pods are spherical (Fig.5.26).

It has a strong tolerance of drought, and can grow on hillside, ditch banks, wilderness, river bank, sand and grassy slope. As a superior forage plant, it is favored by horses, cattle and sheep. The grazing rate is about 80%. Harvesting earlier for high crude protein content in the early stage of growth is advisable.

种子直播或扦插，无性繁殖能力较强。播种量 7.5~15 kg/hm^2，行距 30~60 cm，播深 2 ~ 3 cm。苗期弱，应注意防除杂草。

每公顷产鲜草 15~45 t，调制干草或放牧，适口性好，不会引起臌胀病。亦可用来保持水土。

5.3.4 草木樨状黄芪

学名： *Astragalus meliotoides* Pall

英文名： sweetclover-like milkvetch

多年生草本，主根粗而长，茎直立或斜生，高 30~100 cm（图 5.24），奇数羽状复叶，小叶 3~7，总状花序长，花粉紫色或白色，荚果近球状，椭圆形（图 5.25）。

抗旱，多生于山坡沟旁、荒野、河岸沙地、草坡。为优等饲用植物，马、牛、羊喜食，采食率达 80%，须及早收获，早期蛋白质含量多。

Fig.5.24 Sweetclover-like milkvetch
图 5.24 草木樨状黄芪

Fig.5.25 Sweetclover-like milkvetch—pod
图 5.25 草木樨状黄芪的荚果

5.4 *Onobrychis* Mill. — sainfoin 红豆草属—红豆草

Sainfoin

Scientific name: *Onobrychis viciaefolia*

Sainfoin is distributed mainly in Europe, northern Africa, western and southern Asia. Wild species appear in the north slope of Tianshan Mountain in Xinjiang at latitudes of 1000~2000 m. The cultivation areas are primarily in Inner Mongolia, Xinjiang, Shaanxi, Ningxia and Qinghai provinces. With gorgeous pink flowers and high nutrient value, sainfoin is regarded as the "queen of forages". Ordinary sainfoin and caucasus sainfoin have been introduced from other countries. They have been planted in large areas in Europe, Africa and Asia at present.

红豆草

学名：*Onobrychis viciaefolia*

英文名：sainfoin

主要分布在欧洲、北非、亚洲西部和南部，我国新疆天山北坡海拔1 000~2 000 m半阴坡有野生种。国内种植较多的省区有内蒙古、新疆、陕西、宁夏、青海。红豆草花色粉红艳丽，饲用价值可与紫花苜蓿媲美，故有“牧草皇后”之称。国内栽培的全是引进种，主要是普通红豆草和高加索红豆草。现在欧洲、非洲和亚洲都有大面积的栽培。

5.4.1 Botanical characteristics

This perennial legume has deep, branched taproot with the depth of 1~3 m. Lateral roots are symbiotic with a large quantity of rhizobium bacteria. Stems are erect, hollow, green or purple red reaching heights of 70~150 cm with 5~15 branches (Fig.5.26). The first true leaf is solitary. Additional leaves are odd-pinnately compound with 6~14 pairs of leaflets. Leaflets are ovoid, oblong or elliptic with short hairs on the edge of leaf back. Stipules are triangular. The raceme is 15~30 cm long with 40~75 flowers (Fig.5.27). Flowers are papilionaceous, pink or red or dark red. Pods are flat, yellowish-brown with serrated edges. The pod is rough with convex reticulated veins (Fig.5.28). It is difficult to crack when the pod is mature. Each pod contains one kidney-shaped, green-brown seed. The seed, with a smooth coat, is 2.5~4.5 mm long, 2.0~3.5 mm wide, and 1.5~2.0 mm thick. Thousand seed weight is 13~16 g. The hard seed rate is less than 20%.

5.4.1 植物学特征

红豆草是豆科红豆草属多年生草本植物。深根型，根系强大，主根粗壮，入土1~3 m或更深，侧根随土壤加厚而增多，着生大量根瘤。茎直立，中空，绿色或紫红色，高70~150 cm，分枝5~15个（图5.26）。第一片真叶单生，其余为奇数羽状复叶，小叶6~14对或更多，卵圆形、长圆形或椭圆形，叶背边缘有短茸毛，托叶三角形。总状花序，长15~30 cm，有小花40~75朵（图5.27），蝶形，粉红色、红色或深红色，十分美丽。荚果扁平，黄褐色，果皮粗糙有凸形网状脉纹，边缘有锯齿，成熟后不易开裂，内含肾形绿褐色种子一粒（图5.28）。种皮光滑，长2.5~4.5 mm，宽2.0~3.5 mm，厚1.5~2.0 mm，千粒重13~16 g，带壳千荚重18~26 g，硬实率不超过20%。

Fig.5.26 Sainfoin
图 5.26 红豆草

Fig.5.27 Sainfoin—flower
图 5.27 红豆草的花

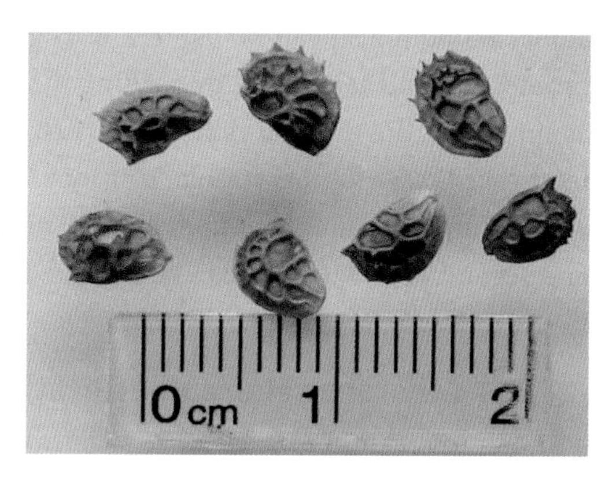

Fig.5.28 Sainfoin—pod
图 5.28 红豆草荚果

5.4.2 Biological characteristics

Sainfoin prefers a cold, dry climate and tolerates drought, but not waterlogging. It has better drought tolerance but poorer cold tolerance than alfalfa. The optimal temperature for growth is 20~25℃. Therefore, it is best adapted where the average annual temperature is 3~8℃, the frost-free period is about 140 days and annual rainfall is about 400 mm. Sainfoin can grow on a variety of soils, even dry, barren soils, gravel soils and silty loam and chalky soils, but it prefers dry calcareous soils, loose carbonate soils and fertile farmland. It is not adapted to acid soils, marsh or areas with a high water table. It can be harvested in the seeding year when sown in the spring. If planted in the fall, green up the next spring is one week earlier than alfalfa but regrowth ability is poorer. Sainfoin is cross-pollinated with a low natural seed setting

5.4.2 生物学特性

红豆草性喜温凉、干燥气候，耐干旱、忌水淹（5 d 即死亡），与苜蓿比，抗旱性强，抗寒性稍弱。生长最适温度为 20~25℃，适应栽培在年均气温 3~8℃，无霜期 140 d 左右，年降水量 400 mm 上下的地区。红豆草对土壤要求不严格，可在干燥瘠薄、土粒粗大的沙砾、沙壤土和白垩土上生长；在富含石灰质的土壤、疏松的碳酸盐土壤和肥沃的田间生长极好；不适宜在酸性土、沼泽地和地下水位高的地方栽培。春播当年即有产量；秋播次年返青较苜蓿早一周，再生性不及苜蓿。红豆草属严格的异花授粉植物，在自然状态下，结实

rate of about 50%.

5.4.3 Cultivation technology

Before seeding, weeds should be removed and adequate phosphate and potassium fertilizer applied. It should be planted in rotations every 5~6 years. Shells covering seeds would not hinder seed germination and the embryo may be injured by decorticating shells. The hard seed rate is 4%~20%. The best time for planting are early spring and August. The seeding rate is 60~75 kg/hm^2 for forage production and 37~53 kg/hm^2 for seed production. Seeding depth should be 3~4 cm and the row spacing is 30~60 cm when seeding alone. Sainfoin also can be planted with grasses like smooth bromegrass and tall fescue by seeding in alternative rows or as mixtures. The optimal time for harvesting is from flowering to seed set. After harvesting, soil should be fertilized with 750 kg/hm^2 phosphorus and 90 kg/hm^2 nitrogen for the first and second year. Nitrogen can be provided by biological fixation afterwards.

5.4.4 Forage value and utilization

Sainfoin can be harvested 2~4 times every year and the yield of hay is about 12~15 t/hm^2. The yield of first cutting is the highest, providing about 50% of the total annual production. Second cutting yield is only half as much as the production of the first cutting. After the first cutting for forage, the sainfoin stand usually can be used for grazing. Cutting at the soil surface does not affect branching and regrowth. However, the stubble height of Caucasus sainfoin should be 5~6 cm to allow regeneration from buds growing from axillae. Seed yield reaches the highest level about 450~750 kg/hm^2 in the second and third year. Seeds shatter easily, so harvest should not be delayed. Seed yield can be increased by fertilizing with phosphate or spraying phosphate around roots.

Fresh forage and hay are high in protein, vitamins and minerals and is highly digestible. Digestibility is 75% when it is harvested in the seedling stage, and 65% when harvested after flowering. Sainfoin is a nonbloating legume for it contains high levels of tannin.

率较低，一般只在 50% 左右。

5.4.3 栽培技术

清除杂草，施足磷、钾肥，不宜连作，隔 5~6 年再种。带壳不影响种子发芽率，去壳会损伤胚，硬实率为 4%~20%；播期以春季（早春）或 8 月为宜，草地建植播种量为 60~75 kg/hm^2，种子田播种量为 37~53 kg/hm^2；单播行距 30~60 cm，播深 3~4 cm。混播时可与无芒雀麦、苇状羊茅等隔行或混播，开花至结荚收获最好。可施磷肥 750 kg/hm^2，前 1~2 年氮肥以 90 kg/hm^2 为宜，之后可靠自身固氮。

5.4.4 饲用价值与利用

红豆草年刈 2~4 次，年产干草 12~15 t/hm^2，均以第一次产量最高（50%），二茬只有头茬的一半。多为一茬收草，二茬放牧，齐地刈割不影响分枝，而高加索红豆草再生芽从叶腋产生，应留茬 5~6 cm。种子产量以第 2~3 年最高，达 450~750 kg/hm^2，落粒性严重，采种不能过迟，施磷肥或根外喷磷可增加种子产量。

无论干、鲜草，均蛋白质含量丰富，维生素及矿物质含量多。消化率高，幼苗期达 75%，花后降至 65%。最大优点为含有较高浓缩单宁，不发生臌胀病。

青饲：茎中空脆嫩，气味芳香，适口性仅次于苜蓿，花后适口性变差。

放牧：刈割两次后，再生草放牧优于苜蓿，含 1%~1.5% 浓缩单宁可沉淀可溶性蛋白质，不会引起臌胀病。

Sainfoin is high in palatability, second only to alfalfa, due to its hollow and crisp stem and fragrant smell. Palatability may decrease after flowering.

After cutting twice, regrowth is more satiable for grazing than alfalfa. Because it contains 1%~1.5% tannin which combines with soluble protein. There is no risk of bloat. It does not produce seed by self-pollination. Yield of honey is 120 kg/hm^2.

5.5 *Coronilla* L.— crownvetch 小冠花属—小冠花

Crownvetch

Scientific name: *Coronilla varia* L.

5.5.1 Botanical characteristics

The plant produces a well-developed taproot with many branches. The root system is mainly concentrated in the 0~40 cm soil layer. The taproot has a diameter of more than 2 cm and can penetrate to a depth of 80cm after only one year. Lateral roots can spread up to 1 m. Crownvetch spreads by rhizomes, which can generate new plants or produce new subterraneous stems. The irregular nodules on the root support strong symbiotic N-fixation. Stems are hollow, angular, upright or inclined, soft and crisp with a height of 90~150 cm. Plant layer is about 50~60 cm(Fig.5.29). Leaves are odd-pinnately compound with 9~27 leaflets. Leaflets are long oblong or obovate with entire edges. The petiole is 15~25 cm long and the stipule is sharp tongue-shaped. The inflorescence from the axillae has 8~22 flowers closely arranged on the top of the pedicel. The pedicel is 8~12 cm. Flowering may last 2~3 months and flowers have variegated color from pink to red to purple-blue. Lathy and sectionalized pods are easily broken on the nodes (Fig.5.30). Seeds are clavate, purple-red or purple-brown and 3.5 mm long and 1 mm wide (Fig.5.31). Thousand seed weight is 3.1~4.1 g. Hard seed rate is 19%~30%.

小冠花

学名： *Coronilla varia* L.

英文名： crownvetch

5.5.1 植物学特征

主根粗壮，直径达 2 cm 以上，侧根发达，主要分布于 0~40 cm 土层，质嫩，色黄白。主根一年后入土深达 80 cm，侧根横向穿走达 1 m 以外，根上有许多不定芽，可形成新植株和新地下茎，根上均有不规则块状或棒状根瘤，较大，固氮效果好。茎中空有条棱，柔软而脆嫩，上升倾卧或铺展生长，着地处不生根，长 90~150 cm，草层高 50~60 cm（图 5.29）。奇数羽状复叶，小叶互生，9~27 片，长椭圆或倒卵形，全缘，总叶柄长 15~25 cm，托叶尖舌状。腋生伞形花序，花梗 8~12 cm，由 8~22 朵小花呈杯状紧密排列于花梗顶端，形似冠，故名小冠花。花期长达 2~3 个月，花色为粉红 – 红色 – 蓝紫色，故名多变小冠花。荚果细长分节，节处易断（图 5.30）。种子棒状，紫红色或紫褐色，长约 3.5 mm，宽 1 mm，千粒重 3.1~4.1 g（图 5.31），种皮坚硬，硬实率达 19%~30%。

Fig.5.29 Crownvetch—stem, leaf and flower
图 5.29 小冠花的茎、叶和花

Fig.5.30 Crownvetch—pod
图 5.30 小冠花的荚果

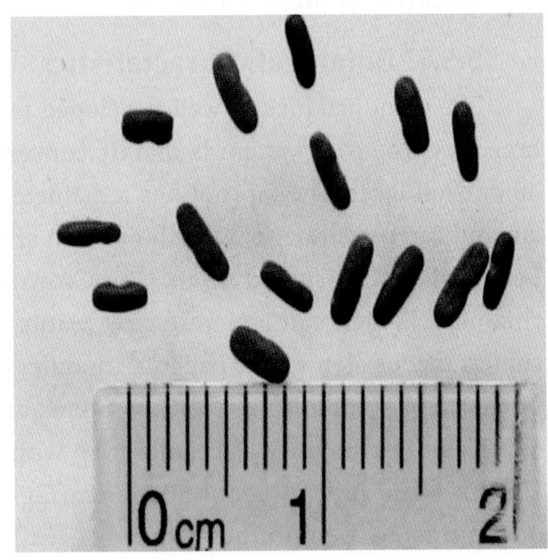

Fig.5.31 Crownvetch—seed
图 5.31 小冠花的种子

5.5.2 Biological characteristics

Crownvetch prefers a warm climate and tolerates cold temperature. The optimal temperature for growth is 20~30℃. The best temperature for germination is 25℃. It has good tolerance of high temperature and does not wither. It can survive winter temperatures of -34℃ in northern Shanxi province and can tolerate cold environment as low as -22℃ in the seedling stage. It has good drought tolerance but not humid and waterlogged soil. Slight shade is best for growth.

5.5.2 生物学特性

温度：喜温耐寒，生长适温为20~30℃，25℃时发芽最快，抗炎热，高温不枯萎，这是与苜蓿等豆科牧草相比的优势之一，晋北 -34℃能越冬，苗期耐 -22℃低温。

水分：抗旱怕涝不耐湿。成株淹水3~4 d 根烂死亡。

光照：轻度遮荫对生长有利，生长

Two years after planting, it can be intercropped with maize or sorghum.It can grow in a wide range of soil, even infertile or waste slag soil. In the seedling stage, crownvetch tolerates soil with 0.5% salt. It grows well on soils with pH 6.8~7.5.

5.5.3 Cultivation technology

The hard seed fraction is high. Procedures to overcome hard seeds include: soaking the seeds with 70~80℃ water and pouring water while stirring until the temperature drops to 30℃; scratching the seed coat; scarifying seeds with concentrated sulfuric acid.

The proper time for sowing is spring and summer. The 1~1.5 m of spacing with a sowing depth of 1~2 cm is recommended. The seeding rate is 7.5 kg/hm^2.

5.5.4 Forage value and utilization

Crownvetch is suitable for grazing and feeding as fresh forage because it has strong trampling resistance. Yield of fresh forage is 60~90 t/hm^2, and leaves comprise 65.6% of the total dry matter of forage production. It should be harvested promptly when the plant height is 50~60 cm. Otherwise, lower leaves mature and fall off if harvested too late. Crownvetch can be cut 3~4 times every year. Cutting should be done from bud stage to early bloom stage for hay production and the hay yield is 15~23 t/hm^2. Cutting for haylage should beat early bloom stage. Crownvech contains β-nitro propionic acid, which is toxic to monogastric livestock. It may inhibit their growth and development if cownvetch accounts for more than 2.5% or 5% of chicken or pig diets, respectively. However, it is not toxic to ruminants. Seeds can be harvested when pods are yellow, dry and shrunken. Seed yield varies among cultivars.

Crownvetch can cover the ground rapidly and can be used for water and soil conservation. It also can be used to protect soil on hillsides and embankments and as a landscaping plant.

两年后宜与玉米、高粱等高秆作物间作套种，可充分利用地力等资源。

土壤：要求不严，瘠薄地、废矿渣地都能生长，苗期耐盐 0.5% 以内，适于 pH 6.8~7.5 的土壤。

5.5.3 栽培技术

硬实种子较多，处理方法为用温水浸种，或用 70~80℃热水浸种，边倒边搅，待温度降至 30℃时止；擦伤种皮；浓硫酸拌种。

春夏秋皆可播种，条播行距 1~1.5 m，播种量为 7.5 kg/hm^2，覆土 1~2 cm。

5.5.4 饲用价值与利用

青饲耐践踏，适宜放牧，鲜草产量 60~90 t/hm^2，干草叶占干重的 65.6%，草丛高 50~60 cm 时刈割，刈割过晚下部叶片会因荫蔽脱落，年刈 3~4 次。做青干草现蕾至初花期刈割，收干草 15~23 t/hm^2。或初花期刈割半干青贮。小冠花含有 β－硝基丙酸，对单胃家畜有毒害作用，占鸡日粮的 2.5%，猪日粮的 5% 时，会使其生长发育受阻，但对反刍家畜无害。当荚果呈黄色干缩状时种子成熟，应及时分批采收，种子产量因品种而异。

小冠花能迅速覆盖地面，是优良的水土保持植物，可用作护坡、护堤和美化环境。

5.6 *Lotus* L. — birdsfoot trefoil 百脉根属—百脉根

Birdsfoot trefoil

Scientific name: *Lotus corniculatus* L.

5.6.1 Botanical characteristics

This perennial legume forms a well-developed taproot with many branches. Distribution of the root system is shallower than alfalfa and concentrates mainly in the 30~60 cm soil layer. Stems are prostrate and well branched (Fig.5.32). The main stem is not distinct. Stems are smooth, without hairs, 60~90 cm high, and develop roots if they are separated from the main plant. Leaves are trifoliolate, with two stipules near the base of the petiole which is near the same size of leaf (Fig.5.33). The umbel has 4 to 8 flowers that are generally bright yellow (Fig.5.34). Pods are long and round and resemble the toes of a bird, and contain 10~15 seeds (Fig.5.35). Seeds are brown and thousand seed weight is 1.2 g (Fig.5.36).

百脉根

学名： *Lotus corniculatus* L.

英文名： birdsfoot trefoil

5.6.1 植物学特征

豆科百脉根属多年生牧草。主根强壮，侧根多而发达，分布较苜蓿浅，脉状分布于 30~60 cm 土层，故名百脉根。茎丛生，分枝性强，无明显主茎，匍匐，长 60~90 cm（图 5.32），光滑无毛，切断能生根。三出复叶，托叶位于总叶柄基部，大小与小叶相近，故又名五叶草（图 5.33）。伞形花序，蝶形花冠，黄色艳丽，小花 4~8 朵（图 5.34）。荚果长而圆，角状，似鸟趾，故名鸟趾豆（图 5.35）。每荚 10~15 粒种子，棕色，千粒重 1.2 g（图 5.36）。

Fig.5.32 Multiple branching birdsfoot trefoil
图 5.32 分枝性极强的百脉根

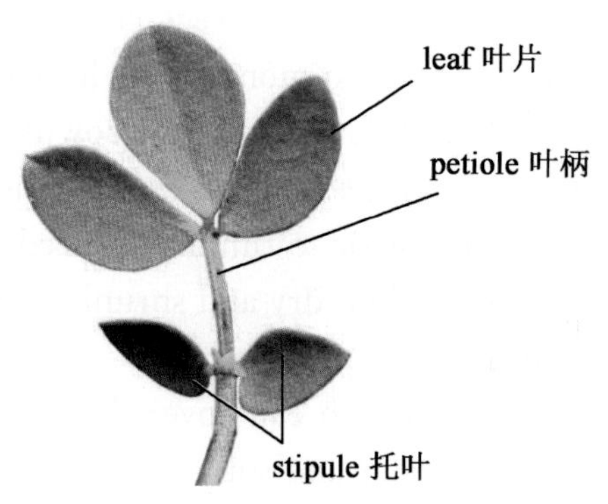

Fig.5.33 Birdsfoot trefoil—leaf
图 5.33 百脉根的叶

Fig.5.34 Birdsfoot trefoil—flower
图 5.34 百脉根的花

Fig.5.35 Birdsfoot trefoil—pod
图 5.35 百脉根的荚果

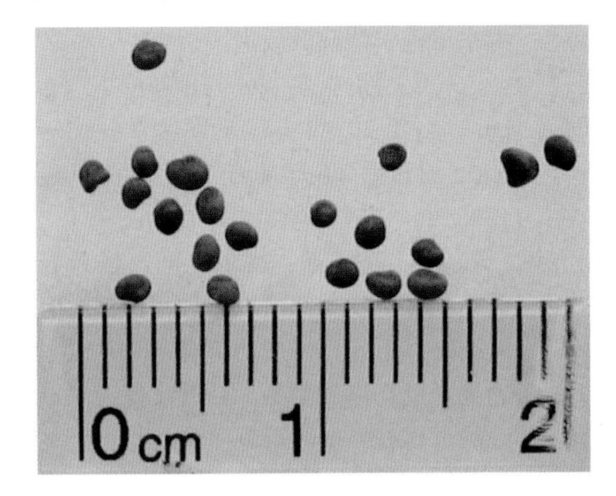

Fig.5.36 Birdsfoot trefoil—seed
图 5.36 百脉根的种子

5.6.2 Biological characteristics

Birdsfoot trefoil prefers a warm and humid climate. Tolerance of cold temperature is poor, but heat tolerance is superior to alfalfa.Drought tolerance is better than clover, but poorer than alfalfa.Birdsfoot trefoil is a heliophile and long-day plant.It is adapted to fertile and moist clay and silty loam, acid and slightly alkaline soils. Optimum soil pH is 6.2~6.5. Birdsfoot trefoil has good tolerance of trampling and grazing and moderate regrowth ability.

5.6.3 Cultivation technology

5.6.3.1 Toil and tillage

Birdsfoot trefoil is adapted to a range of soils, including silty loam, clay loam, infertile, poorly drained, slightly acid

5.6.2 生物学特性

温度：喜温暖湿润气候，不耐寒，耐热力强于苜蓿。

水分：耐旱强于三叶草，但比苜蓿差。

光照：喜光，长日照。

土壤：喜肥沃能灌溉的黏土、沙壤土、酸性及微碱土，pH 6.2~6.5 最好。

耐践踏、耐牧，再生力中等。

5.6.3 栽培技术

5.6.3.1 土壤与耕作

百脉根对土壤要求不严，在沙壤土，

and alkaline soils.

5.6.3.2 Fertilization

Stand of Birdsfoot trefoil can persist for several years and benefits from adequate fertilization. Deep ploughing in fall with adequate base fertilizer before sowing, 22~38 t/hm^2 organic fertilizer, 75~150 kg/hm^2 ammonium nitrate and 375~750 kg/hm^2 calcium superphosphate is recommended. Calcium superphosphate should be mixed with organic fertilizer, and moistened by spraying water and fermented for 20~30 d before application.

5.6.3.3 Sowing

(1) Hard seed treatment: Scratching the seed coat, soaking seeds in warm water or scarifying seeds with concentrated sulfuric acid can improve germination.

(2) Seeding rate: 6~7.5 kg/hm^2 for seed production field with a row spacing of 30~40 cm, 7.5~9 kg/hm^2 for forage production with row spacing 20~30 cm, 9.5~12 kg/hm^2 for broadcast seeding are recommended. Depth of sowing is 1~2 cm.

(3) Time for seeding: Birdsfoot trefoil is appropriately sown in spring and autumn, depending on the regional climate. When planted in autumn, it should be earlier than August to allow plants to grow for more than one month in order to survive the winter. With pastureland renovation, seed can be planted in early winter, allowing germination in the next spring.

(4) Methods of seeding: Birdsfoot trefoil can be planted alone or mixed with grass. When planted in mixture with grasses such as kentucky bluegrass, smooth bromegrass, orchardgrass, perennial ryegrass and meadow fescue, birdsfoot trefoil should occupy 40%~50% of the mixed stand. Plants can be propagated vegetatively using root and stem cuttings to maintain genotype.

5.6.3.4 Field management

In the seedling stage, it is necessary to till and control weeds. Grazing should be avoided to protect seedlings. It is inadvisable to graze heavily or graze after irrigation or rainfall. Two weeks should be allowed for plants to regrow after each grazing. After cutting and grazing the amount of fertilization should be 150~225 kg/hm^2 calcium superphos-

黏壤土，瘠薄地，排水不良的低湿地，微酸性或微碱性土壤均可种植。

5.6.3.2 施肥

百脉根播种后，利用年限较长，耗肥量较大，要结合秋耕深翻一次施足底肥，每公顷施有机肥 22~38 t，硝酸铵 75~150 kg，标准过磷酸钙 375~750 kg，过磷酸钙须事先与有机肥混合，洒水使其湿润，堆积腐熟发酵 20~30 d 再施用。

5.6.3.3 播种

（1）播前硬实处理：机械擦伤法；温水浸泡；浓硫酸浸泡。

（2）播种量：种子田为 6~7.5 kg/hm^2，行距 30~40 cm；收草为 7.5~9 kg/hm^2，行距 20~30 cm；撒播为 9.5~12 kg/hm^2。播种深度为 1~2 cm。

（3）播种期：视地区间气候差异而不同，春夏秋都可以播种，秋播不应迟于 8 月下旬，要使幼苗有一个月以上的生长期，以利越冬。补播改良天然草地的可以在入冬时寄籽播种。

（4）播种方式：单播、混播，常用适宜混播组合的草种有草地早熟禾、无芒雀麦、鸡脚草、多年生黑麦草、高牛尾草等。混播比例以百脉根占 40% ~ 50%为宜。在进行纯种繁育时，可用根或茎切段扦插繁殖。在补播改良天然草地时，可调整控制利用，以利种子充分成熟，便于藉种子自然落粒自行繁衍改良。

5.6.3.4 田间管理

（1）幼苗期应及时中耕，松土锄草，严禁放牧践踏毁坏幼苗。

（2）草地建植后不能过牧，不要在灌水、降雨后放牧，免损生机。每次放牧间隔半月，以利再生。

（3）每次放牧或刈割后要利用降雨或灌水追肥。每公顷施过磷酸钙 150~225 kg，磷二铵或硝酸铵 60~

phate, 60~75 kg/hm^2 phosphate diammonium or ammonium nitrate. Stubble height should be 5~8 cm after cutting and 10~15 cm following grazing.

75 kg，刈割留茬 5~8 cm，放牧留茬 10~15 cm。

5.6.4 Forage value and utilization

5.6.4.1 Harvesting and utilization

Seed should be harvested when 70% of the pods turn black brown or immediately when pods are mature. For forage production, harvest should be from early bloom to full-bloom stage. Grazing should begin from branching stage to bud stage or when canopy height is about 20 cm. It is inadvisable to graze heavily or graze after irrigation or rainfall.

5.6.4.2 Forage value

(1) Birdsfoot trefoil is high in protein with very palatable because of small stems and leaves, soft texture and a large quantity of leaves which comprise half of the total plant weight.

(2) This procumbent plant is suitable for grazing because it is nonbloating due in part to tannin content. Fresh yield is 30~75 t/hm^2 and it can be harvested 2~3 times every year. It also can be used as fresh forage, silage and hay. During storage, accumulation of free hydrogen cyanide can be problematic.

(3) It can be planted in rotation with cereal crops, grasses and other crops. It is also a nectariferous plant.

5.6.4 饲用价值与利用

5.6.4.1 收获利用

种子：70%的果荚变为黑褐色时收获，有条件可随熟随收，分期进行；

收草：初花期至盛花期收获；

放牧：从分枝期至孕蕾期，或草层高度达 20 cm 时适度放牧，切忌重牧或降雨、灌水后放牧。

5.6.4.2 饲用价值

（1）茎叶细小，草质细软，叶重为全株重的一半以上，适口性好；蛋白质含量高，与苜蓿、三叶草相近，是饲用价值很高的优良牧草之一。

（2）匍匐性强，适合于放牧；含单宁不会发生臌胀病，鲜草产量高达 30~75 t/hm^2，年刈 2~3 次。利用时应防止因堆积陈腐产生游离的氢氰酸使动物中毒，也可用于青饲、青贮、调制干草。

（3）常与禾谷类粮草作物和经济作物轮作倒茬利用。是良好的蜜源植物。

5.7 *Melilotus* Mill.—white sweetclover and yellow sweetclover 草木樨属—白花草木樨和黄花草木樨

The genus of *Melilotus* has 20 species, distributed in the temperate and subtropical zones of the northern hemisphere. It was introduced to Shandong province in 1922, and then gradually spread to other locations in China. At present, 9 species occur in China. White and yellow sweetclover are cultivated in large areas.

White sweetclover (Fig.5.37)

Scientific name: *Melilotus albus* (L.)Desr.

Yellow sweetclover (Fig.5.38)

Scientific name: *Melilotus officinalis* (L.)Lam.

此属全世界约 20 种，分布于北半球的温带和亚热带。我国 1922 年在山东进行引种试验，后逐渐推广至其他地区，现有 9 种，栽培较多的为白花草木樨和黄花草木樨两种。

白花草木樨（图 5.37）

学名： *Melilotus albus* (L.)Desr

英文名： white sweetclover

黄花草木樨（图 5.38）

学名： *Melilotus officinalis* (L.)Lam.

Fig.5.37 White sweetclover
图 5.37 白花草木樨

Fig.5.38 Yellow sweetclover
图 5.38 黄花草木樨

Origin of the genus *Melilotus* is controversial. However, it is generally believed to originate in Asia Minor, and spread to countries of Europe in the temperate zone. Cultivars planted in northern China are mainly biennial white and yellow sweetclover. The only differences between them are that the former is taller with white flowers, while the later is shorter with yellow flowers. White sweetclover matures 2 weeks later than yellow sweetclover.

英文名： yellow sweetclover

起源地说法不一，一般认为草木樨属植物起源于小亚细亚，后来传播到整个欧洲温带各国。我国北方栽培的主要是二年生白花草木樨和黄花草木樨，两者的不同是前者花冠白色，株体较高大，后者花冠黄色，株体较低，前者比后者晚熟约两周，其他性状则完全一致。

5.7.1 Botanical characteristics

Genus *Melilotus* plants are annual or biennial with a well-developed taproot. It can penetrate the soil as deep as 1.5 m. The upper portion of the taproot develops into a fleshy root, which is an vital organ to over wintering and is useful for green mature. Stems are upright, thick, round and hollow and 1~4 m high with about 5 branches (Fig.5.39). Leaves are pinnately compound with 3 leaflets per leaf and small stipules at the base of the petiole. The oblong leaflets are smooth on both surfaces and serrated around the entire margin(Fig.5.40). The inflorescence is a raceme from the axillae with many small flowers (40~80) (Fig. 5.41). Pods are oval or

5.7.1 植物学特征

草木樨属一年或两年生豆科草本植物，主根发达，入土 1.5 m 以上，根瘤众多；主根上部发育成肉质根，是越冬的重要器官，也是组成绿肥的重要部分。茎粗直立，圆而中空，高 1~4 m，每株分枝 5 个左右（图 5.39）。三出羽状复叶，叶缘有稀疏锯齿，中间小叶具短柄（图 5.40），托叶小而尖细，披针形，叶厚有苦味。总状花序，腋生花梗长，花小而多，40~80 个（图 5.41）。荚果卵圆或椭圆，表面有网纹，每荚含 1~2

oblong with netlike veins on the surface (Fig.5.42). There are one or two seeds in each pod. Seeds are kidney-shaped and brownish yellow. Thousand seed weight is 1.9~2.5g (Fig.5.43).

粒种子（图 5.42）。种子肾形，棕黄色，千粒重 1.9~2.5 g（图 5.43）。

Fig.5.39 Sweetclover—stem
图 5.39 草木樨的茎

Fig.5.40 Sweetclover—leaf
图 5.40 草木樨的叶片

Fig.5.41 Sweetclover—inflorescence (left:yellow sweetclover; right:white sweetclover)
图 5.41 草木樨的花序（左：黄花草木樨 右：白花草木樨）

Fig.5.42 Sweetclover—pod
图 5.42 草木樨的荚果

Fig.5.43 Sweetclover—seed
图 5.43 草木樨种子

5.7.2 Biological characteristics

Sweetclovers have better tolerance to drought, infertility, salt and alkaline soils than alfalfa. It is well adapted to a warm and humid climate and performs well where annual rainfall is 400~500 mm and annual temperature is 6~8℃. Cold tolerance is not so good. It can grow in a wide range of soils but not acid soil. Tolerance of waterlogging is better than alfalfa and red clover. Sweetclover can grow on soil with pH 7.0~9.0. It is able to improve salty and alkaline soils. Sweetclover has good capacity for reproduction. The developmental period is about 10~15 months. White sweetclover flowers 2 weeks later than yellow sweetclover. Average self-pollination rates are 86% and 26% for white and yellow sweetclover, respectively. Mature pods fall off easily. Seeds often fall and subsequently germinate. The hard seed rate is 30% ~60% .

5.7.3 Cultivation technology

5.7.3.1 Sowing

Methods to reduce hard seed generally include scratching the seed coat, soaking seeds in warm water, scarifying seeds with concentrated sulfuric acid and variable temperature. Sowing can be done in all seasons, but spring is the best. Seeding rate is 15~22.5 kg/hm^2. Sowing method is: seeding in rows at a depth of 1~3 cm. Firming the soil promptly is needed after sowing seeds. Seeding with nurse crops are such as barely, wheat, oat and sudangrass. It is appropriate to sow

5.7.2 生物学特性

草木樨抗旱、耐盐碱、耐瘠薄，抗逆性优于紫花苜蓿。最适宜温暖湿润气候，在年降水量 400~500 mm、年平均气温 6~8℃的地区生长最好，耐寒性不强。对土壤要求不严，但不能适应酸性土壤。比苜蓿和红三叶耐涝；草木樨能改良盐碱地，适应土壤 pH 为 7.0~9.0。草木樨繁殖能力很强，生育期持续 10~15 个月，白花草木樨比黄花草木樨约迟 2 周开花。白花草木樨自花授粉率平均为 86%，黄花草木樨为 26%。成熟的荚果容易脱落，有自落自生的习性。硬实率一般占 30% ~60%。

5.7.3 栽培技术

5.7.3.1 播种

硬实种子常见处理方法有：机械擦种法、热水浸泡、硫酸处理和变温处理法。

播种期：四季均可，但以春播为宜。

播种量：15~22.5 kg/hm^2。

播种方法：条播，覆土深 1~3 cm，播后及时镇压，可用大麦、小麦、燕麦、苏丹草等进行保护播种，保护作

sweetclover when the nurse crop has developed 2~4 true leaves.

物出现 2~4 片叶子时再播种草木樨效果较好。

5.7.3.2 Field management

After turning green in the spring, plants should be protected from trampling, cutting, grazing, pest and disease. In dry and barren regions, tillage, weed control, fertilization and irrigation can promote plant growth and improve the yield of seeds or forage.

5.7.3.2 田间管理

返青后，禁止人畜践踏、割草放牧，要开始防治病虫害。

旱薄地中耕除草保墒，若有条件施肥灌水促进生长，可增加鲜草和种子产量。

5.7.4 Forge value and utilization

(1) Sweetclover should be cut before blooming. Yield of fresh forage is 30~52.5 t/hm^2, up to 67.5 t/hm^2 in some northwest areas. Seed yield is 750~1 500 kg/hm^2. It should be cut when the content of coumarin is low in the morning or evening. Mould forage is harmful to livestock.

(2) N-fixation of sweetclover is very efficient, which can improve soil fertility. When used as green manure, spring and summer are the best time to harvest and incorporate vegetation into soil.

(3) Sweetclover has an indeterminate inflorescence. Seeds at different maturity stages are present on the same plant. They can be directly harvested when one-third are mature or harvested repeatedly.

(4) Sweetclover can be used for soil and water conservation. It can also be planted under trees in arid areas to maintain soil moisture and provide nutrient for trees.

(5) It can be used as fuel and is a nectariferous plant.

5.7.4 饲用价值与利用

（1）花前刈用。鲜草产量达 30~52.5 t/hm^2，西北地区可高达 67.5 t/hm^2。种子产量 750~1 500 kg/hm^2。选择香豆素含量低的时候(中午含量最高，早晨和傍晚含量较低）刈割利用。忌喂牲畜霉变草。

（2）草木樨固氮肥田效果显著。压青时间一般在夏秋。耕翻质量上要做到“耕得深，犁断根，耕得细，不透气。”

（3）收种。草木樨为无限花序，种子成熟时间先后不一，必须适时采收。一般在有 1/3 种子成熟时即收。

（4）草木樨可用于保持水土、旱地造林，与树苗混种可遮挡日光，供给必要养料利于树苗成活生长。

（5）燃料、蜜源植物。

5.8 *Vicia* L. 野豌豆属

The genus *Vicia* consists of 200 species worldwide and most is concentrated in the temperate zone. There are 25 species in China, distributed in both northern and southern areas. At present, many species have found application in agriculture production such as broad bean, hairy vetch, smooth vetch and common vetch. They had been planted in large areas for forage and green manure 2400 years ago in China.

野豌豆属也称巢菜属，全世界 200 多种，多生于温带，我国约有 25 种，南北均有分布。目前有许多栽培种应用于生产，如蚕豆、毛苕子、光叶苕子、箭筈豌豆。2 400 多年以前我国已大量栽培，作为牧草或绿肥已经广为应用。

5.8.1 Common vetch

Scientific name: *Vicia sativa* L.

Common vetch is native to northern Europe and western Asia and is presently planted worldwide. It was introduced to our country in the 1950s and subsequently had been planted throughout our country.

5.8.1.1 Botanical characteristics

This annual herbaceous plant has a well-developed, but shallow taproot, which supports symbiosis with rhizobial bacteria. Erect or prostrate stems are thin, soft, prismatic, multi-branched, reaching a height of 80~120 cm(Fig.5.44). Leaves are even-pinnately compound with 8~16 leaflets and a tendril at the tip of each leaf(Fig.5.45). Leaflets are lanceolate or obround, and the concave tip is truncate with a small sharp point in the middle. Stipules are half arrow-shaped with an entire margin on one side and the other side with 1~3 serrations. Flowers (generally 1 to 3) from the axillae are usually purple or red, seldom white, and papilionaceous with short pedicels (Fig.5.46). Pods are strip-shaped, slightly flat and 4~6 cm long. Mature pods are brown and crisp with 7~12 seeds. Seeds are spherical or oblate. Thousand seed weight is 50~60 g (Fig.5.47).

5.8.1 箭筈豌豆

学名： *Vicia sativa* L.

英文名： common vetch

原产欧洲南部、亚洲西部，目前世界各地都有种植。我国于 20 世纪 50 年代引入，随后推广至全国。

5.8.1.1 植物学特征

一年生草本，主根肥大，入土不深，根瘤多。茎细，柔软，有条棱，多分枝，斜生或攀缘，长 80~120 cm（图 5.44）。偶数羽状复叶，具小叶 8~16 枚，顶端具卷须（图 5.45）。小叶倒披针形或长圆形，先端截形凹入并有小尖头。托叶半箭头形，一边全缘，一边有 1~3 锯齿。基部有明显腺点。花 1~3 朵生于叶腋，花梗短，蝶形，紫红或红色，个别白色。花柱背面顶端有一簇黄色冉毛（图 5.46）。荚果条形，稍扁，4~6 cm，成熟荚果褐色，含种子 7~12 粒，易碎。种子球形或扁圆形，色泽因品种而异。千粒重 50~60 g（图 5.47）。

Fig.5.44 Common vetch
图 5.44 箭筈豌豆

Fig.5.45 Common vetch—leaf
图 5.45 箭筈豌豆的叶片

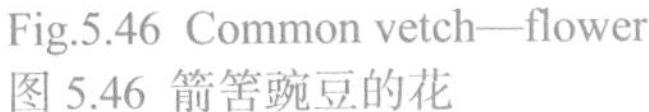

Fig.5.46 Common vetch—flower
图 5.46 箭筈豌豆的花

Fig.5.47 Common vetch—seed
图 5.47 箭筈豌豆种子

5.8.1.2 Biological characteristics

Common vetch prefers a warm climate and has good cold tolerance, but is not heat tolerant. Accumulated temperature above 0 ℃ is 1 700~2 000 ℃ for plant growth and development. It is quite drought tolerant but is sensitive to soil water content. Common vetch can grow on a wide range of soils, even on acid and infertile soils except for saline soil. It performs well on gravel and clay soil with the pH of 5.0~6.8.

5.8.1.3 Cultivation technology

Seeds should be vernalized before sowing. It is appropriate to seed in all seasons except for winter in the northern areas. Seeding rate is 60~75 kg/hm^2 for forage production and 45~60 kg/hm^2 for seed fields with a seeding depth of 3~4 cm and row spacing of 20~30 cm. When plants germinate, cotyledons remain in the soil. It can be planted in mixture with grasses.

5.8.1.4 Utilization

Fresh forage yield is 11~19 t/hm^2, and may reach 33 t/hm^2 in favorable conditions. It can be used as fresh forage, grazing and hay. Stubble height after harvest should be 5~6 cm.Seeds of common vetch contain alkaloids and hydrogen glycosides. Hydrogen glycoside can release hydrocyanic acid after hydrolysis. Hydrocyanic acid is toxic to livestock if excessive common vetch seed is fed. Hydrogen glycosides content is highest at the green pod stage. Common vetch seed can be detoxified by soaking, washing, squashing, frying or

5.8.1.2 生物学特性

喜温凉气候，抗寒能力强，但不耐炎热，生长发育需大于 0 ℃，积温 1 700~2 000 ℃。较耐旱，但对水分敏感。对土壤要求不严，耐酸耐瘠薄能力强，而耐盐能力差。能在 pH 5.0~6.8 的沙砾质至黏质土壤上生长良好。

5.8.1.3 栽培技术

种子应进行春化处理，北方春、夏、秋播种，南方四季均可播种。播种量 60~75 kg/hm^2，种子田播种量为 45~60 kg/hm^2，行距 20~30 cm，播深 3~4 cm，子叶不出土。可与禾本科混播。

5.8.1.4 利用方式

产鲜草 11~19 t/hm^2，高者可达 33 t/hm^2。放牧、刈割或调制干草均可，留茬 5~6 cm。

籽实利用：野豌豆种子中含有生物碱和氢甙，氢甙经水分解后释放出氢氰酸，食用过量会使人畜中毒，青荚期含量高应避开。

去毒处理：食用、饲用前应浸泡、淘洗、磨碎、炒熟、蒸煮等，应避免大量、长期、连续使用。

此外，亦可用作绿肥。

stewing before feeding animals. Feeding excessive common vetch seed for long periods should be avoided. It can be used as green manure also.

5.8.2 Hairy vetch (Russian vetch)

Scientific name: *Vicia villosa* Roth

5.8.2.1 Botanical characteristics

Hairy vetch is an annual or biennial herbaceous plant. Climbing stems are thin, 2~3 m long with 20~30 branches. Stems and leaves are covered with long hairs. Canopy height is about 40 cm (Fig. 5.48). Leaves are even-pinnately compound with 5~10 pairs, narrow leaflets and a tendril at the tip of each leaf. Racemes occur singly from the axillaae with numerous blue-violet or purple florets (Fig. 5.49). Pods are roundly rhombus and smooth without hairs. Each pod contains 2~8 seeds(Fig. 5.50). Seeds are round and dark. Thousand seed weight is 25~30 g (Fig. 5.51).

5.8.2 毛苕子

学名： *Vicia villosa* Roth

英文名： hairy vetch，Russian vetch

5.8.2.1 植物学特征

野豌豆属一年生或越年生豆科草本植物。攀缘茎，细长，达 2~3 m，全株密被长柔毛，草丛高 40 cm（图 5.48）。每株 20~30 分枝。偶数羽状复叶，小叶狭长 5~10 对，顶端有卷须（为变态小叶）。总状花序腋生，长毛梗，10~30 朵小花聚生于花梗的一侧，花冠蓝紫色、紫色（图 5.49）。荚果圆状菱形，光滑无毛，每荚含种子 2~8 粒（图 5.50）。种子圆形黑色，千粒重 25~30 g（图 5.51）。

Fig.5.48 Hairy vetch
图 5.48 毛苕子

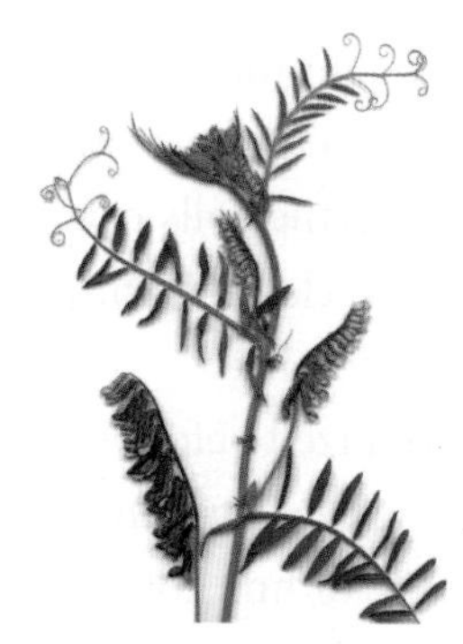

Fig.5.49 Hairy vetch—stem, leaf and inflorescence
图 5.49 毛苕子的茎、叶和花序

Fig.5.50 Hairy vetch—pod
图 5.50 毛苕子的荚果

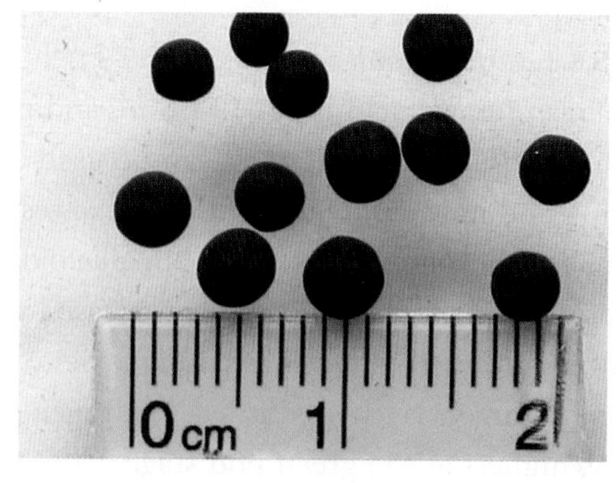

Fig.5.51 Hairy vetch—seed
图 5.51 毛苕子种子

5.8.2.2 Biological characteristics

Hairy vetch is a transition plant from winterness to springness. Its growth period is longer than smooth vetch and flowering and maturity are later. It prefers a warm and humid climate. Optimal temperature for growth is 20 ℃ . It does not tolerate high temperatures (≥ 30℃). However, it is winter hardy and able to survive in the temperatures as low as -30℃ . It tolerates drought, but not waterlogging. The appropriate annual rainfall for growth is 500~800 mm. Hairy vetch is shade tolerant. It grows well below trees and tall crops. Hairy vetch is well suited to sandy soils with pH 5~8.5. It is very salt tolerant. Adequate phosphorus and potassium fertilizers are important to hairy vetch yield and microelements like molybdenum, boron and manganese can significantly improve yield.

5.8.2.3 Cultivation technology

Hairy vetch seeding rate is 45~75 kg/hm^2 for forage production and 30~37.5 kg/hm^2 for seed production. Spring and autumn are generally suitable for planting. Planting with other crops is advisable because it can climb upwards using tendril to facilitate branching and reduce leaf rot.

5.8.2.4 Utilization

Hairy vetch has slender shoots and higher nutrient value than smooth vetch. The nutrient value of 500 kg of green hairy vetch hay is equivalent to 250 kg soy-bean cake. The best time to cut is from bud to early bloom stage or when canopy height reaches 40~50 cm. Cutting frequency is 2~3 times per year. Poor regrowth ability requires that the stubble should be left higher. Forage yield is 26~43 t/hm^2. It can be used as fresh forage, silage, hay and for grazing.

5.8.2.2 生物学特性

属冬性向春性过渡的植物，生长期较光叶苕子长，开花成熟较晚。

温度：喜温暖湿润气候，最适合在20℃生长，不耐高温（≥ 30℃）、抗寒性强，-30℃仍能生存。

水分：抗干旱，不耐水淹，适合年降水量为500~800 mm。

光照：耐阴，在果园或高秆作物行间生存良好。

土壤：喜沙质土壤，耐盐碱，pH 5~8.5可生长良好。

肥力：对磷、钾敏感，钼、硼、锰等微量元素的利用能显著增产。

5.8.2.3 栽培技术

播种量收草用为45~75 kg/hm^2，采种用为30~37.5 kg/hm^2，春、秋皆可播种。与其他作物间作套种，能借卷须攀缘向上，增加分枝，促进生长，减少基叶腐烂。

5.8.2.4 利用方式

茎叶较细，营养价值比光叶苕子高（500 kg优质绿色干草的营养价值相当于250 kg豆饼），现蕾至初花期收获最好，也可在40~50 cm高时开始刈割，每年2~3次。毛苕子再生性差，应该注意留茬高度。鲜草产量为26~43 t/hm^2。可青饲、放牧或青贮，或调制干草。

5.9 *Lespedeza* Mich.—shrub lespedeza, Chinese lespedeza and dahurian bushclover 胡枝子属—二色胡枝子、截叶胡枝子和达乌里胡枝子

Shrub lespedeza

Scientific name: *Lespedeza bicolor*

Chinese lespedeza (sericea lespedeza)

二色胡枝子

学名：*Lespedeza bicolor*

英文名：shrub lespedeza

Scientific name: *Lespedeza cuneata*

Dahurian bushclover

Scientific name: *Lespedeza davurica*

5.9.1 Botanical characteristics

The genus Lespedeza are subshrubs or shrubs with a height of 70~200 cm(Fig. 5.52). Leaves are normally pinnately trifoliolate. Leaflets are entirely edged with a small sharp thorn in the tip. The racemes from leaf axils have numerous purple or white florets (Fig. 5.53). Pods are obovate with brown or green-yellow seeds. The growing period is 90~170 d(Fig. 5.54).

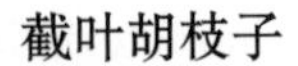

截叶胡枝子

学名： *Lespedeza cuneata*

英文名： Chinese lespedeza, sericea lespedeza

达乌里胡枝子

学名： *Lespedeza davurica*

英文名： dahurian bushclover

5.9.1 植物学特征

多年生半灌木或灌木，高 70~200 cm（图 5.52），三出复叶，小叶全缘尖端有小尖刺，总状花序腋生，花紫色或白色（图 5.53），荚果倒卵形，种子褐色、绿黄色等。生育期 90~170 d，生长期 150~230 d（图 5.54）。

Fig.5.52 Lespedeza
图 5.52 胡枝子

Fig.5.53 Lespedeza—inflorescence
图 5.53 胡枝子的花序

Fig.5.54 Lespedeza—seed
图 5.54 胡枝子的种子

5.9.2 Biological characteristics

Lespedeza plants are mainly distributed on hilly areas without trees in the temperate zone, northern hillsides and in the Yangtze valley where annual rainfall exceeds 300 mm. It has a good tolerance to drought, infertile soil and cold and some salt tolerance.

5.9.2 生物学特性

多生长在年降水量 300 mm 以上的温带无林山地或中山阴坡及长江流域。耐干旱、耐瘠薄、耐寒、耐轻度盐碱。

5.9.3 Cultivation and utilization

Hard seeds should be treated through physical or chemical methods to overcome dormancy. Row spacing is about 30~100 cm depending on the size of the species. Seeding depth is 1~3 cm. Seeding rate is 6~9 kg/hm^2 depending on seed weight. For some species, seeding rate may be as high as 22 kg/hm^2.

Seed yield of *Lespedeza* is 220~1 280 kg/hm^2. Hay production is 2 250~7 500 kg/hm^2. Cutting or grazing should be before the bloom stage. *Lespedeza* is palatable for cattle, horses and sheep. It can be used soil and water conservation and as green manure. It is a nectariferous plant with a long flowering period. It can be used as fuel and fiber board. Seed oil content is 11.34%. Oil can be extracted for and used to make soap.

5.9.3 栽培与利用

种子处理硬实后，根据植株大小定行距，一般为 30~100 cm，播深 1~3 cm，播量根据千粒重，每公顷 6~9 kg，多者可达 22 kg/hm^2。

胡枝子（属）产种子 220~1 280 kg/hm^2，产干草 2 250~7 500 kg/hm^2。牛、马、羊喜食，放牧或刈割，花前利用。

故枝子还是水土保持、绿肥植物。无异味，营养丰富，花期长，可做很好的蜜源植物。亦可做农村薪材、纤维板。种子含油 11.34%，可榨油制皂。

5.10 *Hedysarum* L. 岩黄芪属

5.10.1 Taluo shrubby sweetvetch

Scientific name: *Hedysarum laeve* Maxim.

Taluo shrubby sweetvetch is long-lived psammophyte with a height of 90~200 cm (Fig. 5.55). A conoid taproot can be as deep as 2 m and lateral roots are distributed in the soil. Rhizomes are thick, with as many as 33 branches. Flowers are red and pods have 1~3 nodes(Fig. 5.56). Thousand seed weight is about 11 g. The developmental period is 100 days. It has a good tolerance to drought, cold, high temperature and infertile soil. It can even perform well where annual rainfall is 250 mm. Hard seeds should be treated through physical or chemical methods to break dormancy. Taluo shrubby sweet-

5.10.1 羊柴

学名： *Hedysarum laeve* Maxim.

英文名： taluo shrubby sweetvetch

长寿命沙生植物，株高 90~200 cm（图 5.55），主根圆锥达 2 m 深，侧根分布在沙层中，根茎粗大，分枝多达 33 个，花红色，荚果 1~3 节（图 5.56），多发育 1 节，千粒重约 11 g，生育期 100 d。耐旱、耐寒、耐热、耐瘠薄，在年降水量 250 mm 的地区仍能正常生长。播种需处理硬实种子，行距 30~45 cm，播深 2 cm，去壳种子播种量为

vetch should be planted using a raw spacing of 30~45 cm , with a seeding depth of 2 cm, and a seeding rate of 30~45 kg/hm^2 without shells. In the seeding stage, it is necessary to control weeds and irrigate. Hay and seed yields are 2 250~3 750 and 150 kg/hm^2, respectively. It can be used soil and water conservation and as forage and fuel.

30~45 kg/hm^2，苗期注意给水除杂草等管理。每公顷产干草 2 250~3 750 kg，种子 150 kg。可做饲料、水土保持、燃料用。

Fig.5.55 Taluo shrubby sweetvetch
图 5.55 羊柴

Fig.5.56 Taluo shrubby sweetvetch—pod
图 5.56 羊柴的荚果

5.10.2 Slender branch sweetvetch

Scientific name: *Hedysarum Scoparium* Fish. et Mey.

Slender branch sweetvetch is perennial psammophytes subshrub with a height of 90~200 cm(Fig. 5.57). The tap-root is shallow and lateral roots are mainly distributed in the 20~80 cm soil layer with as many as 80 branches. The odd-pinnately leaf has 7~11 leaflets. The raceme from the axillae has 5~7 purplish red florets (Fig. 5.58). Pods have 2~4 nodes with reticular veins and white hairs (Fig. 5.59).

5.10.2 花棒

学 名： *Hedysarum Scoparium* Fish. et Mey.

英文名： slender branch sweetvetch

多年生沙生半灌木，高 90~200 cm（图 5.57），主根不长，侧根发达分布于 20~80 cm 沙层，分枝多达 80 个，奇数羽状复叶，小叶 7~11 个。总状花序腋生，小花 5~7 朵，花紫红（图 5.58）。荚果 2~4 节，有网状纹，密生白色绒毛（图 5.59）。

Fig.5.57 Slender branch sweetvetch
图 5.57 花棒

Fig.5.58 Slender branch sweetvetch—flower
图 5.58 花棒的花

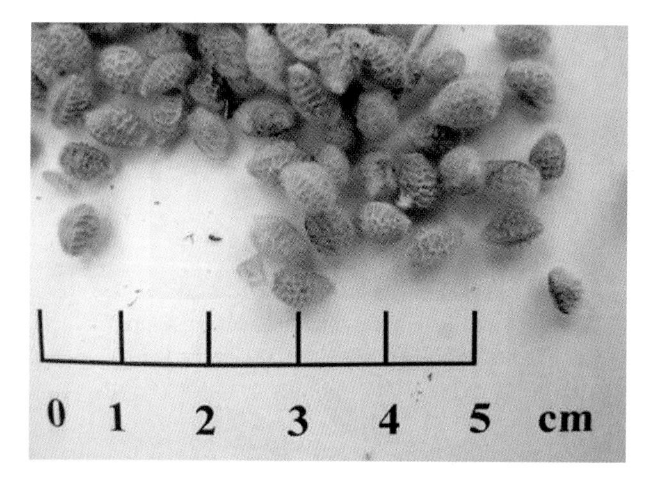

Fig.5.59 Slender branch sweetvetch—pod
图 5.59 花棒的荚果

5.11 *Amorpha* L. — *Amorpha fruticosa* 紫穗槐属—紫穗槐

Indigobush amorpha

Scientific Name: *Amorpha fruticosa*

Indigobush amorpha is a perennial deciduous shrub native to North America. It was introduced from America to Shanghai in the 1920s as an ornamental plant, and from Japan in the 1930s to protect soil of slope and as an ornamental plants and in the beginning of 1950s it was used for green manure and forage.

5.11.1 Botanical characteristics

The well-developed root system is mainly distributed in the of 0~45 cm soil layer. The height of the stem is 3~4 m and young stems are covered with hairs(Fig. 5.60). The leaf is odd-pinnately and alternatively compound with 11~25 leaflets. Leaflets are oval-round with an obtuse, rounded or slightly concave apex and a small, sharp tip. Leaflet margins are entire with glandular spots. The terminal raceme is intensively compound with purplish blue florets(Fig. 5.61). Flowers have obcordate vexils without petals and keels. Crooked pods are brown with warty glandular dots. Pods containing 1~2 seeds are indehiscent when they are ripe(Fig. 5.62). Thousand seed weight is about 10.5 g.

紫穗槐

学名： *Amorpha fruticosa*

英文名： indigobush amorpha

为多年生紫穗槐属落叶灌木，原产于北美，我国 20 世纪 20 年代从美国引入上海做观赏植物，30 年代从日本引入做公路、铁路护坡和观赏用，50 年代才开始做绿肥和饲料。

5.11.1 植物学特征

根系发达，主要分布在 0~45 cm 土层，茎高 3~4 m（图 5.60），幼枝密被绒毛，互生奇数羽状复叶，小叶 11~25 枚，卵状距圆形先端钝圆或微凹，具小尖头。全缘，具黑褐色腺点。花紫蓝色，密集总状花序，顶生，旗瓣倒心形，无翼瓣及龙骨瓣（图 5.61）。荚果弯曲，棕褐色，有瘤状腺点，果荚不开裂，含种子 1~2 粒，千粒重 10.5 g（图 5.62）。

Fig.5.60 Indigobush amorpha
图 5.60 紫穗槐

Fig.5.61 Indigobush amorpha—inflorescence
图 5.61 紫穗槐的花序

Fig.5.62 Indigobush amorpha—pod and seed
图 5.62 紫穗槐的荚果和种子

5.11.2 Biological characteristics

Indigobush amorpha has good tolerance of saline and alkaline soils, waterlogging and cold temperature. It prefers light, but tolerates shade and can survive in a wide range of soils.

5.11.3 Cultivation technology

It is best to sow seeds in spring. Shells of pods contain grease which impedes water imbibition by seeds. Pods should be broken by grinding or soaking in warm water. Seeds should be sown with pods using a seeding rate of 45~60 kg/hm^2, a row spacing of 1~2 m and a seeding depth of 4 cm. The appropriate plant density is 450~600 thousand per hm^2.

5.11.4 Utilization

Indigobush amorpha can be used as fresh forage and grazing. It can be used to produce leaf powder for it is high in leaf yield and nutrient value. Leaf powder is high in protein (22.84%), vitamin E and fat. Vitamin E content is 14 times that of maize, and crude protein content is higher than alfalfa leaf powder. It is also a nectariferous plant and can be used

5.11.2 生物学特性

适应性强，耐盐碱、耐水淹，耐寒、喜光耐阴，对土壤要求不严。

5.11.3 栽培技术

种子繁殖，春播较好，种荚皮含有油脂，难吸水膨胀，应碾压种子破碎种皮或温水浸种。

每公顷播带荚种子 45~60 kg，1~2 m 宽条播或穴播，播深 4 cm，每公顷 45~60 万株。

5.11.4 利用方式

可用于放牧或青饲、调制叶粉等。叶量丰富，营养价值高，叶粉含蛋白质 22.84%，维生素 E 含量是玉米的 14 倍，粗蛋白质含量比苜蓿粉还高，脂肪含量也较高。可作为良好的蜜源植物。枝条可编筐篓、造纸、生产人造纤维。种子可生产油漆、甘油、润滑油。可作水土保持植物。

for soil and water conservation. Stems can be made into baskets, paper and artificial fibers. Seeds can be used to produce oil paint, glycerinum and lubricating oil.

5.12 *Leucaena glauca* (L.) Benth. — horse tamarind 银合欢属—银合欢

Horse Tamarind

Scientific name: *Leucaena glauca* (L.) Benth. or *Leucaena leucocephala* (Lam.) de Wit.

Horse Tamarind is native to Yucatan in Mexico and was introduced to Taiwan 300 years ago. In 1961, Tropical Cops Research Institute in south China introduced seeds from Central America to plant experimentally in Guangxi Animal Husbandry Institute. Currently, it has been successfully established in Guangxi, Guangdong, Fujian, Hainan, Taiwan, Yunnan, Zhejiang and Hubei provinces in large areas. The range of introduction is from the Xisha Islands to Wuchang in Hubei province.

5.12.1 Botanical characteristics

Horse tamarind is a perennial evergreen shrub or tree. The upright stem is 2~10 m or more high(Fig. 5.63). Branches have brown, corky pores and are thornless, hairy when young, and brown and hairless when old. Leaves are even bi-pinnately compound. Petioles are 12~19 cm long, split into 5~15 pairs of minor petioles. Leaflets are small, hairless, narrowly oblong and point-tipped, 1.2~1.7 cm long and 3~4 mm wide. The capitulum from the axillae is a round, white, 2~3 cm diameter puffball of numerous white stamens with hairy, pale yellow anthers(Fig. 5.64). Flowers are mature to form clusters of linear, 10~18 cm long, flat, thin, drooping, green pods which dry to papery brown containing a necklace-like row of seeds. Every pod contains 6~25 seeds(Fig. 5.65). Seeds are flat, oval, and a rich glossy brown color. Thousand seed weight is about 35 g (Fig. 5.66).

银合欢

学 名： *Leucaena glauca* (L.) Benth., *Leucaena leucocephala* (Lam.) de Wit.

英文名： horse tamarind

银合欢原产于墨西哥的尤卡坦（Yucatan）半岛一带，我国台湾早在300多年前就已进行引种。我国华南热带作物研究院于1961年从中美洲引进少量种子，1964年在广西畜牧所试种。目前，我国的广西、广东、福建，海南、台湾、云南、浙江、湖北已有较大面积栽培。银合欢引种范围从西沙群岛到湖北武昌。

5.12.1 植物学特征

常绿灌木或小乔木。树干直立，高2~10 m或更高（图5.63），幼枝被短柔毛，老枝无毛，具褐色皮孔。偶数二回羽状复叶，有羽片4~8对，长6~9 cm，叶轴长12~19 cm，基部膨大，在第一对羽片及最顶端一对羽片着生处各有1枚黑色腺体，每个羽片具小叶5~15对，条状长椭圆形，长1.2~1.7 cm，宽3~4 mm，先端钝或锐尖，无毛。头状花序1~2个腋生，直径2~3 cm，有长花序梗（图5.64）；每花序有小花10~160余朵，花小，白色，5数，萼筒状，外面有毛，花瓣极狭，分离，长约为雄蕊的1/3，雄蕊10，通常疏被柔毛，子房有短柄，柱头凹下呈杯状。荚果条形，长10~18 cm，无毛，有网纹，顶端具凸尖，纵裂，每个荚果含有种子

Fig.5.63 Horse tamarind
图 5.63 银合欢

Fig.5.64 Horse tamarind—flower
图 5.64 银合欢的花

Fig.5.65 Horse tamarind—pod
图 5.65 银合欢的荚果

Fig.5.66 Horse tamarind—seed
图 5.66 银合欢的种子

5.12.2 Biological characteristics

Horse tamarind is native to the tropical zone and prefers a warm, humid climate. Optimal temperature is 25~30℃. It grows slowly when the temperature approaches 12℃ or lower and growth may stop at temperatures below 10℃ or above 35℃. Leaves fall off due to cold injury when temperature is below 0℃. As a heliophilous specie, horse tamarind is slightly tolerant to shade but can perform well in full sunlight. It grows very well in the region of 1 000~3 000 mm annual rainfall and can also survive in areas of 250 mm annual rainfall and in the southern areas with limited dry season rainfall or without rainfall for several months. However, it may fail to thrive with prolonged waterlogged conditions. Horse tam-

6~25 粒（图 5.65），种子卵形，扁平，有光泽，千粒重约 35 g（图 5.66）。

5.12.2 生物学特性

银合欢原产于热带地区，喜温暖湿润的气候条件，生长最适温度为 25~30℃，12℃生长缓慢，低于 10℃，高于 35℃，将停止生长，0℃以下叶片受害脱落。属阳性树种，稍耐阴，在无荫蔽条件下生长最好，对日照要求不太严格。在年降水量 1 000~3 000 mm 地区生长良好，但也能在 250 mm 地区生长。能耐南方旱季少雨条件，数月无雨仍能存活。不耐水渍，长时间积水生长

arind has a wide range of soil adaption, and prefers neutral and slightly alkaline soil with pH 6.0~7.7. It can even survive in cracks of rocks.

5.12.3 Cultivation technology

Seeds of horse tamarind must be treated by scratching the seed coat, soaking the seeds with warm or boiling water or scarifying seeds with concentrated sulfuric acid because of its hard and waxy seed coat, poor water imbibition and hard seed rate of 80%~90%. It can be planted in row with 40~80 seeds per meter, at a seeding rate of 15~30 kg/hm^2, using a row spacing of 60~80 cm at a sowing depth of 2~3 cm. Thinning plants is needed after germination, leaving a healthy plant every 20cm. It can also be planted with hole sowing with 4~5 seeds in each hole, 2cm deep and covering with 2~3 cm soil. When sowing directly on the soil surface, it should be planted at the beginning of the rainy season in the fertile soil with a pH above 5.5. The seedbed should be tilled and well preparation. Weeds and pest controls must be timely.

5.12.4 Utilization

Foliage can be used as fresh forage, for grazing and to produce leaf powder. Horse tamarind grows extremely fast, produces new branches after cutting, and produce high stem and leave yields of 45~60 t/hm^2 in Hainan province. Regrowth capacity is also strong when mature plants are cut down. For example, a three-years-old horse tamarind plant can produce 15~25 branches during the year following cutting. Foliage and seeds contain the toxic amino acid, mimosine, which can cause health problems for livestock if ingested regularly. Feed should be cooked and prepared to remove most of the mimosine and only consumed occasionally in limited amounts. The proportion of horse tamarind should be less than 25% of a ruminant diet and less than 15% for non-ruminant animals. When used to grazing, average daily dry matter intake should be 1.7%~2.7% of body weight.

不良。银合欢对土壤要求不严，可在中性或微碱性（pH 6.0~7.7）的土壤生长，亦可在岩石缝隙中生长。

5.12.3 栽培技术

银合欢种皮坚硬，表面有蜡质，吸水能力差，硬实率达 80%~90%，播种前必须进行处理。可采用擦破种皮、热水浸泡、沸水浸种、硫酸浸种等方法处理。条播可用人工或机械进行，每米播种 40~80 粒，每公顷用种 15~30 kg，行距 60~80 cm，播深 2~3 cm。出苗后间苗，每隔 20 cm 留一壮苗。也可穴播，深 5 cm，盖土 2~3 cm，每穴播种 4~5 粒。用种子直播，以雨季开始时为宜，并要选土层深厚、肥沃、pH 5.5 以上的地方种植，需全垦，整地要精细。注意除草防虫。

5.12.4 利用方式

可直接饲喂、放牧及制作叶粉。银合欢速生，刈割后萌芽抽枝多，鲜茎叶产量高，在海南省一般年产鲜茎叶 45~60 t/hm^2。成龄银合欢树被砍伐后，萌发再生能力也很强，3 年生银合欢，砍伐后当年可萌芽抽枝 15~25 条。叶和种子有含羞草素，可引起中毒，应控制采食量。反刍动物日粮不超 25%，非反刍动物日粮不超 15%。放牧家畜，干物质的日采食量应控制在体重 1.7%~2.7% 以内。

5.13 *Caragana* L.—korshinski peashrub 锦鸡儿属—柠条

Korshinski peashrub

Scientific name: *Caragana korshinskii*

Korshinski peashrub is perennial deciduous shrub which grows to a height of 1~3 m and always grows as several individuals clumped together(Fig. 5.67). Its well-developed taproot can penetrate soil as deep as 5~6 m and even below 9 m under some conditions. Young stems with rectangular corners are grey white, turning green and becoming glossy and grey yellow with maturity. Leaves are pinnately compound with 6~8 pairs of leaflets(Fig. 5.68). Leaflets are obovate or oval with thick, silky hairs. Stipules are hardened into thorns. Flower from the axillae are usually one or two with yellow crowns (Fig. 5.69). Pods are lanceolate or short, round lanceolate, slightly flat, keratinized and dark red brown (Fig. 5.70). Seeds are irregularly kidney-shaped or elliptical spherical, brown or tawny yellow. Thousand seed weight is 55 g.

柠条

学名： *Caragana korshinskii*

英文名： korshinski peashrub

落叶多年生灌木，高 1~3 m，常多数丛生（图 5.67）。根系发达，入土 5~6 m，深者达 9 m 以上。幼枝具棱，灰白色，后变绿色，木质化后灰黄色，有光泽。羽状复叶 6~8 对，倒卵形或长圆形，密生绢毛，托叶硬化成刺状（图 5.68）。腋生花 1~2 朵，花冠黄色（图 5.69）。荚果披针形或短圆状披针形，稍扁，革质，深红褐色（图 5.70），种子呈不规则肾形或椭圆状球形，褐色或黄褐色，千粒重 55 g。

Fig.5.67 Korshinski peashrub
图 5.67 柠条

Fig.5.68 Korshinski peashrub—leaf
图 5.68 柠条的叶

Fig.5.69 Korshinski peashrub—flower
图 5.69 柠条的花

Fig.5.70 Korshinski peashrub—pod and seed
图 5.70 柠条的荚果和种子

Korshinski peashrub prefers light and tolerates cold and high temperature. It is extremely tolerant to drought and can develop into flourishing and dense shrubbery on the dry, barren hillsides with annual rainfall of 300 mm. Korshinski peashrub is well adapted to infertile soil and reproduces easily from a well-developed root system and with strong tillering ability. It is excellent for use in soil and water conservation. Twigs and green leaves can be used as fodder. Firm branches can be used as fuel.Branches and leaf yield is 2 250~3 000 kg/hm^2. Seed yield is 240~300 kg/hm^2. It is palatable for sheep, goats and camels but not horses and cattle. It also can be processed into green-hay powder. Pods and seeds can be used to fatten livestock. It can provide wind protection and stabilize sands due to its well-developed and dense root system. It is also a nectariferous and medical plant.

In addition, the other two commonly used species are middle peashrub (scientific name: *Caragana intermedia*) (Fig. 5.71, Fig. 5.72, Fig. 5.73) and little-leaf peashrub (Fig. 5.74, Fig. 5.75). (scientific name: *Caragana microphylla*).

喜光、耐寒，耐高温。极耐干旱，能在年降水量 350 mm 左右的干旱荒山上形成茂密的灌木林。适应性强，耐瘠薄，易繁殖，根系发达，萌蘖性强，保土性能好；嫩枝绿叶可作饲料；枝条坚实，发火力旺，湿材也能燃烧，是深受群众欢迎的水保林、薪炭林树种。

枝叶干物质产量为 2 250~3 000 kg/hm^2，种子 240~300 kg/hm^2。可全年放牧，绵羊、山羊和骆驼喜食，马牛采食较少，可加工成草粉利用。荚果和种子可用于催肥家畜。根系强大，可用于防风固沙。亦是蜜源和药用植物。

此外，生产中常见的还有中间锦鸡儿（学名：*Caragana intermedia*，英文名：middle peashrub）（图 5.71，图 5.72，图 5.73）和小叶锦鸡儿（学名：*Caragana microphylla*，英文名：little-leaf peashrub）（图 5.74，图 5.75）。

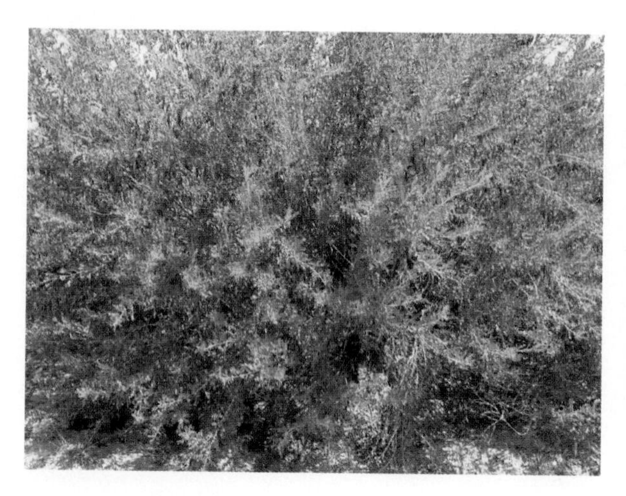

Fig.5.71 Middle peashrub
图 5.71 中间锦鸡儿

Fig.5.72 Middle peashrub—pod
图 5.72 中间锦鸡儿的荚果

Fig.5.73 Middle peashrub—seed
图 5.73 中间锦鸡儿的种子

Fig.5.74 Little-leaf peashrub
图 5.74 小叶锦鸡儿

Fig.5.75 Little-leaf peashrub—seed
图 5.75 小叶锦鸡儿的种子

5.14 *Stylosanthes* Sw.—brazilian stylo 柱花草属—柱花草

Brazilian stylo (brazilian lucerne, common stylo)

Scientific Name: *Stylosanthes gracilis* H.B.K, *Stylosanthes guianensis* (Aubl.) Sw. var. *guianensis, Stylosanthes gracilis* Kunth., *Stylosanthes guyanensis* (Aublet) Sw.

Brazilian stylo is native to South America and is distributed in tropical and subtropical zones. It was introduced to China in 1962.

5.14.1 Botanical characteristics

Brazilian stylo is a tropical perennial(Fig. 5.76). Covered by fuzzy hairs, the plant has an obvious taproot and well-developed lateral roots. The root system is 2 m deep. Tender stems have many oblique branches and reach a height of 100~150 cm(Fig. 5.77). Leaves are pinnately trifoliolate. Leaflets are lanceolate. Central leaflet is slightly larger(Fig. 5.78). The spike from the axillae is compound with small, yellow or purple florets. Pods are small with only one seed. Seeds are kidney-shaped and tawny. Thousand seed weight is 2.5 g (Fig. 5.79).

柱花草

学 名： *Stylosanthes gracilis* H.B.K, *Stylosanthes guianensis* (Aubl.) Sw. var. *guianensis, Stylosanthes gracilis* Kunth., *Stylosanthes guyanensis* (Aublet) Sw.

英 文 名： brazilian stylo, brazilian lucerne, common stylo

原产于南美洲。分布于热带、亚热带地区，我国 1962 年引入。

5.14.1 植物学特征

热带多年生草本，全株被茸毛（图 5.76），主根明显，侧根发达，入土深 2 m，茎细嫩，分枝多，斜卧，长 100~150 cm（图 5.77）。三出复叶，小叶披针形，中间叶稍大，两侧叶小（图 5.78）。复穗状花序，腋生，花小，黄色或紫色。荚果小，内含 1 粒种子，种子肾形，黄褐色，千粒重 2.5 g（图 5.79）。

Fig.5.76 Brazilian stylo
图 5.76 柱花草

Fig.5.77 Brazilian stylo—stem
图 5.77 柱花草的茎

Fig.5.78 Brazilian stylo—leaf
图 5.78 柱花草的叶

Fig.5.79 Brazilian stylo—seed
图 5.79 柱花草的种子

5.14.2 Biological characteristics

Brazilian stylo is adapted to well-drained sandy and loam soils with an annual rainfall of 600~1 800 mm without freezing. It has good tolerance of heat, low phosphorus soils and drought and has good pest resistance. Tolerance of cold and waterlogging is poor. Shrubby stylo can tolerate hot temperatures and prefers a humid climate, but not cold or saline soils.

5.14.3 Cultivation technology

Even though Brazilian stylo has wide soil adaptation, it requires certain moisture and nutrient levels. It can be sown a 60 cm row spacing, a seeding depth of 1~2 cm and a seeding rate of 7.5 kg/hm^2. It can be planted by broadcasting, hole sowing, seedling transplants or cuttings.

5.14.4 Utilization

Brazilian stylo can be cut when the canopy height is 80~90 cm leaving a stubble height of 30 cm. Fresh forage yield is 30~60 t/hm^2. It can be used for fresh forage, silage, hay and grazing.

5.14.2 生物学特性

适于在年降水量 600~1 800 mm，无霜冻、排水良好的沙壤和壤土中生长。耐热，耐低磷，耐干旱，抗虫害，不耐低温和浸渍。灌木状柱花草（shrubby stylo）耐热、喜湿、怕冻，耐盐性差。

5.14.3 栽培技术

虽对土壤要求不严，但仍需要一定水肥条件。种子直播行距 60 cm，播深 1~2 cm，播种量 7.5 kg/hm^2，还可撒播、穴播等。育苗可移栽、扦插。

5.14.4 利用方式

草层高度 80~90 cm 即可刈割，留茬 30 cm，产鲜草 30~60 t/hm^2。可青饲、青贮、调制干草、放牧。

5.15 *Pocokia ruthenical* (L.) Boisb.— ruthenian medic(alfalfa) 扁蓿豆属—扁蓿豆

Ruthenian medic (alfalfa)

Scientific name: *Pocokia ruthenica, Trigonella ruthenica, Medicago ruthenica*

5.15.1 Botanical characteristics

Ruthenian medic is a perennial herbaceous plant. Cultivars reach a height of 70~80 cm(Fig. 5.80). Well-developed taproot and lateral roots can penetrate soil as deep as 80~110 cm and support many nodules. Upright stems are smooth and clumped. Leaves are pinnately trifoliate. Leaflets are long oval with serrate margins and their tips are round or trun-

扁蓿豆

学名： *Pocokia ruthenica*，*Trigonella ruthenica*，*Medicago ruthenica*

英文名： ruthenian medic（alfalfa）

5.15.1 植物学特征

豆科多年生草本，栽培种高 70~80 cm，主根侧根粗大，入土 80~110 cm，根瘤较多。茎直立或上升，光滑，丛生（图 5.80）。羽状三出复叶，狭长而椭圆形，先端圆形或截形，微凹或有

Fig.5.80 Ruthenian medic
图 5.80 扁蓿豆

Fig.5.81 Ruthenian medic—leaf
图 5.81 扁蓿豆的叶

Fig.5.82 Ruthenian medic—flower
图 5.82 扁蓿豆的花

Fig.5.83 Ruthenian medic—seed
图 5.83 扁蓿豆的种子（标尺刻度为厘米）

cate with slightly concave or small, sharp points (Fig. 5.81). Leaflets are rough and covered with white pubescence on the under surface. Stipules are lanceolate. The raceme from the axillae has 4~12 florets and flowers are yellow with deep purple markers (Fig. 5.82). The pod is flat with 2~4 seeds. Seeds are oval and light yellow. Thousand seed weight t is 3.1~3.4 g (Fig. 5.83).

5.15.2 Biological characteristics

Ruthenian medic prefers a warm climate and has good tolerance of drought and alkaline soil. It is high in protein and is palatable to all livestock. It is not tolerant to high temperature but can grow in infertile soil. The developmental period is 120~140 d.

5.15.3 Cultivation technology

Ruthenian medic can be sown singly or in mixture with other plants using a row spacing of 30 cm and a seeding depth of 1~1.5 cm and seeding rate of 15~20 kg/hm^2. It grows very slowly in the seedling stage.

5.15.4 Utilization

Fresh forage yield is 25~30 t/hm^2 and seed yield is 150~220 kg/hm^2. It can be used as fresh forage or hay and is best harvested in the bud stage. The best time to graze is when the plant is 30 cm high with 3~4 branches. It can be used for soil and water conservation.

小尖头，边缘有锯齿，叶面粗糙，下有白色柔毛（图 5.81）。托叶披针形。总状花序，自叶腋生出，4~12 朵小花，花黄色带深紫斑（图 5.82），荚果扁平，含 2~4 粒种子。种子距圆状椭圆形，淡黄色，千粒重 3.1~3.4 g（图 5.83）。

5.15.2 生物学特性

喜温耐寒，抗旱性强，抗碱性强，蛋白质含量高，适口性好，各种家畜均喜食。不耐热，耐瘠薄，生育期 120~140 d。

5.15.3 栽培技术

单播或混播，行距 30 cm，播种量 15~20 kg/hm^2，播深 1~1.5 cm，苗期细弱，生长缓慢。

5.15.4 利用方式

可产鲜草 25~30 t/hm^2，产种子 150~220 kg/hm^2。可作青饲或调制干草，现蕾至开花期刈割。株高 30 cm，具 3~4 分枝时，可放牧利用。亦可用作水土保持植物。

Chapter 6

Grasses

禾本科牧草

6.1 *Bromus* L. 雀麦属

Bromus species are annual or perennial grasses. This genera consists of more than 100 species, widely distributed in the temperate zone, northern hemisphere. Approximately 16 species are distributed in the warmer areas of northern China. In China, the primary cultivated species at present are smooth bromegrass (*Bromus inermis*), brome(*Bromus japonicus*), rescue brome(*Bromus catharticus*) and mountain brome(*Bromus marginatus*).

6.1.1 Smooth bromegrass

Scientific name: *Bromus inermis* Leyss.

Origin and geographical distribution: The main cultivated grass of *Bromus* is smooth bromegrass which originated in Europe. The wild species of smooth brome spread to the temperate zones of Asia, Europe and North America, mainly growing on the hillsides, the side of the road, and riverbanks. In China, smooth bromegrass is distributed in the northeast, north, and northwest. It has become a significant cultivated grass in the dry and cold places of Europe and Asia. In the northeast areas of China it was introduced in 1923. Now it plays an important role in the north region of China as a cultivated grass.

6.1.1.1 Botanical characteristics

A perennial cool season grass, reproduces vegetatively through horizontal stems growing below the soil surface, called rhizomes, forming roots and producing new plants (Fig.6.1).Hairless erect stem, cylindrical, 50~120 cm high (Fig.6.2). Leaf sheaths are closed (Fig.6.3). Ligules are membranous, with no auricles. 4~6 leaves are flat, long, narrow and lanceolate. Leaf blade has conspicuous “M”- or “W”-shaped constriction, blade is 7~16 cm long and 5~8 mm wide (Fig.6.4). Open panicle (main axis with subdivided branching) is 10~30 cm long, 2~8 branches, erect with ascending branches (Fig.6.5). Each branch bears 1 or 2 spikelets. Spikelets are long and narrow, ovate, with 4~8 florets inside.

雀麦属植物为一年生或多年生草本。本属有 100 多种，广布于地球北温带。我国约有 16 种，主要分布于北方较温暖地区。目前在我国主要栽培的有无芒雀麦、雀麦、扁穗雀麦、高山雀麦等。

6.1.1 无芒雀麦

学名： *Bromus inermis* Leyss.

英文名： smooth bromegrass

别名： 禾萱草、无芒草

起源与地理分布： 雀麦属栽培牧草主要为无芒雀麦。无芒雀麦原产于欧洲，其野生种分布于亚洲、欧洲和北美洲的温带地区，多分布于山坡、路旁、河岸。我国东北、华北、西北等地都有野生种。该草现已成为欧洲和亚洲干旱、寒冷地区的重要栽培牧草。我国东北 1923 年开始引种栽培，现在是北方地区很有栽培价值的禾本科牧草。

6.1.1.1 植物学特征

无芒雀麦为禾本科雀麦属多年生冷季型牧草，根系发达，具繁殖能力极强的地下短根茎（图 6.1）。茎秆光滑直立，圆形，高 50 ~ 120 cm（图 6.2）。叶鞘闭合（图 6.3）；叶舌膜质，无叶耳；叶片 4 ~ 6，狭长披针形，长 7 ~ 16 cm，宽 5 ~ 8 mm，通常无毛，叶片中部有一个很明显皱缩的“M”形或“W”形标志（图 6.4）。开散式圆锥花序，长 10 ~ 30 cm，穗轴每节轮生 2 ~ 8 个枝梗，每枝梗着生 1 ~ 2 个小穗（图 6.5）；小穗狭长卵形，内有小花 4 ~ 8 个；颖披针形，边缘膜质；外稃宽披针形，具 5 ~ 7 脉，通常无芒或背部近顶端具有

Glumes are lanceolate, with membranous margins. Lemmas are oblong-lanceolate, 5~7 veined and generally awnless or 1~2 mm long near the top of the lemma. The palea is slightly shorter than the lemma. The caryopsis narrow, oval and 9~12 mm long. Thousand seed weight is 3.2-4.0g (Fig.6.6).

6.1.1.2 Biological characteristics

Smooth bromegrass has a well developed root system, with strong rhizomes, which expand vigorously. It is useful in sand stabilization and preventing soil erosion. It is adapted to diverse climates, especially in cold and dry zones. It has some saline alkaline tolerance and can tolerate water logging for as long as 50 d.

长 1 ~ 2 mm 的短芒；内稃较外稃短。颖果狭长卵形，长 9 ~ 12 mm，千粒重 3.2 ~ 4.0 g（图 6.6）。

6.1.1.2 生物学特性

根系发达，地下茎强壮，蔓延能力极强，可防沙固土，对气候条件适应性广，特别适于寒冷干燥地区，较耐盐碱，耐水淹时间可长达 50 d 左右。

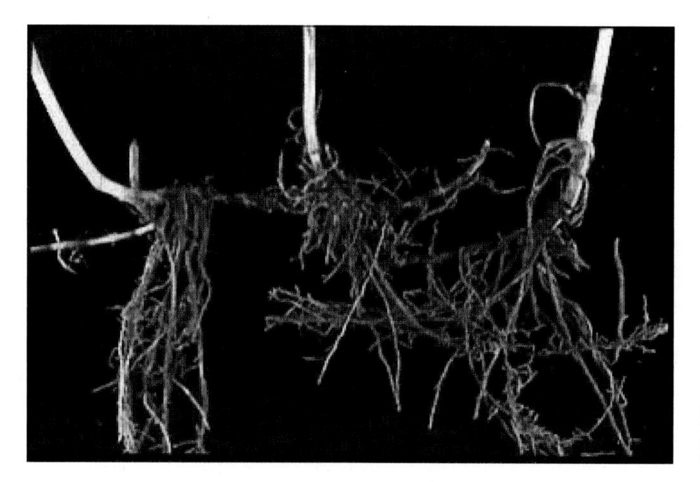

Fig. 6.1 The rhizomes of smooth bromegrass
图 6.1 无芒雀麦的地下根茎

Fig. 6.2 The blooming stand of smooth bromegrass
图 6.2 开花期的无芒雀麦

Fig. 6.3 The closed sheath of smooth bromegrass
图 6.3 无芒雀麦的闭合叶鞘

Fig. 6.4 The "M" or "W"-shaped constriction on leaf blade of smooth bromegrass
图 6.4 无芒雀麦叶片的"M"形或"W"形皱缩特征

Fig. 6.5 The open panicle of smooth bromegrass
图 6.5 无芒雀麦的圆锥花序

Fig. 6.6 The seeds of smooth bromegrass
图 6.6 无芒雀麦的种子

6.1.1.3 Cultivation and utilization

In crop rotation, smooth bromegrass can be mixed with alfalfa (*Medicago sativa* L.), sainfoin (*Onobrychis viciifolia*), red clover (*Trifolium pratense* L.), sweet clover (*Melilotus officinalis*) and grasses like timothy (*Uraria crinita* L.). Careful soil preparation is important to enhance establishment and increase yield. Fall plowing, accompanied with proper fertilization, is beneficial for smooth bromegrass growth and development. Sowing in spring, summer or early autumn all can

6.1.1.3 栽培利用技术

在轮作中可与紫花苜蓿、红豆草、红三叶和草木樨等牧草混播，也可与其他禾本科牧草如猫尾草等混播。精细整地是保苗和提高产量的重要措施。秋翻结合施肥对无芒雀麦的生长发育有良好的作用。无芒雀麦的播种期因地制宜，春播、夏播或早秋播均可。播种方法上，条播、撒播均

be adopted. Proper sowing time can be decided according to local circumstances. It can be sown by drilling or broadcasting. When drilling, the row spacing is 15~30 cm. For seed production row spacing should widen to 45 cm. Seeding rate is 22.5~30 kg/hm^2 when drilled alone. For seed production seeding rate can be reduced to 15~22.5 kg/ hm^2.For broadcasting, seeding rate should be increased to about 45.0 kg/hm^2. Sowing depth is usually 2~4 cm, 2~3 cm in clay, and 3~4 cm in sandy soils. Sowing depth should be increased to 4~5 cm in arid and windy locations that are easily dessicated. Smooth bromegrass has a high nitrogen demand. Sufficient nitrogen must be provided to smooth bromegrass, especially when it is drilled alone. Adequate P fertilizer and K fertilizer are also needed.

Smooth brome is a long-lived perennial grass. It's easy to be established. However, it grows slowly during the seeding year, and is easily damaged by weeds. Effective weed control is needed in the first year. Proper hay harvest time is at anthesis. Spring sowing may allow one harvest the first year. Seed produced the year of sowing is generally a limited yield and poor quality, not suitable for harvest. By the second year and third year, smooth bromegrass exhibits vigorous development and growth with highest seed yields, and the most appropriate time to harvest. Seed is usually harvested when 50%~60% of spikelets turn yellow. Seed yield is 600~750 kg/hm^2.

6.1.2 Rescue brome

Scientific name: *Bromus catharticus*

Origin and geographical distribution: Rescue brome (*Bromus catharticus*) occurs naturally in Argentina. It has been widely cultivated in Australia, and New Zealand. It is also cultivated in China.

6.1.2.1 Botanical characteristics

Rescue brome (*Bromus catharticus*) has dense, but thin and delicate roots. Plant height reaches 60 to 100 cm. Culms are erect and tufted. Leaf sheaths are pubescent. Leaves are 20~30 cm long and 4~7 cm wide. The panicle is lax, and branched, bearing 2~5 spikelets at the apex. Spikelets are

可。一般条播行距 15 ~ 30 cm，种子田可加宽行距到 45 cm，播种量单播时 22.5 ~ 30 kg/hm^2，种子田可减少到 15 ~ 22.5 kg/hm^2。如采用撒播，播种量可增至 45.0 kg/hm^2 左右。播时覆土深度一般为 2 ~ 4 cm，黏性土壤为 2 ~ 3cm，沙性土壤为 3 ~ 4 cm，春季干旱多风的地区由于土壤水分蒸发较快，覆土深度可增至 4 ~ 5 cm。无芒雀麦需氮甚多，须充分施氮肥，尤以单播时为甚。同时还要适当地施磷、钾肥。

无芒雀麦是长寿禾本科牧草，较易抓苗，但播种当年生长缓慢，易受杂草危害。因此，播种当年要注意中耕除草。干草的适当收获时间为开花期，春播当年可收一次干草。播种当年结籽量少，种子质量差，一般不宜采种。第 2 ~ 3 年生长发育最旺盛，种子产量高，适宜收种，在 50% ~ 60% 的小穗变为黄色时收种，种子产量为 600 ~ 750 kg/hm^2。

常见品种：‘公农’无芒雀麦、‘奇台’无芒雀麦、‘锡林郭勒’无芒雀麦、‘新雀 1 号’无芒雀麦、‘乌苏 1 号’无芒雀麦、‘林肯’无芒雀麦、‘卡尔顿’无芒雀麦等。

6.1.2 扁穗雀麦

学名：*Bromus catharticus*

英文名：rescue brome

别名：野麦子

起源与地理分布：扁穗雀麦原产于南美洲的阿根廷，在澳大利亚、新西兰等国已广为种植。我国也引种栽培。

6.1.2.1 植物学特征

须根细弱，较稠密。株高 60 ~ 100 cm，直立，丛生，叶鞘被柔毛；叶片长 20 ~ 30 cm，宽 4 ~ 7 mm；圆锥花序疏松，分枝，顶端着生 2 ~ 5 个小穗，小穗通常扁平，含 6 ~ 12 个小花；

generally flat, containing 6~12 florets. The caryopsis is tightly covered by the seed coat. There are trichoma at the top of caryopsis. Panicle length is about 20 cm. Thousand seed weight is 10~13 g.

6.1.2.2 Biological characteristics

Rescue brome is a short-lived perennial grass. It is adapted to a moist climate and warm winters. Drought tolerance of rescue brome is good, but not prefer to waterlogging and hot summers are detrimental. Rescue brome has poor cold resistance. So it does not overwinter in Beijing, Inner Mongolia, or Gansu. Rescue brome requires fertile soil. It prefers fecund clay loam. It can tolerant to some extent of saline alkaline.

6.1.2.3 Cultivation and utilization

Rescue brome has large seed and is easily established. Generally, spring or fall seeding can be successful. In north areas of China, spring seeding is typical; In the south of Yangtze River basin, where weather in winter is warm, seeding is generally successful in fall. Rescue brome is rarely, seeded alone except for seed harvest. Ordinarily, it is sown in mixtures with ryegrass (*Lolium perenne* L.), white clover (*Trifolium repens* L.), common orchardgrass (*Dactylis glomerata* L.) and red clover. When sown alone, seeding rate is 22.5~30 kg/hm^2. The rate with white clover (*Trifolium repens* L.) is 22.5 kg/hm^2, and with common orchardgrass (*Dactylis glomerata* L.), rye grass (*Lolium perenne* L.) and red clover is 11.25 kg/hm^2. When drilled, row spacing should be 15~30 cm, the sowing depth is 3~4 cm. Seed yield is about 1 500-2 250 kg/hm^2.

颖果贴于稃内，顶端具毛茸，穗长约 20 cm，千粒重 10 ~ 13 g。

6.1.2.2 生物学特性

扁穗雀麦属为短期多年生草。适于湿润而冬季温暖的气候，耐旱力尚强，但不爱积水，亦不喜炎热的夏季。耐寒性较差，在北京、内蒙古、甘肃等地不能越冬，对土壤肥力的要求较高，喜肥沃的黏质壤土，能在盐碱地生长，耐碱力颇强。

6.1.2.3 栽培利用技术

扁穗雀麦种子大，较易建植，春播或秋播。北方地区宜春播，长江流域以南冬季温暖地区，宜秋播。除收种外，很少单独播种，常与黑麦草、白三叶、鸭茅和红三叶草混种。单播时播种量为 22.5 ~ 30 kg/hm^2，与白三叶混播时，播种量为 22.5 kg/ hm^2，与鸭茅、黑麦草和红三叶草混种时，其播种量为 11.25 kg/hm^2。条播行距 15 ~ 30 cm，播深 3 ~ 4 cm，种子产量为 1 500 ~ 2 250 kg/hm^2。

6.2 *Leymus* Hochst.
赖草属

The genus of *Leymus* contains about 30 species of perennial grasses, occurring in the cold temperate zone of the Northern Hemisphere. The majority of species occur natural-

赖草属牧草为多年生草本。本属约有 30 种，分布于北半球寒温带，多数种产于中亚，少数产于欧美；我国有 9

ly in Central Asia, a minority occur in Europe and America. There are 9 species in China. They distribute in the northeast and northwest provinces. Most of them grow in the steppe.

6.2.1 Chinese wildrye

Scientific name: *Leymus chinensis*(Trin.) Tzvel.

Origin and geographical distribution: Chinese wildrye has a wide ecological amplitude. It is a xerophyte. The primary distribution is in eastern of Eurasian steppe. According to estimates, the total area of Chinese wildrye is 420 000 km^2, about half growing in China. Most of them are found in the Northeast Plain and the eastern Inner Mongolia Plateau.

6.2.1.1 Botanical characteristics

Chinese wildrye has a long thin rhizome. Its culm height is 30~60 cm. Culms are thin and round, with stripe and no branches, solitary or laxly tufted. Basal residue of the leaf sheath is fibrous. Leaf blades are narrow and lanceolate. The texture is hard and thick. Leaves are erect, pale turquoise, 10~30 cm long, and 4~8 mm wide. The abaxial surface is smooth, and the adaxial surface is scabrous, but pubescent. Leaf blades are flat or involute after wilting. Usually, the leaf sheath is longer than the internode, clasping, slide and glabrous. There is a sharp-tooth on the each side of the oral area. The ligule is plump, very short, cyclic annular, and yellow-white. The spike acrogeneous, is erect, bearing lax spikelets. The lower spikelet discontinuous. Spike length is 12~18 cm, and spikelets are 8~20 mm, with 5~10 florets inside. Spikelets are usually 2 per node or 1 per node at the base or apex of the spike, pink green, and yellow when mature. Glumes are conical, and 5~9 mm long, 1-veined, with a ciliolate margin and not covering the lemma. Lemmas are lanceolate. The apex is acute. The margin is awnless without serration, 5-veined, smooth and glabrous. The first lemma is 8~11 mm. The palea is narrow and thin, slightly shorter than lemma, with 2 ridge, so the margin is involute and covered by the palea, with some hair growing on the ridge. The apex is 2-toothed. Botanical characteristics of Chinese wildrye are shown in Fig.6.7.

种，分布于东北和西北诸省（区），多生长于草原和草甸草原地带。

6.2.1 羊草

学名： *Leymus chinensis*(Trin.) Tzvel.

英文名： Chinese wildrye

别名： 碱草

起源与地理分布： 羊草生态幅度很广，属于广域性旱生植物，主要分布区为欧亚大陆草原东部。据估算，羊草草原的总面积达 42 万 km^2，其中我国境内占一半左右，且集中分布于东北平原和内蒙古高原东部。

6.2.1.1 植物学特征

株高 30 ~ 60 cm，直立，有细长根茎。茎秆细而圆，有条纹，不分枝，单生或少数丛生，基部叶鞘残留呈纤维状。叶片狭披针形，质地较硬而厚，直立，苍绿色，长 10 ~ 30 cm，宽 4 ~ 8 mm，下面平滑，上面粗涩，具有细毛，扁平或干后内卷。叶鞘常较节间长，颇紧密，光滑无毛，口部两侧各有一尖齿。叶舌较肥厚，极短，环状，黄白色。穗状花序顶生，直立，生稀疏之小穗，下部小穗间断，长 12 ~ 18 cm，小穗长 8 ~ 20 mm，含 5 ~ 10 朵小花，通常每节 2 枚，或在花序基部及顶端单生，粉绿色，成熟时变黄。颖锥形，长 5 ~ 9 mm，具 1 脉，边缘有微纤毛，不正覆盖外稃；外稃披针形，先端尖锐，全缘无芒，具 5 脉，平滑无毛，长 8 ~ 11 mm；内稃狭而薄，较外稃稍短，具 2 脊，边缘内卷，包于外稃之内，脊有毛，先端具 2 齿。羊草植物学特征如图 6.7。

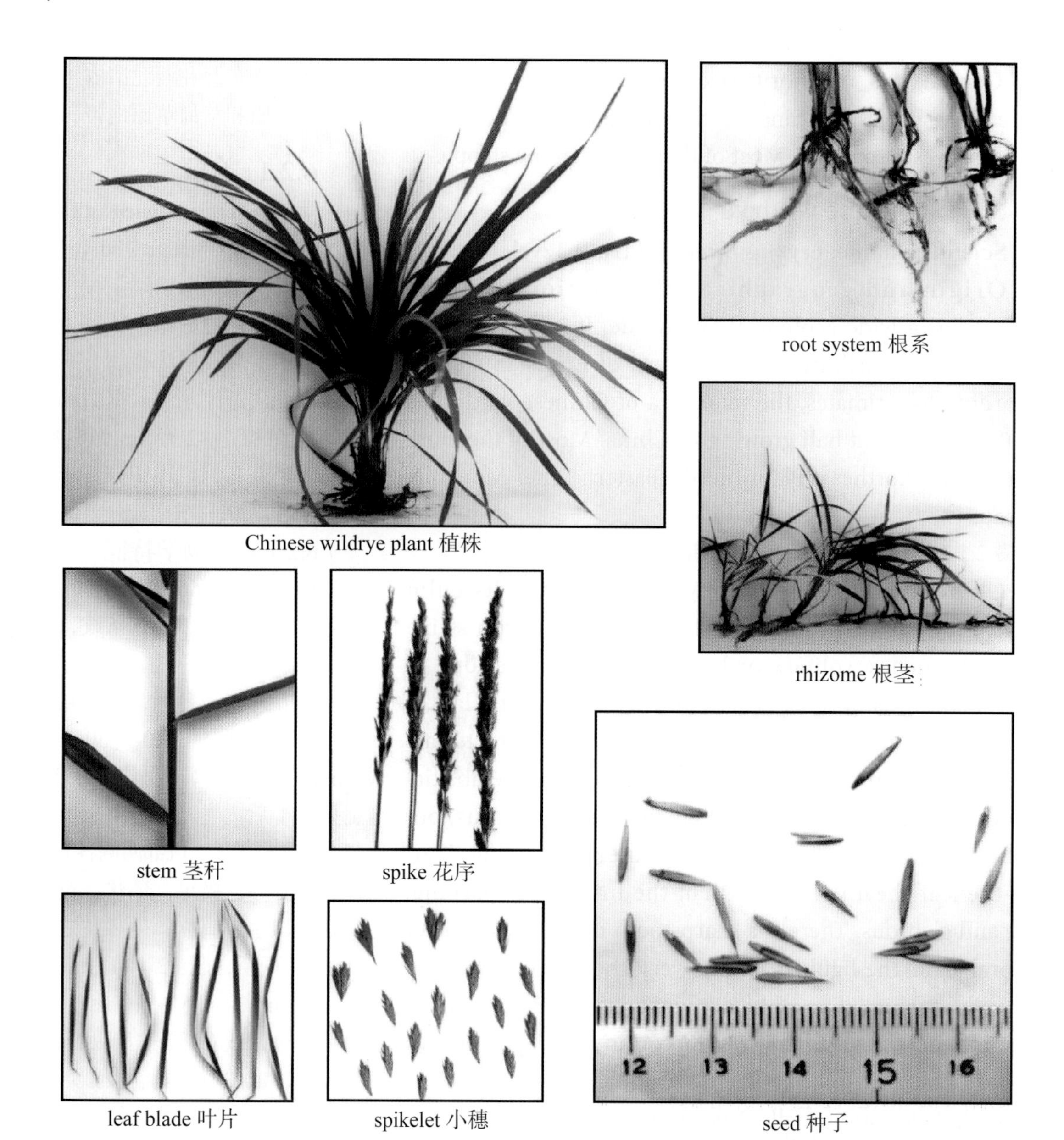

Fig. 6.7 The botanical characteristics of Chinese wildrye
图 6.7 羊草植物学特征

6.2.1.2 Biological characteristics

Chinese wildrye is a perennial grass with excellent drought tolerance, and its well developed rhizome confers trampling tolerance. Resistance to cold is good. Tolerance of saline alkaline soils is excellent, so it is also named *nutans*. Chinese wildrye reproduces vapidly via rhizomes. It is aggressive, which limits invasion of other plants in communi-

6.2.1.2 生物学特性

羊草系多年生草本，因根茎发达，故既耐干旱，亦耐践踏，对寒冷的抵抗力也很强，且能耐碱，可生长于盐碱土，故有碱草之名。羊草具有强大的根茎繁殖能力，侵占性很强，在群落中能排挤其他植物的入侵，能使群落种类组成相

ty, and the components of community would be is relative simple. Because of the widely spread of Chinese wildrye, the composition of community of Chinese wildrye has various types because its widely spread diversity.

Chinese wildrye turns green in April. Vegetative growth is vigorous from June to early July, during the boot and head stage. Growth rate gradually slows. Heading and flowering of Chinese wildrye are indeterminate, extending from late June to early August. Anthesis lasts for 50~60 d. Seeds reach milk stage 15~20 d after pollination, dough stage in 25~30 d, full ripe stage in 30~35 d. The time from green-up to fully mature seed is about 110 d. Vegetative growth still continues after seed maturation. The entire growth period can extend for 200 d.

6.2.1.3 Cultivation and utilization

Chinese wildrye grows slowly during the seedling stage. A well prepared seedbed and suitable moisture are needed. Both spring and fall seeding are acceptable as well as summer if the seedbed is weed free. Suitable sowing time varies with the district and climate because the germination of Chinese wildrye requires warm temperature and adequate moisture. The germination rate of Chinese wildrye is low. Seedling survival is poor. Careful tillage is important and generous seeding rates are beneficial. Acceptable seeding rates are 30~45 kg/hm^2. Chinese wildrye is best sown with a row spacing of 30 cm and sowing depth 2~3 cm. Compacting after sowing is recommended.

Chinese wildrye may be established using rhizomes rather than seed. Rhizomes are divided into pieces containing 2~3 nodes which are buried in a well prepared ditch. This method is an efficient way to establish Chinese wildrye because of a high survival rate, rapid growth, and potential use during the first year. Chinese wildrye grows slowly at the seedling stage requiring finely weed control.

Sowing Chinese wildrye is mixture with millet (*Setaria italica*), flax (*Linum usitatissimum* L.) or some other crops, can limit weed invasion and increase economic benefit the first year. Chinese wildrye rarely heads during the sowing year. Most plants overwinter as vegetative shoots. Lush grass

对比较单纯。由于羊草分布广，组成的群落类型多样。

羊草 4 月返青，而营养体的旺盛生长在 6 月至 7 月上旬，处于孕穗、抽穗期；随后生长速度又逐渐减慢。羊草抽穗、开花十分不齐，从 6 月下旬至 8 月上旬开花期达 50 ~ 60 d，开花授粉后 15 ~ 20 d 种子达乳熟期，25 ~ 30 d 蜡熟，30 ~ 35 d 完熟。从返青到种子成熟约 110 d。结实后营养生长仍在继续，整个生长期可达 200 d。

6.2.1.3 栽培利用技术

羊草苗期生长缓慢，要求苗床土壤深厚细碎，墒情适宜。播种羊草春秋季均可，如播前进行除草，夏季播种也可。由于羊草种子发芽时要求较高的温度和充足的水分，适宜播期可因地区、气候特点而异。羊草种子发芽率较低，且幼苗容易死亡，必须精细整地，同时要适当加大播种量，以 30 ~ 45 kg/ hm^2 为宜，羊草宜于单播，行距 30 cm，播深 2 ~ 3 cm，播后镇压。

羊草除直接用种子播种建植人工草地外，也可用根茎建植。方法是将羊草根茎切分成 5 ~ 10 cm 的小段，每小段保留 2 ~ 3 个节，按一定行距埋入已开好的播种沟内。该法成活率高，生长快，当年见效，是建立羊草草地的有效途径。羊草苗期生长缓慢，应注意清除田间杂草。

为保护幼苗、抑制杂草，羊草可以与谷子、胡麻等作物混播，也可提高当年经济效益。播种当年，羊草极少抽穗，绝大多数以营养枝越冬，第 2 年后才能充分发育，长成茂盛的草群。羊草为刈牧兼用型牧草，栽培的羊草主要用于刈割调制干草，孕穗期至始花期刈割为宜。羊草种子收获量低，采种较困难。可在穗头变黄、籽实变硬时分批采收，也可在 50% ~ 60% 穗变黄时集中采收。

results at full development the second year. Chinese wildrye can be used for cutting or grazing. Chinese wildrye is most often used to cut for hay. Hay is best cut between the boot and heading stage. Seed yield is always low and harvesting seed is difficult. It may be collected several times when spikelets become yellow and the seed become stiff. A single harvest when about 50%-60% of spikelets become yellow may be adequate.

6.2.2 Common aneurolepidium

Scientific name: *Leymus secalinus*(Georgi) Tzvel.

Origin and geographical distribution: Common aneurolepidium(*Leymus secalinus*) is spread through arid and semiarid areas of north China and Qinghai-Tibet Plateau. It occurs widely within a small area. It is most often found on mild-saline low-land. It is the dominant species in communities found in saline marshy grasslands. It sometimes occurs accompanying species comprising communities at low mountain, hill and mountain meadows.

6.2.2.1 Botanical characteristics

Culm height is 40~90 cm, with long rhizomes. Culms are erect and strong, without branches and densely ciliate on the upper part portions. Culms are solitary or tufted. The base of the sheath is fibrous. Leaf blades are thin and long, relatively thick and hard. The abaxial surface is smooth, and adaxial surface is scabrous and always rolled, and viridescent, the leaves 8~30 cm long, and 4~7 mm wide. The spike is acrogenous, terete, 8~20 cm long, and 1~2 cm wide. Spikelets arranged closely, with lower spikelets often discontinuous. It could be distinguished from Chinese wildrye because each node of the rachilla bears 2~4 spikelets, which are 10-15 mm long, containing 3~7 florets. Glumes are shorter than the spikelets.

6.2.2.2 Biological characteristics

Common aneurolepidium is a mid-xerophyte. Not only can it tolerant drought and resist cold, but it also has good tolerance to saline soils. It is well adapted to a wide range of soils. Its area of adaption compared with Chinese wildrye is much wider. Common aneurolepidium emerges early in the spring, generally turning green in late March or early April,

常见品种：‘吉生 1 号’‘吉生 2 号’‘吉生 3 号’‘吉生 4 号’‘农牧 1 号’。

6.2.2 赖草

学名：*Leymus secalinus*(Georgi) Tzvel.

英文名：common aneurolepidium

别名：阔穗碱草、老披碱草、宾草

起源与地理分布：赖草分布于我国北方和青藏高原干旱、半干旱地区，分布较广，但面积不大。常出现于轻度盐渍化低地上，是盐化草甸的建群种。在低山丘陵和山地草原中，有时作为群落的主要伴生种出现。

6.2.2.1 植物学特征

株高 40 ~ 90 cm，具长根茎，秆直立粗壮，不分枝，上部密生柔毛，单生或少数丛生。基部叶鞘残留，呈纤维状；叶片细长，较厚硬，下部平滑，上面粗涩，常旋卷，呈淡绿色，长 8 ~ 30 cm，宽 4 ~ 7 mm。穗状花序顶生，圆柱形，长 8 ~ 20 cm，宽 1 ~ 2 cm，小穗排列紧密，下部小穗常呈间断状。其与羊草的区别在于：小穗常 2 ~ 4 枚着生于每节穗轴，小穗长 10 ~ 15 mm，含 3 ~ 7 小花，颖短于小穗。

6.2.2.2 生物学特性

赖草属中旱生植物，耐旱抗寒，也具有较强的耐盐性，对土壤适应范围极广，比羊草有更广泛的生态适应区域。赖草春季萌发早，一般 3 月下旬至 4 月上旬返青，5 月下旬抽穗，6—7 月开花，7—8 月种子成熟。其生长形态随生境条件的变化而有较大变化，在干旱或盐渍化较重的地方，生长低矮，有时仅有 3 ~ 4 片基生叶。在水肥条件好时，株丛生长繁茂，根茎迅速繁衍，能形成独立的优势群落。

Heading in late May, and flowering in June and July. Seed matures in July and August. Morphology can vary considerably changes with differing growing conditions. Growth is low in dry or heavily saline areas, sometimes with only 3~4 leaves. In well a fertilized, well irrigated area, it grows lushly, and the rhizomes reproduce rapidly, sometimes resulting in a solitary community.

6.2.2.3 Cultivation and utilization

Through introduction and domestication, cultivars adapted to the arid and saline areas of the northwest have been selected. It can be used for cutting and grazing. Good results have been achieved by transplanting rhizomes in desert grassland in the east of Helan mountain in Ningxia. Normally, after seeding, germination occurs in 9 d, tillering in 23 d, elongation in 35 d, heading in 65 d, flowering in 70 d, and plants mature in 100 d. During the seeding year, vegetation can be cut three times, in June, August and September. Seed yield is 41.5 kg/hm^2. Common aneurolepidium can be used medicinally as well as forage. Roots have properties which reduce favor and as a diuretic. Common aneurolepidium can be used for windbreaks, for sand stabilization and soil and water conservation.

6.2.2.3 栽培利用技术

通过引种驯化，可培育为适应我国西北干旱地区轻盐渍化土壤的刈牧兼用的栽培草种，在宁夏贺兰山东麓荒漠草原地区栽培（实行根茎移栽）试验，结果良好。正常情况下，播种后9 d出苗，23 d分蘖，35 d拔节，65 d抽穗，70 d开花，100 d成熟。当年于6月、8月、9月刈割3次，种子产量为41.5 kg/hm^2。赖草除作饲用外，根可入药，具有清热、止血、利尿作用；又可用作防风固沙或水土保持草种。

6.3 *Agropyron* J. Gaertn. 冰草属

The genus *Agropyron* is comprised of perennial grasses. This genus contains about fifteen species. It is widely distributed in Eurasia, the prairies of America, Canada and Mongolia.It occurs in grassland and desert grassland, primarily in the northeast, north, northwest and southwest over China.

冰草属系多年生草本。本属共约有15种，在欧亚大陆、北美大草原等有大面积分布，我国分布于草原及荒漠草原地区，主要是东北、华北、西北、西南等地。

6.3.1 Crested wheatgrass

Scientific name: *Agropyron cristatum*(Gaertn.)

Origin and geographical distribution: Crested wheatgrass occurs in Eurasia, and throughout in Europe. In China, it occurs primarily in the arid grassland zones of the northeast, Inner Mongolia, Hebei, Shanxi, Shaanxi, Gansu,

6.3.1 扁穗冰草

学名： *Agropyron cristatum*(Gaertn.)

英文名： crested wheat grass

别名： 野麦子、冰草、麦穗草、山麦草

Qinghai, Ningxia, Xinjiang and neighboring provinces. It is an important forage for improvement at arid and semiarid grasslands.

6.3.1.1 Botanical characteristics

Crested wheatgrass has a fibrous root system. Culms are tufted, erect, geniculate at the base. The upper portion is pubescent, reaching a height is 45~60 cm. Leaf blades are 5~10 cm long and 2~5 mm wide. Sheaths are closed and smooth. The auricle is short and acute. Spikes are 6~8cm long. Spikelets are sessile, growing on both sides of the rachilla. Spikelets are imbricate, and closely arranged. Each spikelet includes 3~4 florets. The lemma is ciliate, with a short awn at the apex. Thousand seed weight is 2 g. Botanical characteristics of crested wheatgrass are shown in Fig.6.8.

6.3.1.2 Biological characteristics

Crested wheatgrass is a typical, perennial, herbaceous xerophyte. It prefers a dry and cold climate, has excellent tolerance of both cold and drought stress, and can grow in the semi-desert zone. During drought, growth is reduced. When moisture returns, it recover rapidly. Alkaline tolerance is also good. It prefers chestnut soils of grassland. Crested wheatgrass (*Agropyron cristatum*) may persist for 10~15 years. In China, it occurs in the northeast, northwest, Inner Mongolia, and Qinghai province. It is a primary pasture forage. In the Xilingol League areas of Inner Mongolia, it emerges in April. Heading occurs in June and July, and mature seed in August. By September, it is withered and yellow.

Crested wheatgrass tillers profusely. In the seeding year, each plant may produce up to 25~55 tillers, which may flower rapidly, shed seed and reseed. The tillering node is located 2 cm beneath the soil surface. Crested wheatgrass is cross-pollinated and wind pollinated. Self-pollination causes dysgenesis. Flowering begins in the upper third of the spikelet followed by opening of the upper and lower opening in order. The lowest spikelet flowers last. After pollination, seed reaches the milk stage in 8~10 d, dough stage in 20~25 d, and fully ripe in 27~30 d. In hot, dry weather, seed matures faster. Seed maturity is delayed by cool, rainy weather.

起源与地理分布：扁穗冰草起源于欧亚大陆，遍布于欧洲。我国主要分布在东北地区，内蒙古、河北、山西、陕西、甘肃、青海、宁夏和新疆等省区的干旱草原地带；是改良我国干旱、半干旱草原的重要牧草。

6.3.1.1 植物学特征

须根系，外具沙套。茎秆丛生，直立，基部膝状弯曲，上被短柔毛，株高45 ~ 60 cm，叶长5 ~ 10 cm，宽2 ~ 5 mm；叶鞘紧密而光滑，叶耳短而光，尖锐；穗状花序，长6 ~ 8 cm，小穗无柄，着生于茎轴两侧，紧密排列整齐成羽状；每穗含3 ~ 4花；外稃有毛，尖端常具短芒。种子千粒重为2 g。扁穗冰草植物学特征如图6.8。

6.3.1.2 生物学特性

扁穗冰草为典型的多年生旱生草本植物，性喜干燥冷凉气候，抗旱抗寒性都较强，能在半沙漠地带生长。干旱时生长虽见停滞，一有水分供给，即又恢复生长。耐碱性也很强，最适于在草原栗钙土上生长。生长期可达10 ~ 15年，我国东北、西北、内蒙古、青海等地均有分布，为牧区的主要牧草之一。在内蒙古锡林郭勒盟4月开始发芽，6—7月抽穗，8月结实，9月枯黄。

扁穗冰草分蘖力很强，播种当年单株分蘖可达25 ~ 55个，并很快形成丛状。种子自然脱落，可以自生。分蘖节位于土表以下2 cm处。扁穗冰草是异花授粉植物，靠风力传粉，自花授粉大部不孕。首先由花序的上1/3处小穗开始开放，然后向下、向上开放，花序最下部小穗最后开放。冰草授粉后8 ~ 10 d达乳熟期，20 ~ 25 d达蜡熟期，27 ~ 30 d达完熟期。在炎热、干燥的天气下，种子成熟较快，凉爽、多雨时种子成熟时间将延长。

Fig. 6.8 The botanical characteristics of crested wheatgrass
图 6.8 扁穗冰草的植物学特征

6.3.1.3 Cultivation and utilization

Crested wheatgrass seed is generally large with a high and consistent germination rate. Seedbed preparation should include ploughing, repeated leveling and fully reducing clods. Harrowing land before sowing in newly reclaimed wasteland, should be followed with leveling. In

6.3.1.3 栽培利用技术

冰草种子较大，纯净度高，发芽较好，出苗整齐，但整地仍要精细，土地耕好后，要反复耕耱，充分粉碎土块。在新垦荒地播前要耙碎草皮、地面整平。最好对于新垦的荒地种过 1 ～ 2 年作物

newly reclaimed wasteland, planting a crop for 1~2 years in order for the soil mellow, and then planting grass, is often successful.

Crested wheatgrass can be sown in spring, summer or fall in the North of China, but most often in spring, crested wheatgrass is best sown in summer in areas of poor soil moisture, less rain and more wind during spring. Sowing after rain is preferred. Fall sowing also can be practiced, but earlier is better, especially in relative cold areas. Sowing time influences overwintering success. Sowing in April and May is preferred in the cold, high mountain areas of Gansu and Qinghai. It can also be sown in June and July. In the middle part of Shaanxi province, it can be sown in July and August. Drill spacing is 20~30 cm. Seeding rate is 15~22.5 kg/hm^2 and sowing depth is 3~4 cm. Management should be careful following germination. Crested wheatgrass can be sown in mixture with grasses, such as, redtop, bluegrass or forages including roegneria (*Roegneria kamoji* ohwi), alfalfa (*Medicago sativa*) and sainfoin(*Onobrychis viciaefolia*). Mixtures with legumes can increase forage quality and yield.

If crested wheatgrass is invaded by weeds during establishment, intertillage for weeding control should be implemented. Crested wheatgrass has excellent tolerance of drought and barren soil, but during dry periods timely irrigation and N fertilization will markedly increase seed and grass yield. Crested wheatgrass can produce high seed yields that are easily harvested with high germination.

The best time for harvesting is at head stage. After flowering, crude protein and palatability will decline markedly. Since the regrowth is not good, it can only be cut one time per year. Grass of regrowth can be used for grazing. Seed production requires excellent management. Harvesting must be timely. Seed harvest should be at late dough stage or at full maturity. Timely harvest should limit yield loss due to seed shedding. Common varieties include **'fairway' 'parkway' 'ruff' 'ephraim' 'kirk'** and **'douglus'.**

后，待土壤熟化，再栽种冰草，将会获得成功。

冰草春、夏、秋均可播种，我国北方地区多春播，在春季少雨多风、土壤墒情较差的地区夏播为宜，夏季降雨后抢墒播种。秋播也可，但宜早不宜迟，特别在较寒冷地区影响其安全过冬。在甘肃、青海高山寒冷地区播种时间以4—5月份为宜，也可在6—7月份播种，陕西省渭北为7—8月。条播行距为20 ~ 30 cm，播种量为15 ~ 22.5 kg/hm^2，播深3 ~ 4 cm，出苗后应加强管理。亦可与小糠草、早熟禾、鹅观草、苜蓿、红豆草等混播，以提高产量与品质。

冰草在生长期间如田间杂草滋生，要加强中耕除草。冰草虽抗旱耐瘠薄能力强，但在干旱地区或遇干旱年份，如能适时灌溉，追施氮肥，则可显著提高青草和种子产量，扁穗冰草结籽较多，易于收集，发芽力颇强。

冰草的适宜刈割期为抽穗期，开花后蛋白质含量和适口性明显下降。它的再生能力差，一年只能刈割一次，再生草可用作放牧。种子田要加强田间管理，收获要及时，一般应在蜡熟末期或完熟期收获。以免种子脱落，影响产量。

常见品种：'fairway' 'parkway' 'ruff' 'ephraim' 'kirk' 'douglus'。

6.3.2 Desert wheatgrass

Scientific name: *Agropyron desertorum* (Fisch.) Schul.

Origin and geographical distribution: Desert wheatgrass occurs in the temperature steppes of Eurasia. It is sourced from Jilin, western Liaoning, Inner Mongolia, Shanxi, Gansu and Xinjiang province. It also occurs in Mongolia and the Soviet Union. Native to Siberia and central Asia desert wheatgrass is cold resistant, drought resistant, well adapted to dry and cold climate conditions. It is a typical xerophytic grass. It has minimal soil requirements, but prefers loam and sand soil. It is suitable for cultivation in arid, desert steppe.

6.3.2.1 Botanical characteristics

Desert wheatgrass is a perennial grass. It possesses horizonal or downward oriented rhizomes. The exterior sheath is fibrous. Culms are erect as high as 30~50 cm, loose tufted, glabrous, or pubescent just below the spike. Leaf blades are shorter than internodes. Culms are tightly wrapped. The ligule is short and small. Leaf blades are 5~10 cm long and 1~1.5 mm wide, always rolled into a cone. The spike is erect, cylindrical, 2~5(9) cm long and 5~7(9) mm wide, with 4~6 florets. Glumes are boat shaped. The first glume is 2~3 mm long. The second is 3~4 mm long. Awns are 2 mm long. The lemma is also boat like 5~6 mm long, the base is blunt round, and awns are 1~1.5 mm long. Length of the palea is equal to or a little longer than lemma. Caryopsis sticks with lemmas, about 3 mm long, bronzing, ciliate on the apex. Botanical characteristics of desert wheatgrass are shown in Fig.6.9.

6.3.2.2 Biological characteristics

Desert wheatgrass has a well developed root system, extending to a depth of 0~15 cm.Tolerance of drought and cold are excellent. Precipitation requirement is 150~400 mm. It is a typical pasture xerophyte. Desert wheatgrass greens up in very early spring. Regrowth capacity is good. It is suitable for grazing. In winter, some overground leaves and culm which can be left well, gradually withered leaves can be securely left in stems. Desert wheatgrass not demands more on soil, but usually prefer grow in sandy soil, sand, sandy slopes and inter

6.3.2 沙生冰草

学名: *Agropyron desertorum* (Fisch.) Schul.

英文名: desert wheatgrass

别名: 荒漠冰草

起源与地理分布: 沙生冰草分布于欧亚大陆的温带草原区。产于我国吉林、辽宁西部、内蒙古、山西、甘肃、新疆等地；蒙古及原苏联也有分布。原产于西伯利亚和中亚。抗寒、耐旱，适应干燥寒冷气候条件，是典型的旱生牧草。对土壤要求不严，喜生于沙质壤土、沙地。适宜在干旱荒漠草原种植。

6.3.2.1 植物学特征

多年生草本。具横走或下伸的根状茎，须根外具沙套。秆直立，高 30 ~ 50 cm，呈疏丛型，光滑或在花序下被柔毛。叶稍短于节间，紧密裹茎，叶舌短小；叶片长 5 ~ 10 cm，宽 1 ~ 1.5 mm，多内卷成锥状。穗状花序直立，圆柱形，长 2 ~ 5（9）cm，宽 5 ~ 7（9）mm；小穗长 4 ~ 9 mm，含 4 ~ 6 小花；颖舟形，第一颖长 2 ~ 3 mm，第二颖长 3 ~ 4 mm，芒长达 2 mm；外稃舟形，长 5 ~ 6 mm，基盘钝圆，芒长 1 ~ 1.5 mm；内稃等长或微长于外稃。颖果与稃片黏合，长约 3 mm，红褐色，顶端有毛。沙生冰草植物学特征如图 6.9。

6.3.2.2 生物学特性

沙生冰草的根系较发达，主要分布于 0 ~ 15 cm 的土层中。耐旱和耐寒性强，对自然年降水量要求为 150 ~ 400 mm 之间，是一种比较典型的草原旱生植物。沙生冰草返青早，再生性也较好，适于放牧利用。到冬季地上部分茎叶能较好地残留下来，渐干枯的叶子也能牢固地残留在茎上。沙生冰草对土

Fig. 6.9 The botanical characteristics of desert wheatgrass
图 6.9 沙生冰草植物学特征

dune lowland. Desert wheatgrass occurs with other species in sandy areas. At times it becomes the dominant species in sandy areas and formats desert wheatgrass grassland.

6.3.2.3 Cultivation and utilization

Desert wheatgrass seeds are relatively large. Thousand seed weight is 2.5 g. Seeds generally have high germination rates. Seedlings grow slowly so careful weed control is necessary to protect the seedling against invasion. Generally grass can not be harvested during the sowing year. The livestock carrying capacity should be carefully managed when desert

壤不苛求，但通常喜生于沙质土壤、沙地、沙质坡地及沙丘间低地。沙生冰草在沙地植被中主要作为伴生种出现，有时在局部覆沙地段或沙质土壤上可成优势种，形成沙生冰草草原。

6.3.2.3 栽培利用技术

沙生冰草的种子比较大，千粒重为 2.5 g，发芽率较高，出苗整齐，但幼苗期生长缓慢，应加强除草，以免杂草欺苗。一般播种当年不能利用，利用沙

wheatgrass is used for pasture. Excessive grazing will result in degeneration of desert wheatgrass. Also grazing should not be started too early. Sowing in spring or fall, using a seeding rate of 11~22 kg/hm^2 is suitable. It can be sowed with alfalfa, erect milkvetch or other legume forages to establish mixed grassland for grazing. During the second year, desert wheatgrass expresses vigorous growth, growing lush culms and blades and provides good quality forage with excellent palatability. It can be harvested 2~3 times each year. Fresh yield is 15~22 t/hm^2. Common cultivars include **'norden' 'desert'** and **'summit'**.

生冰草改良的草场，应注意载畜量，过高则使沙生冰草退化，同时始牧期也不宜过早。宜在夏秋两季播种，播种量为 11~22 kg/hm^2，可与苜蓿、沙打旺等豆科牧草混播，建成混播草地放牧利用。建植第二年生长势强，茎叶繁茂，草质优良，适口性好。一年可刈割 2 ~ 3 次，鲜草产量为 15 ~ 22 t/hm^2。

常见品种：'norden' 'desert' 'summit'。

6.4 *Poa* L. 早熟禾属

The genus *Poa* L. is comprised of annual and perennial grasses. The genus includes hundreds of species, distributed in the temperate and frigid zones, rarely in tropical zone. There are 78 species in China. Most species of this genus provide excellent pasture, but the majority is wild species. Most commonly used species are Kentucky bluegrass, common bluegrass and Canada bluegrass.

早熟禾属为一年生或多年生草本。约有数百种，多分布于温带及寒带，热带稀少。我国主要有 78 种。本属多数种类均为优良牧草。但多数为野生种，常见栽培的主要有草地早熟禾、普通早熟禾和加拿大早熟禾。

6.4.1 Kentucky bluegrass

Scientific name: *Poa pratensis* L.

Origin and geographic distribution: Kentucky bluegrass originated in Europe, northern Asia and North Africa, and then spread to America. It has spread over all the temperate regions of the earth. It occurs in Northeast China, Shandong, Jiangxi, Hebei, Shanxi, Gansu, Sichuan and Inner Mongolia.

6.4.1.1 Botanical characteristics

Kentucky bluegrass has creeping rhizomes. It roots to a depth of about 1 m, most roots are near the surface within a depth of 15 cm, rarely over 40 cm. Culms grow to a of height with 30~60 cm, smooth, erect and solitary or clustered (Fig.6.10A). Leaves are above the narrowly linear, 5~10 cm long, 1~6 mm wide, soft, blue-green with obvious veins at the base of the ridge. The leaf tip is boat-shaped.

6.4.1 草地早熟禾

学名：*Poa pratensis* L.

英文名：Kentucky bluegrass

别名：蓝草、肯塔基蓝草

起源与地理分布：草地早熟禾原产于欧洲各地、亚洲北部及非洲北部，后来传至美洲。现已遍及全球温带地区。我国东北、山东、江西、河北、山西、甘肃、四川、内蒙古等省(区)都有野生。

6.4.1.1 植物学特征

具匍匐根茎，为根茎疏丛型下繁草，根深达 1 m 左右，根茎在表层 15 cm 内为最多，在 40 cm 以下的根量很少；株高 30 ~ 60 cm，光滑，直立，单生或丛生（图 6.10A）；叶狭线形，长 5 ~ 10 cm，宽 1 ~ 6 mm，质软，蓝绿色，基部有明显的脊脉，

Sheaths are smooth or rough (Fig.6.10B, C). The ligule is membranous, short and blunt, about 2 mm and sometimes degraded. The panicle is 5~20 cm long branching upward. Spikelets are dense at the top, 3~6 mm long, flat, containing 3~5 flowers (Fig.6.10D, E). The top is oval or ovate-lanceolate. Lemma length is 2.5~3.5 mm and sharp, raw textured with 5 veins at the base, following the ridge and side veins central health filamentous pubescence. The palea is shorter than lemma with a rough ridge to small cilia. The caryopsis is fusiform, triangular, about 2 mm long. Thousand seed weight is 0.4 g (Fig.6.10D, E).

叶尖呈舟形（图 6.10B，C）；叶鞘光滑或粗糙，叶舌膜质，短而钝，长约 2 mm，有时退化；圆锥花序，长 5 ~ 20 cm，分枝向上或散开（图 6.10D，E）；小穗密生顶端，长 3 ~ 6 mm，扁平，含小花 3 ~ 5 朵，顶端卵圆形或卵状披针形，光滑或脊上粗糙，第一颖具 1 脉，第二颖具 3 脉；外稃长 2.5 ~ 3.5 mm，尖锐，基部生网纹，有脉 5 条，脊及边脉中部以下生丝状柔毛；内稃较外稃为短，脊粗糙至具小纤毛；颖果纺锤形，具三棱，长约 2 mm，千粒重 0.4 g（图 6.10D，E）。

Fig. 6.10 Kentucky bluegrass (A. stand; B. stem; C. blade; D. panicle; E. seeds)
图 6.10 草地早熟禾（A. 早熟禾草地；B. 茎秆；C. 叶片；D. 圆锥花序；E. 种子）

6.4.1.2 Biological characteristics

Kentucky bluegrass grows early in spring, flowering in May and June. Growth shows slowly during dry-hot summer, resumes during the rainy autumn until late fall. Bluegrass is suited to warm, humid climate, with good winter hardy and poor drought tolerance. In arid areas it can grow well if irrigated. It also has good tolerance to long-term water lodging. It grows well in well-drained, loose textured soils, especially soils rich in humus. It flourishes in calcareous soils. Although the grass begins to grow early in the spring, it flowers and produces seed once a year. It seldom flowers and produces seed after cutting, even in a suitable environment for growth.

6.4.1.3 Cultivation and utilization

Bluegrass grows slowly, but may persist for more than 10 years. Therefore, it is better used for long-term grassland. It is suitable for grazing because of its low height and tolerance of animal use, simplifying grassland management. It is not well suited to the grass field rotation system. Because of small seed size, seedlings grow slowly. Seedbed preparation should include land leveling and fine crushing to enhance emergence. Sowing period varies with location. In warm regions the appropriate sowing time is autumn, while in alpine pastoral areas sowing immediately after the spring thaw appropriate either drilling or broadcasting may be used. Seeding rate is 7.5~15.0 kg/hm^2 with a sowing depth of 1~2 cm. Drill spacing is about 30 cm. It is intolerant of weeds which should be controlled after full stand. Bluegrass established is often sows alone, but it can also be mixed sow with white clover, trefoil, tall fescue, orchardgrass, alfalfa and other grasses to increase production and improve quality.

Kentucky bluegrass responds well to fertilizer, such as nitrogen, phosphorus and potassium which can increase production. Heading, flowering and early spring growth can be increased with fertilizer. Root distribution is shallow, so irrigation may be necessary during drought conditions. A combination of irrigation and fertilizer can not only improve forage production and nutrition but also extend persistence.

6.4.1.2 生物学特性

早春生长很早，5—6月开花，如遇夏季干热天气，则生长停滞，秋后遇雨又继续生长，直到晚秋。草地早熟禾宜于温暖湿润的气候生长，耐寒性极强，而耐旱性较差。在干旱地区如进行灌溉，也能良好生长。但在酷热的夏季，即使灌溉也生长不良。能耐长期水淹，喜排水良好、质地疏松的土壤，尤以富含腐殖质之土为宜。在石灰质较多的土壤生长更好。此草虽春季发育很早，但每年只开花结籽一次，即使在生长环境适宜之处，再生草亦很少开花结籽。

6.4.1.3 栽培利用技术

草地早熟禾生长较慢，而生长年限达10年以上，因而适宜种在长期草地，又因其株体较低而耐牧性较强，故适于种在以牧为主的草地，便于草地的管理。在草田轮作制中如草地年限较短，则不宜使用。种子细小，故苗期生长缓慢。整地要精细，做到土地平整细碎，以利出苗。播种时期因地而异。在温暖地区以秋播为宜，而在高寒牧区春季土壤解冻后，即行播种。播种采用条播和撒播，播种量为7.5 ~ 15.0 kg/hm^2，播种深度1 ~ 2 cm，条播行距30 cm左右，它不耐杂草，齐苗后要清除。早熟禾一般单播，也可与白三叶、百脉根、苇状羊茅、鸭茅、紫花苜蓿等混播，以提高产量和改进品质。

草地早熟禾对肥料的反应良好，如施氮、磷、钾或全价肥料能显著增产。在抽穗开花和早春萌发时均可追肥。它的根系在土壤中分布较浅，故干旱时期有条件者应多次灌溉。结合施肥进行灌溉，既提高草料产量，又增加营养，也能延长草地寿命。

Kentucky bluegrass seeds are subject to exfoliation after maturing. Seed should be harvested when spikes change from green to yellow, and the cob turns yellow. Common varieties include **'wabash' 'kentucky' 'connie'** and **'midnight'.**

6.4.2 Roughstalk bluegrass

Scientific name: *Poa trivalis* L.

Origin and geographic distribution: Roughstalk bluegrass is native to Europe. It is cultivated in many areas of China.

6.4.2.1 Botanical characteristics

Stems and leaves appear apple green, turning purple after frost. Stems are tufted, either erect or prostrate creeping on the soil and taking root 30~60 cm high, with 3~4 nodes. Stems below the spike are rough. The leaf sheath is complete and the ridge is significant. The ligule is membranous, and the bottom ligule is short and the upper ligule is longer. Panicles are erect and whorled on the rough branches. The spikelet is oval with 2~3 florets. The glume is lanceolate. The first glume is narrower with a pulse while the second glume has 3 veins with thorns on the edge. A ridge is obvious, also with thorns. The top of lemma is sharp with 5 veins. Those in the middle and both sides are obvious. The lower half of the middle veins is pilose. Remaining veins are hairless. The base row has long, cottony hair. Seed length is 1.8~2.5 mm, which is shorter than Kentucky bluegrass.

6.4.2.2 Biological characteristics

Roughstalk bluegrass is a perennial tufted grass with dense stolons. Underground stems also spread as rhizomes. New branches occur on their nodes, growing quickly into a dense clump. Roughstalk bluegrass prefers moist and fertile clay. It can survive under the shade, but it intolerant of heat but has good cold resistance. Roots are shallow and topsoil must have adequate moisture to keep grass growing well.It is intolerant of drought. Irrigation is required under drought condition, plants are short, stems and leaves turn red. Growth will resume after rain. Roughstalk bluegrass should not be planted in loose, barren sand. If mild, humid winter, roughstalk bluegrass growth roughstalk bluegrass will recover in early spring. The fastest growth occurs in May and June,

草地早熟禾种子成熟后易脱落，为减少落粒损失，当穗由绿变黄，穗轴已见枯黄时适收。

常见品种：‘瓦巴斯’‘肯塔基’‘康尼’和‘午夜’。

6.4.2 普通早熟禾

学名：*Poa trivalis* L.

英文名：roughstalk bluegrass

别名：粗茎兰草

起源与地理分布：普通早熟禾原产于欧洲，我国多有栽培。

6.4.2.1 植物学特征

茎叶呈苹果绿色，经霜后带紫色。茎秆丛生，直立或茎部倾斜匍匐而落土生根，高 30 ~ 60 cm，具 3 ~ 4 节。叶鞘完整，脊显著；叶舌膜质，下部的叶舌较短，上部的叶舌长而有光。圆锥花序直立，轮生于粗糙的枝上。小穗卵形，有小花 2 ~ 3 朵；颖披针形，第一颖较窄，具 1 脉，第二颖具 3 脉，边有刺，脊明显，脊上亦有刺；外稃顶端尖锐，具 5 脉，而以中间及两边者为明显，中间脉纹的下半部具柔毛，其余脉纹无毛。种子长 1.8 ~ 2.5 mm，较草地早熟禾为狭。

6.4.2.2 生物学特性

普通早熟禾系多年生丛生禾草，密生匍匐茎，亦有地下茎蔓延根际，其节上发生新枝，可迅速长成丛密的草皮，且持久。普通早熟禾喜湿润而肥沃的黏土，耐阴不耐酷热，耐寒性强，其根入土较浅，表土须墒情充足，不耐干旱。遇旱时须进行灌溉，否则生长不旺，植株低矮，茎叶变红，雨后即能恢复生长。因此普通早熟禾不宜栽种于疏松贫瘠的沙土。

栽种普通早熟禾的地区，如冬季温和潮湿，则春季生长很早，5—6 月间

heading in early summer. Growth become slowly in summer, slow down later, and becoming lush again after rain in late summer.

6.4.2.3 Cultivation and utilization

Roughstalk bluegrass is suitable for long-term grassland establishment; it is easily mixed with legumes and the other grasses. Because of its spreading rhizomes, it can take full advantage of gaps when mixed with other forage species, minimizing the ground, suitable for conserving moisture, and preventing weeds invasion. It is an excellent bottom grass. In addition to the arid regions, it is suitable for mixed ordinary blue¬grass grassland. Grasses like timothy, orchard grass, ryegrass and legume forages like red clover, white clover mused are appropriate in mixture with ordinary blue-grass. Seeds of or¬dinary bluegrass are small, maturate soil is good for seeding more than on newly cultivated soil. Seed-bed preparation must be fine and shallow seeding. Unless the soil is moist, it should be packed before planting in order to facilitate emergence. Sowing can be done in either spring or autumn. Autumn is more appropriate if autumn rainfall is abundant. Seeding rate is about 11.2~15.0 kg/hm^2. Only 25%~50% seeds are needed if mixed with other forage species.

When roughstalk bluegrass was grown 2~4 years, rhizomes spread over forming a dense sod which can limit development of new growth; suitable renovation may increase soil aeration and promote new growth. This can be done during late autumn when the pasture stop growing, the growth of the grass will recover after contact with the soil through a whole winter season. Raking can be carried out in early spring or after each mowing, but should be combined with fertilization, irrigation, then grasses will grow flourish. Seed yield is 450 kg/hm^2.

Roughstalk bluegrass grows faster than Kentucky blue¬grass and can get higher yield in the second year if growth conditions are favorable. Forage has high nutrient value and good palatability. A variety of livestock find the forage palatable, especially cattle.

Roughstalk bluegrass is very tolerant of grazing. It

生长最盛，夏初抽穗，其后生长减退，夏后得雨，生长又盛。

6.4.2.3 栽培利用技术

普通早熟禾适于栽种在长期草地，并易于与豆科和禾本科牧草混播。因蔓生匍匐茎，与其他牧草混种时能充分利用空隙，保蓄地面的水分，防止杂草的侵入，为优良的下繁草。除干旱地区外，混播普通早熟禾最为有利。禾本科牧草中的猫尾草、鸭茅、黑麦草以及豆科牧草中的红三叶、杂三叶、白三叶等皆宜与普通早熟禾混播。

普通早熟禾种子细小，宜于播种在熟化的土壤，不宜播种在新垦的土上。整地务必细碎，播种宜浅，除非土壤湿润，否则须于播种前先行镇压，以利于出苗。播种时期春秋皆可，秋季如雨水充足，以秋播为宜。播种量为 11.25 ~ 15.0 kg/hm^2，如与其他牧草混种，每公顷只需用单播量的25% ~ 50%。

普通早熟禾生长 2 ~ 4 年后，如地下茎蔓延过度，草皮絮结，有碍新株发育时，须切碎草皮，增加土壤通气，以促进新株的生长。在晚秋牧草基本停止生长时进行，经过一个冬季，可使切断的根茎很好地与土壤接合，翌年春季回暖时期牧草会很快恢复生长。耘耙时间可于早春或每次刈割后进行，同时应结合施肥、灌溉，则其生长复盛，种子产量约为 450 kg/hm^2。

普通早熟禾生长比草地早熟禾为快，如生长条件适宜，第二年即能获得较高产量。其草料营养价值很高，适口性好，各种家畜皆很嗜食，尤以牛所喜食。

普通早熟禾耐牧性甚强，虽经践踏和不断地嗜食，牧后生机易于恢复。普通早熟禾与草地早熟禾同属下繁草，前

recovers well after grazing despite constant trampling and grazing. Roughstalk bluegrass and Kentucky bluegrass are both bottom grass. The former grows taller and is suitable for mowing for hay. With enough rainfall, appropriate raking and fertilizing after grazing or cutting, it still can produce abundant forage.

者植株较高，故亦适于刈割，可调制优质干草。普通早熟禾再生性稍差，但如雨量充沛，且于刈牧后能加以耘耙、施肥等复壮处理，则仍能获得更高产量。

6.5 *Festuca* L. 羊茅属

The genus includes hundreds of species in the worldwide, widely distributed in cold temperate regions. There are 23 species in China; most can be used for forage. The species cultivated in China primarily are meadow fescue and tall fescue.

羊茅属植物全世界有百余种，广布于寒温带地区。我国有 23 种，大都可以饲用。我国栽培的主要有草地羊茅和苇状羊茅。

6.5.1 Meadow fescue

Scientific name: *Festuca pratensis* Huds.

Origin and geographic distribution: Meadow fescue originated in the warm zones of Eurasia. It is cultivated in warm and humid region of the world or irrigated areas. China has wild species, but those being cultivated are introduced species. It was introduced in the 1920s. Now it is cultivating in the northeast, north, northwest, Shandong and Jiangsu provinces, especially in the warm, humid regions of northern temperate zone or southern subtropical areas of high altitude

6.5.1.1 Botanical characteristics

The stem is smooth, erect and 50~140 cm tall. Leaves are unfurled, 10~60 cm long, 4~8 mm wide and dark green. The based ridge is significant; sheaths are mostly shorter than internodes. The auricle is highlight. Ligules are small and blunt, light green, with eyelash appearance on both edges. The inflorescence is erect or dropping on top, 10~20 cm long, spreading, branching near the base into spikelets with 3~13 flowers in each. Spikelets are mostly 6~8, 8~12 mm long, 2 with membranous glumes of different sizes ridged and sharp. The first glume has 1 pulse in general. The second glume has 3 veins, 3~4 mm long. The lemma 5~7 mm long with 3~7

6.5.1 草地羊茅

学名： *Festuca pratensis* Huds.

英文名： meadow fescue

别名： 牛尾草、草地狐茅

起源与地理分布： 草地羊茅原产于欧亚温暖地带，在世界温暖湿润地区或有灌溉条件的地方均有栽培。我国也有野生种，但栽培的均为引入种，自 20 世纪 20 年代引入以来，现在东北、华北、西北及山东、江苏等地均有栽培，尤其适于北方温暖带或南方亚热带高海拔温暖湿润地区种植。

6.5.1.1 植物学特征

茎秆光滑，直立，高 50 ~ 140 cm；叶片平展，长 10 ~ 60 cm，宽 4 ~ 8 mm，深绿色，基部之脊颇显；叶鞘大多较节间为短；叶耳突出，叶舌小而钝，淡绿色，边缘均具睫毛。圆锥花序直立或顶端下垂，长 10 ~ 20 cm，散开，近基部处分枝生小穗，小穗有花 3 ~ 13 朵，而以 6 ~ 8 朵花为多，长 8 ~ 12 mm，颖 2，膜状，大小不一，有脊，尖锐，第一颖多生 1 脉，第二颖 3 脉，长 3 ~

veins, back round, without mans on the top. The palea is slightly shorter than the lemma with two ridges.

6.5.1.2 Biological characteristics

Meadow fescue is a perennial thinning grass, sometimes producing short, creeping underground stems or rhizomes. Rhizomes grow into short rhizome thinning brushwood. It is deep-rooted, to 160 cm under suitable conditions. The majority of fibrous roots are concentrated in the top 20 cm of soil. Viability is good at it grows in shade and barren mountain. It is tolerant of drought and humidity, so it can grow on poorly drained clay, lowland swamp, and low and wet sides which are not suitable to most other general grasses. It tolerates some flooding. Because of deep roots, it can grow in loose droughts soil if the soil is deep. In the meadow fescue is a reliable forage producer, with a shorter growth period during drought. In fertile, moist, clay loam on the loose, or dewatered wetlands, growth is better. After sowing, meadow fescue grows slowly becoming vigorous during the second or third year. It grows quite early in spring, resume growth when temperatures reach 2~5 ℃ and growth until autumn. Meadow fescue produces many tillers, those times more regulative than unproductive, lead to a high leaf: stem radio. Heading occurs in May and June and plant mature in June and July. The time from green up to seed maturation is about 100~110 d.

6.5.1.3 Cultivation and utilization

Meadow fescue seed is small and seedbed preparation should be fine. Seed depth is 2~3 cm. It is suitable for long-term stand establishment and often mixed with timo¬thy, orchard grass, ryegrass, white clover etc. except for seed production purpose. The row spacing is 30 cm. Seeding rate is 15 kg/hm2. Usually it should be weeded 1~2 times in seedling stage. Do not utilize it if seeded in summer or autumn, especially no grazing. Mow¬ing and seed harvest may start from the second year. The growth period of tall fescue is long, with good regrowth habit. It can be mowed 3~5 times annually under good soil moisture and fertility conditions. Tall fescue is very tolerant of grazing. The first grazing in the seeding year should be started at the jointing stage. Frequent rota-

4 mm，外稃 5 ~ 7 mm 长，3 ~ 7 脉，背圆，顶端无芒；内稃较外稃稍短，有 2 脊。

6.5.1.2 生物学特性

草地羊茅是多年生疏丛型草本植物，具匍匐短根茎，根系发达，在适宜条件下，根深可达 160 cm，但绝大多数须根集中于 0 ~ 20 cm 土层内。生活力颇强，能生长于荫下及瘠薄的山地。既耐旱，又耐湿，能生长于排水不良的黏土、沼泽低地，及其他地势低湿不适于栽种一般牧草之处，能耐一定时期的水淹；根系深，如土层较厚，即使在疏松干旱地亦能生长。在水旱不调之处，如混种草地羊茅，则草料生产较为可靠，唯在干旱时生长期较短。在肥沃、湿润、疏松的黏壤土，或疏干的沼泽地上，则生长更好。

草地羊茅播种后生长较慢，第二、三年生长始盛，春季生长颇早，当气温达 2 ~ 5℃时即可返青，直至秋季能继续生长。分蘖颇多，特别是营养枝较多，为生殖枝的 3 倍，叶的比例较高。5、6 月抽穗，6、7 月种子即可成熟，从返青至种子成熟为 100 ~ 110 d。

6.5.1.3 栽培利用技术

草地羊茅种子细小，应精细整地，适当覆土，以 2 ~ 3 cm 为宜。宜种于干湿不调的长期草地，除预备收种者外，较少单独播种，常与猫尾草、鸭茅、黑麦草和白三叶等混种。条播行距 30 cm，播种量为 15 kg/hm^2。苗期须除草 1 ~ 2 次，夏播或秋播最好当年不要利用，尤其是放牧。第二年开始正常刈割和收籽利用，生长期长，再生性强，水肥条件好时可年刈割 3 ~ 5 次，以抽穗期刈割为宜，耐牧性很强。年内首牧应在拔节期进行，频繁轮牧既可防止草丛老化，又可形成稀疏草皮。种子落粒

tional grazing will slow maturation and lead to form a sparse sod, but seed should be harvest in the dough stage because of its strong shattering tendency. The germination rate of newly harvested seed is low and reaches normal after 100 d. Seed longevity can be 3 years or longer.

6.5.2 Tall fescue

Scientific name: *Festuca arundinacea Schreb.*

Origin and geographic distribution: Tall fescue originated in Western Europe. Wild tall fescue occurs in the humid area of Xinjiang, Central part of Northeast of China. The cultivation started in Britain and the United States in the 1920s, and it has now become one of the most important cultivated grasses in Europe and the United States. It was first introduced by Northeast of China in 1923, currently it becomes a primary species that is used to the establish of artificial grasslands and for the reseeding of natural pasture in the warm regions of the north. It is not only used widely throughout the world for forage but also as turfgrass.

6.5.2.1 Botanical characteristics

Tall fescue and meadow fescue are quite similar. The difference is that tall fescue is thicker than meadow fescue. The culms are as 80~150 cm as tall. The upper part of the stem, based leaf sheaths and the top of sclerophyllous are all very rough. It is hairless on the edge of ligules and auricles, and there is no awn on top of the lemma or, if present, awns are short 0.7~2.5 mm long, which distinguishes between meadow fescue and tall fescue. Botanical characteristics of tall fescue are shown in Fig.6.11.

6.5.2.2 Biological characteristics

As a perennial bunchgrass, tall fescue has a wide environmental amplitude. It is adapted to humid a climate and fertile, loose soils even those with a high groundwater level. It has strong resistance to cold even at 1℃ , and grows rapidly with temperate above 4℃ . It is also adapted to hot regions. It has no strict soil requirements, and it grows in overflow and poorly drained soil. So it is adapted to both acid and alkaline soils. And it is tolerant of moderate salt levels. Tall fescue can be sow in late March in the northern region and matures in

性强，采种宜在蜡熟期进行。新收获种子发芽率低，100 d 后可达正常，种子寿命可保存 2 ~ 3 年或更长。

6.5.2 苇状羊茅

学名： *Festuca arundinacea* Schreb.

英文名： tall fescue

别名： 苇状狐茅、高狐茅

起源与地理分布： 苇状羊茅原产于欧洲西部，我国新疆、东北中部湿润地区有野生。20 世纪 20 年代初开始在英、美等国栽培，目前是欧美重要栽培牧草之一。在我国东北地区引种，始于 1923 年，现已是北方暖温带地区建立人工草地和补播天然草场的重要草种，尤其是作为草坪草种在全球显示出非常巨大的作用。

6.5.2.1 植物学特征

苇状羊茅与草地羊茅颇相似，不同点在于其较草地羊茅更为粗大，株高 80 ~ 150 cm，且上部的茎秆、下部的叶鞘及硬叶之上面皆甚粗糙。叶舌及叶耳边缘无毛，外稃顶端无芒或具长 0.7 ~ 2.5 mm 的短芒，可与草地羊茅区别，苇状羊茅植物学特征如图 6.11。

6.5.2.2 生物学特性

多年生疏丛草本。生态幅度较广，而适于湿润气候和肥沃疏松的土壤，在地下水位高的情况下也能很好地生长。较耐寒，在 1℃的温度下也能继续生长，温度高于 4℃时，生长速度加快。也适于较炎热的地方。对土壤要求不严格，在水泛地、排水不良的土壤上均能生长，并能适应酸性与碱性土壤，具有一定的耐盐能力。

苇状羊茅在北方地区 3 月下旬播种，8 月上旬成熟，播种当年可完成发育周期，越冬良好。在东北地区，生长

Fig. 6.11 The botanical characteristics of tall fescue
图 6.11 苇状羊茅植物学特征

early August. It completes the development cycle in the sowing year and overwinters well. Green up is in April during the second year in the Northeast, and heading occur from the end of May to early June. Flowering occurs in mid June and plants mature by early July. The time from spring green up to seed maturity is about 90 d.

第二年 4 月初返青，5 月底至 6 月初抽穗，6 月中旬开花，7 月初种子成熟，从返青至种子成熟约需 90 d。

6.5.2.3 Cultivation and utilization

Tall fescue is a deep-rooted, high yielding grass which

6.5.2.3 栽培利用技术

苇状羊茅为深根型高产牧草，要求土层深厚、底肥充足。为此在播种前一年秋季应深翻耕，并按 30 t/hm^2 厩肥施足基肥。播前须耙耱 1 ~ 2 次。苇

needs deep soil and adequate fertilizer. Therefore in the fall before sowing, soil should be deep plowed and 30t/hm^2 animal manure. Land must be raked once or twice before sowing. Tall fescue can be planted in spring, summer or autumn depending on local conditions. The short rhizomes of tall fescue make it suitable as a monoculture, but it can also be mixed with alfalfa, white clover, red clover or milkvetch. Row spacing is 30 cm. Seeding rate is 15~30 kg/hm^2. Tall fescue has very limited resistence to weeds at the seedling stage. Cultivating and weeding is critical. Eliminating weeds before sowing should be through. New stands should be weeded after each mowing as well. Cutting should occur before or at early flowering stage. Seed harvest is best at dough stage. Seed is lost due to shattering if seed harvest is delayed. Losses can be up to 35%~40%. Seed life of tall fescue is very short. Germination rate drops sharply when stored 4~5 years. Seed viability is an important inder before sowing.

Common varieties: Introduced varieties include **'fine'** **'lexus'** **'may keane'** **'jaguar 3'**. Local varieties are **'yancheng'**, improved varieties of the **'yangtze river No.1'** **'beishan No.1'** **'qiancao No.1'**.

状羊茅易于建植，根据各地条件，可春、夏、秋播。苇状羊茅短根茎具侵占性，适于单播，也可与紫花苜蓿、白三叶、红三叶和沙打旺等混播。条播行距30 cm，播量为15 ~ 30 kg/hm^2。苇状羊茅苗期不耐杂草，中耕除草是关键，除播前和苗期加强灭除杂草外，每次刈割后也应进行中耕除草。牧草宜于花前或初花期刈割，采种宜于种子蜡熟初期，延迟收获种子易脱落，损失量可达35% ~ 40%。苇状羊茅种子寿命很短，贮藏4 ~ 5年后发芽率急剧下降，生产上应注意供播种子的活力。

常见品种：主要引进品种有‘法恩’‘凌志’‘可奇恩’‘美洲虎3号’；地方品种有‘盐城’；育成品种有‘长江1号’ ‘北山1号’ ‘黔草1号’。

6.6 *Elytrigia* L.—quackgrass 偃麦草属—偃麦草

The gensus *Elytrigia* L. includes about 50 species worldwide. In the temperate regions of the Western Hemisphere, including America and Canada there are more than 80 cultivated varieties with high yield, good quality and strong pest resistance. There are six species in China, among them *Elytrigia repens*, *Etrichpphora* and *Elytrigia elongata* grow well.

Quackgrass

Scientific name: *Elytrigia repen* (L.) Desv.

Origin and geographic distribution: Quackgrass is distributed in the meadow steppe, especially in northeast of China, Inner Mongolia, Hulun Buir, Xilin Gol Grassland, Ningxia, Gansu, Qinghai, Xinjiang, Tibet and other provinces.

偃麦草属全世界约有50种，产于西半球的温带地区，包括美国和加拿大等北温带国家，已培育了80多个产量高、品质好、抗逆性强的优良栽培品种。我国共有6个种，其中以偃麦草、毛偃麦草和长穗偃麦草等生长良好。

偃麦草

学名：*Elytrigia repen*(L.) Desv.

英文名：quackgrass

别名：速生草

起源与地理分布：偃麦草分布于我国东北草甸草原，内蒙古呼伦贝尔、锡

It also occurs abroad in northern Mongolia, Siberia, Korea, Japan, India and Malaysia.

Botanical characteristics

Quackgrass is a perennial grass. Erect and 60~120 cm tall, sparsely tufted. The sheath is glabrous or hairy. Leaf blade is rough or slightly hairy. Auricles are small and membranous. The ligule is short and blunt, about 0.5 mm long. The spike is erect, 8~18 cm long, 10 mm wide. Every spikelet contains 6~10 flowers. The lanceolate glume is pointed leading to the awn at the apex. The glume is shorter than lemma and there are cilia on the back and hairs on the top of ovary. Thousand seed weight is 4.0 g. Botanical characteristics of quackgrass are shown in Fig.6.12.

Biological characteristics

Quackgrass is a thermophilic grass. It prefers loose and moist soil, common in the river alluvial plain, valleys and lowlands. It can tolerate mild soil salinity and has high resistance to grazing and tolerates trampling. It develops long, thick underground stems, so fecundity is very strong. It flowers in late June to mid-July. Limited quantities of seed mature in August.

Cultivation and utilization

Germination and seedling growth of quackgrass requires good moisture and fertile soil. Good seedbed preparation should proceed before planting, manure application can serve as base as fertilizer. Spring, summer and autumn are appropriate for sowing, either drilling or broadcasting. Sowing depth is 2~3 cm. When drilling the seeding rate is 15~30 kg/hm^2. Quackgrass can be mixed with legumes like alfalfa and sainfoin. Yield can be increased with irrigation and top dressed fertilizer after each cutting.

Quackgrass has well-developed roots and spreads aggressively, so it is often used in water and soil conservation as well as protection of rail and road slopes and embankments. But it is not a suitable the introduction for the field rotation system as it may become a weed which is difficult to eradicate.

Common varieties: There are no registered domestic quackgrass varieties in most areas. Wild materials are generally used.

林郭勒草原，及宁夏、甘肃、青海、新疆和西藏等省区；在蒙古北部、西伯利亚地区、朝鲜、日本、印度和马来西亚也有分布。

植物学特征

多年生草本，具根茎，秆直立，株高 60 ~ 120 cm，秆疏丛生。叶鞘无毛或分蘖叶鞘有毛；叶片上面粗糙或略有毛，背部粗糙；叶耳小膜质；叶舌短而钝，长约 0.5 mm。穗状花序直立，长 8 ~ 18 cm，宽约 10 mm；小穗含 6 ~ 10 朵花；颖披针形，先端具芒状尖头。内稃短于外稃，背生纤毛；子房上端有毛；千粒重 4.0 g。偃麦草植物学特征如图 6.12。

生物学特性

偃麦草是一种喜温牧草。喜疏松湿润的土壤，常见于河流冲积平原，沟谷低地。能忍受轻度的土壤盐渍化。耐牧性很强，不怕践踏；具粗长而发达的地下茎，故其繁殖力甚强，易形成单一群丛，6 月下旬到 7 月中旬花期，8 月间种子成熟，结籽不多。

栽培利用技术

偃麦草种子发芽及幼苗生长要求有较好的土壤水肥条件，故播前必须精细整地，施入厩肥作底肥。播种时期春、夏、秋皆可。条播或撒播，覆土深度 2 ~ 3 cm。条播时播种量为 15 ~ 30 kg/hm^2，偃麦草可与豆科牧草紫花苜蓿、红豆草等间混播。每次刈割利用后，如能进行灌溉和追肥，可获得较好的增产效果。

偃麦草根茎相当发达，侵占性较强，也适于作为水土保持和保护铁路和公路边坡、堤岸的植物。但不宜引入大田轮作中，否则将成为一种恶性杂草，难以根除。

Fig. 6.12 The botanical characteristics of quackgrass
图 6.12 偃麦草植物学特征

6.7 *Hordeum* L. 大麦草属

Many *Hordeum* L. species are found in China. By introduction and acclimatization in recent years, Two main cultivated species are foxtail barley and *Hordeum bog-danii*.

大麦草属牧草在我国分布有数种，近年来引种驯化并具栽培价值的主要有两种，即短芒大麦草和布顿大麦草。

6.7.1 Foxtail barley

Scientific name: *Hordeum brevisubulatum* (Trin.) Link

Origin and geographical distribution: Foxtail barley is distributed in the swales of Northeast China, North China and Xinjiang. There are wild species in the former Soviet republics of the Caspian Sea, Siberia and The People's Republic of Mongolia. Many locations in northern China have domesticated materials in recent years. It provides good grazing.

6.7.1.1 Botanical characteristics

Foxtail barley is perennial, with short rhizomes, dense fibrous roots, and slender culms usually slim, erect or geniculate at lower nodes. Plants are 30~90 cm high. Leaf length is about 30 cm and leaf width is 3~5 mm. Green or gray-green leaves are concentrated in the lower part of the plant. Spikes are 5~10 cm long and turn purple when mature. Spikelets grown three per section of rachis. Lateral spikelets on both sides are usually small, rudimentary or male, petiolate. Glumes are lanceolate and rough. Lemmas are without awns. Glumes are the shape of sessile spikelets resembling the petiolate. The lemma is nearly smooth, with the apex acuminating into a short awn. The caryopsis about 3 mm, with a pileous apex. The ventral sulcus is linear and not obvious. Thousand seed weight is 2~3 g. Botanical characteristics of foxtail barley are shown in Fig.6.13.

6.7.1.2 Biological characteristics

Foxtail barley is a premature grass. It grows well in light saline alkali soil. It has strong tillering and can be divided from 40 to 70 generally, reaching up to 130 superlatively. Foxtail barley matures slowly in the seeding year, and most plants are nutritious. Only individual plants can head and flower, without seeding and mature. In the second year, it turns green in April, heads in June, flowers in July, and matures seed in August. The complete growth period is 110~130 d. Seed matures consistently and has good fecundity but seed shatters easily. Seed harvest must be timing.

6.7.1 短芒大麦草

学名: *Hordeum brevisubulatum* (Trin.) Link

英文名: foxtail barley

别名: 野大麦、野黑麦、莱麦草

起源与地理分布: 分布于我国东北、华北及新疆等地低湿草地上，里海、西伯利亚地区和蒙古均有野生种分布。近年来我国北方各地引种驯化，表现为生长期短、耐盐碱、草质柔软，是一种良好的放牧-刈割兼用的多年生禾本科牧草。

6.7.1.1 植物学特征

短芒大麦草属多年生草本，具短根茎,须根稠密; 秆细,直立或下部节膝曲，株高 30 ~ 90 cm；叶片长约 30 cm，宽 3 ~ 5 mm，绿色或灰绿色，集中于植株下部；穗状花序，长 5 ~ 10 cm，成熟时带紫色，小穗每 3 枚生长于穗轴各节，两侧小穗通常较小或发育不全或为雄性，有柄，中间小。穗颖呈针状，粗糙，外稃无芒，无柄小穗之颖形似有柄者，外稃近于平滑或贴生微毛，先端渐尖成短芒；颖果长约 3 mm，顶端被毛，腹沟不明显，线形，千粒重 2 ~ 3 g。短芒大麦草植物学特征如图 6.13。

6.7.1.2 生物学特性

短芒大麦草是一种早熟性禾草，宜在轻盐碱土壤生长。分蘖力强，一般可分蘖 40 ~ 70 个，最高达 130 多个。短芒大麦草播种当年发育慢，大部分植株体处于营养状态，仅有个别株体抽穗开花，但不能结籽成熟。第二年 4 月返青，6 月抽穗，7 月开花，8 月种子成熟。生育期为 110 ~ 130 d。种子成熟一致，结实性好，但易断穗落粒，应注意收种时期。

Fig. 6.13 The botanical characteristics of foxtail barley
图 6.13 短芒大麦草植物学特征

6.7.1.3 Cultivation and utilization

Foxtail barley needs deep tillage in the autumn, and it also need harrowing again before sowing. Spring, summer, or sowing is possible. With good irrigation and moist soil, spring seeding is recommended, otherwise it shculd be seeded in summer and autumn after rain. Seeding rate is 7.5~15.0 kg/hm^2. Seed depth is 3~4 cm. Row spacing is 15~30 cm, and packing soil after sowing is useful. Seedlings grow slowly in the seeding year, and can be affected by weeds. Cultivating and weeding should be done when seedlings have 3~4 true leaves. Jointing stage and heading stage should be

6.7.1.3 栽培利用技术

短芒大麦草喜欢生长在湿润的轻度盐渍化土壤上。秋季深翻土地，播种前再行耙耕一次。春、夏、秋皆可播种。如有灌溉条件或春墒较好，可进行春播，否则须在夏季或秋季雨后播种。播种量为 7.5 ~ 15.0 kg/hm^2，播种深度 3 ~ 4 cm。通常条播，行距 15 ~ 30 cm，播后镇压。播种当年幼苗生长缓慢，易受杂草抑制。因此，当小苗长出 3 ~ 4 片真叶时，进行中耕除草，拔节及孕穗

irrigated may occur 1~2 times depending on growing conditions. It is better not to cut or graze in the seeding year. In the second year, one irrigation, if available, should coincide with resumption of growth in early spring. A first cutting should be at the beginning of the heading stage to flowering leaving a stubble height of 2~3 cm. A second time cutting at boot or heading stage should be considered in locations with a short frost-free period. The second cutting should be 30 d before growth ends, or it will influence overwintering. Pasture can be grazed after elongation. The grazing period should not be longer than 7~10 d, and strengthening management at intermission.The grazing plot should not be too small. Seed field should be harvested in a timely schedule. When seed of the middle spikelets is mature, seed should be harvested. Seed yield is generally 225~450 kg/hm^2.

可根据条件灌水 1 ~ 2 次。播种当年最好不刈割或放牧。第二年早春有条件地区可灌一次返青水。刈草地可于抽穗–开花初期进行第1次刈割，留茬高度2 ~ 3 cm，第二次刈割一般在孕穗、抽穗期进行，无霜期短的地区，第二次刈割不得迟于停止生长前30 d，否则影响越冬；放牧场可在拔节后放牧，放牧采食期不得多于 7 ~ 10 d，放牧间歇期应加强管理，放牧小区不易太小。种子田采种应及时，一般当穗中部种子成熟时即可采种。短芒大麦草种子落粒性强。一般种子产量为 225 ~ 450 kg/hm^2。

常见品种：野生栽培品种‘察北野大麦’和育成品种‘军需 1 号’。

6.7.2 *Hordeum bogdanii* Wilensky

Scientific name: *Hordeum bogdanii* Wilensky

Origin and geographical distribution: It is mainly distributed in Xinjiang, Gansu, and Qinghai. It also occurs in Mongolia, Russia, Central Asia, Siberia and Europe.

6.7.2.1 Botanical characteristics

Hordeum bogdani is a perennial grass. Erect culms, often geniculate at the base, 40~80 cm tall and the stem has 5~7 nodes. Nodes are slightly prominent and densely gray hairy at the nodes. Leaf sheaths are densely hairy when it is young, with a membranous ligule, about 1 mm wide. Leaf blades are 6~15 cm long, and 4~6 mm wide. Spikes are erect, bent at the apex, usually gray-green, 5~10 cm long, and 5~7 mm wide. Rachis internode length is about 1 mm. Lateral triple spikelets have 1.5 mm. Glumes are 6~7 mm. Central spikelet is sessile and lanceolate. Glumes are setaceous. Lemma herbaceous, setaceous, about 6 mm, with 7 mm aristate at the apex pubescent with fine bristles on the back. The palea is longer than the lemma. Flowering is from June to September. The caryopsis is obovate, 3 mm long, and the thousand seed weight is 1.56 g.

6.7.2.2 Biological characteristics

Cold resistance of *Hordeum bogdanii* is excellent, endur-

6.7.2 布顿大麦草

学名：*Hordeum bogdanii* Wilensky

起源与地理分布：布顿大麦草主要分布在新疆、甘肃、青海。蒙古、俄罗斯、中亚、西伯利亚地区、欧洲也有分布。

6.7.2.1 植物学特征

多年生草本，具根状茎，形成疏丛，秆直立，基部有时膝曲，高 40 ~ 80 cm，茎 5 ~ 7 节，节稍突出，密被灰毛。叶鞘幼嫩时被柔毛；叶舌膜质，长约 1 mm；叶片长 6 ~ 15 cm，宽 4 ~ 6 mm。穗状花序直立，端略弯曲，通常呈灰绿色，长 5 ~ 10 cm，宽 5 ~ 7 mm，穗轴节间长约 1 mm，易断落；三联小穗两侧具长约 1.5 mm 的柄，颖长 6 ~ 7 mm，外稃贴生细毛，连同芒长约 5 mm，中间小穗无柄，颖披针形，外稃草质，披针形，长约 6 mm，先端具长约 7 mm 的芒，背部贴生短柔毛或细刺毛，内稃长于外稃。花期 6—9 月。颖果倒卵形，长约 3 mm，千粒重 1.56 g。

6.7.2.2 生物学特性

布顿大麦草抗寒能力较强，

ing freezing at -35℃. It is a saline-alkali tolerant plant, with poor drought resistance. Plants grow well with moist conditions. Tillering ability is moderate. Plants grow slowly in the seeding year. Rapid growth during the second year produces higher yields.

在 –35℃下不受冻害。耐盐碱，但抗旱性较差，在湿润环境下生长繁茂。布顿大麦草分蘖能力中等，一般当年能达 8 ~ 42 个分蘖。布顿大麦草种植当年生长较慢，第二年即进入生长盛期，产量也较高。

6.7.2.3 Cultivation and utilization

Both spring and autumn seedling is possible, but spring is best. When early spring ground temperatures reach 10 ℃, sowing can begin. Deep tillage in the autumn before sowing and harrowing in the early spring provide a good seedbed. Seeding rate is 15 kg/hm^2. Row spacing is 15~30 cm and seed depth is 3~4 cm. *Hordeum bogdanii* wilensky grows slowly at seedling stage, and is affected by weeds easily. Intertillage and weeding tiller and elongation stage should irrigate in time. Seed shatter easily after maturity. Seed harvest should begin when 50%~60% of the seed is mature. Seed yield is 375~400 kg/hm^2.

6.7.2.3 栽培利用技术

布顿大麦草春播、秋播均可。但以春季为好。当早春地温达到 10℃时即可播种。播前应秋翻，早春耙耱播种。播种量一般为 15 kg/hm^2，条播，行距 15 ~ 30 cm，播深 3 ~ 4 cm。布顿大麦草由于苗期生长缓慢，易受杂草危害，要特别注意中耕除草，分蘖拔节期应及时灌水。种子成熟后，极易脱落，应在种子成熟达 50% ~ 60% 时即进行采种。种子产量为 375 ~ 400 kg/hm^2。

6.8 *Phleum Pratense* L.—timothy 猫尾草属—猫尾草

The *Phleum* L. genus includes both annual and perennial grasses. This genus has about ten species, distribute in the warm and cold zones of two hemispheres. There are 3 wild species, and one introduced species in China. Most species of this genus are excellent forage plants. Herd grass is widely used as a main forage around the world, and is also widely used as the main forage of the crop rotation system. In addition, *P.alpinum* is an excellent wild forage grass.

Timothy

Scientific Name: *Phleum pratense* L.

The origin and geographical distribution: Timothy is native to the temperate areas of Eurasia, and is cultivated in of Europe and the United States. It is an important grass in the Soviet Union. In China, it is grown in the regions of northeast, north and northwest, but occupies a small area. It can be used for haymaking providing high feed value.

猫尾草属，又名梯牧草属，为一年生或多年生草本。本属约有 10 种，分布于两半球温寒地带，我国野生有 3 种，引进有一种。全属多为优良饲用植物，其中猫尾草是世界上最为广泛利用的主要牧草，又是草田轮作的牧草。此外，高山猫尾草（*P. alpinum*）是我国的一种野生优良牧草。

猫尾草

学名： *Phleum pratense* L.

英文名： timothy

别名： 梯牧草、鬼蜡烛

起源与地理分布： 猫尾草原产于欧洲和亚洲的温带，欧美各国均有栽培，在俄罗斯被列为重要牧草。我国东北、华北、西北各省区都有种植，但目前面

Botanical characteristics

Timothy is a perennial bunch grass. Plant height is 80~110 cm, sometimes reaching 120 cm. Fibrous roots are very developed. Erect stem, short internodes, 6~7 nodes. Lower nodes are mostly oblique. The first and second nodes at the base are more developed, inflating spherically. Leaf blade is oblate, smooth and glabrous, acute at the apex. Leaf blade is 10~30 cm long and 0.3~0.8 cm wide. Leaf sheaths are lax and clasping, grow was in at the internode. The ligule is a triangle and auricles are round. Panicles consist of small spikelets in a tight column, 5~10 cm long. There is only one flower per spikelet. Glumes are round with a keel, hairy on the edge with a short awn at the apex. Lemma length is half of the glume, without an awn at the apex. The palea is narrow and thin, shorter than the lemma. Seeds are tiny, nearly round and easily to separated with the glumes. Thousand seed weight is only 0.37 g. Botanical characteristics of timothy are as shown in Fig.6.14.

Biological characteristics

Timothy grows well in a cold humid climate. Plants grow well in areas where rainfall is above 700 mm. Cold resistance is good. Seed germinate when soil temperature is 3~4 ℃. The optimum temperature for heading stage is 18~19℃. The winter hardiness is excellent. When spring temperature is above 5℃, it turns green. When fall temperature is below than 5℃ in fall, it stops growing. Water tolerance is good, but drought resistance is poor. Plants cannot endure summer at with sustained drought and high temperature above 35℃. Plants grow in neutral or weakly acidic soils, better between pH 5.3-7.7. It has poor tolerance of alkali. Timothy is long-lived, generally 6~7 years, with better management; it will persist 10~15 years. Yield of fresh grass is 15 000-45 000 kg/hm^2 and can reach 75 000 kg/hm^2 when mixed with red clover. Seed yield is 300~750 kg/hm^2.

Cultivation and utilization

Seed field can be improved with supplementary pollination or honeybee pollination at the flowering blooming stage. Late pollination is not conducive to nutrient accumulation in roots. Clean cultivation before sowing

积较小。刈制干草，饲用价值极高。

植物学特征

猫尾草是禾本科猫尾草属多年生疏丛型草本植物。株高 80 ~ 110 cm，高者达 120 cm。须根发达。茎直立，节间短，6 ~ 7 节，下部节多斜生，基部 1 ~ 2 节处较发达，膨大呈球形。叶片扁平细长，光滑无毛，尖端锐，长 10 ~ 30 cm，宽 0.3 ~ 0.8 cm，最大宽度为 1 cm 以上。叶鞘松弛抱茎，长于节间；叶舌为三角形；叶耳为圆形。圆锥花序，小穗紧密，呈柱状，长 5 ~ 10 cm；每个小穗仅有一花。颖圆，具龙骨，边缘有茸毛，前端有短芒；外稃为颖长之半，顶端无芒；内稃狭薄，略短于外稃，种子细小，近圆形，易与颖分离。千粒重仅 0.37 g。猫尾草植物学特征如图 6.14。

生物学特性

猫尾草喜冷凉湿润气候。在年降水量 700 mm 以上地区生长良好。抗寒能力较强，地温 3 ~ 4℃时种子发芽，抽穗期适宜温度为 18 ~ 19℃。越冬性好。春季气温高于 5℃时开始返青，秋季气温低于 5℃停止生长。较耐淹浸，不耐干旱。在 35℃以上的持续高温干燥条件下，不能安度夏季。要求中性及弱酸性土壤，以 pH5.3 ~ 7.7 时为最适宜。耐碱性较弱，在石灰质含量多的土壤上生长不良。猫尾草寿命较长，一般生长 6 ~ 7 年，在管理条件较好的情况下，可生活 10 ~ 15 年。一般鲜草产量为 15 000 ~ 45 000 kg/hm^2，种子产量为 300 ~ 750 kg/hm^2，与红三叶混播时鲜草产量可达 75 000 kg/hm^2。

栽培利用技术

种子田在开花期要借助人工辅助措施授粉或利用蜜蜂授粉，以提高结实率。过迟不利于植株根部和根颈部营养物质

Fig. 6.14 The botanical characteristics of timothy
图 6.14 猫尾草植物学特征

improves establishment. Good soil moisture and organic fertilizer application of about 30 t/hm^2 are desirable. Sowing of spring and summer are appropriate in cold areas, fall rainy season. Row spacing is 20~30 cm, and the seeding rate is 12~18 kg/hm^2.For broadcasting the seeding rate should be 30 kg/hm^2. Seed depth should be 0.5~1.2 cm, with packing after sowing. Weed control is needed during the seedling stage. Common varieties include **'climax' 'timothy' 'hokuon' 'topas' 'piccolo'** and **'winmor'.**

积累。猫尾草播前要精细整地，保证良好的土壤墒情，施有机肥量为 30 t/hm^2。在寒冷地区春播或夏播，也可雨季秋播。条播行距 20~30 cm，播量为 12 ~ 18 kg/hm^2，撒播播种量为 30 kg/hm^2，播深 0.5~1.2 cm，播后适当镇压，苗期注意清除杂草。

常见品种: 'climax' 'timothy' 'hokuon' 'topas' 'piccolo' 'winmor'。

6.9 *Dactylis* L.—orchardgrass 鸭茅属—鸭茅

The genus *Dactylis* has about five species of perennial grasses. It is distributed in Eurasian temperate zones and North Africa. Only common orchardgrass have cultivation value.

Orchardgrass

Scientific name: *Dactylis glomerata* L.

Origin and geographical distribution: Orchardgrass is native to Europe, North Africa, and temperate regions of Asia. Now it is distributed worldwide in temperate regions. This plant is world famous excellent forage. There is a long cultivation history. It is currently cultivated in all of Europe, the Africa plateau, northeast of North America, South America, Oceania and Japan. Orchardgrass occurs at the edge of the forest zone of Chinese, wild species distribute in Xinjiang Tianshan Mountains, Emei Mount and Erlang mountain of Sichuan, and scattered in Greater Khingan Range Southeast slope. Common orchardgrass is used in addition to domesticated local wild species and introduction is from Denmark, USA, Australia and other countries. Now there is large areas of cultivation in Qinghai, Gansu, Xinjiang, Shaanxi, Guizhou, Sichuan, Yunnan, Hubei, Jilin, and Jiangsu provinces

Botanical characteristics

Orchardgrass is a perennial grass. Plant height is 60~140 cm (Fig.6.15A), it is flat at the base of stem. Radical leaf is more than stem leaf, they are collapsible when young. Leaf blades are 7~25 cm long and 2~8 mm wide. The leaf surface is rough. The median vein is prominent. The leaf cross-section is “V” shaped (Fig.6.15B). Leaf sheath is closed. Auricles are absent. The ligule is membranous and obvious (Fig.6.15C). The whole plant is glabrous and blue-green. The expanding panicle is 5~15 cm long. Spikelets grow on top of the rachilla. It resembles a chicken foot, is known as chicken feet grass. Each spikelet contains 3~5 flowers, green or slightly purple. It is cross-pollinated. Two glumes are of unequal length. There is bristle on the ridge. Seed is small.

鸭茅属也称鸡脚草属，多年生草本。全属约有5种，分布于欧亚温带和北非，仅鸭茅有栽培价值。

鸭茅

学名： *Dactylis glomerata* L.

英文名： common orchardgrass

别名： 鸡脚草、果园草

起源与地理分布： 鸭茅原产于欧洲、北非及亚洲温带地区，现在全世界温带地区均有分布。该草是世界上著名的优良牧草之一，栽培历史较长，现已在欧洲各地、非洲高原、北美东北部、南美、大洋洲及日本等温带地区栽培。鸭茅野生种在我国分布于新疆天山山脉的森林边缘地带及四川的峨眉山、二郎山等地，并散见于大兴安岭东南坡地。栽培鸭茅除驯化当地野生种外，多引自丹麦、美国、澳大利亚等国。目前，青海、甘肃、新疆、陕西、贵州、四川、云南、湖北、吉林、江苏等地均有大面积栽培。

植物学特征

多年生禾草。根系发达。疏丛型，丛径可达30 cm，株高60~140 cm，茎基部扁平（图6.15A）。基叶多，茎叶少，幼时折叠。叶片长7~25 cm，宽2~8 mm。叶面及边缘粗糙，中脉突出，断面呈“V”形（图6.15B）。叶鞘封闭。无叶耳。叶舌膜质，明显（图6.15C）。全株光滑无毛，蓝绿色。圆锥花序开展，长5~15 cm，小穗着生在穗轴的一侧，密集成球状，簇生于穗轴的顶端，形似鸡脚，故名鸡脚草。每小穗含小花3~5朵，绿色或稍带紫色，异花授粉，两颖不等长，脊上有硬毛。种子较小，外稃

The lemma is prominent and keeled. Apex aristate, 1mm. It is bent to the side slightly. Thousand seed weight is about 1.0 g(Fig.6.15D).

背部突起成龙骨状，顶端具长 1 mm 的短芒，稍向一侧弯斜，千粒重 1 g 左右(图 6.15 D)。鸭茅植物学特征如图 6.15。

Fig. 6.15 The botanical characteristics of orchard grass
(A. stand; B. membranous ligule; C. leaf and panicle; D. seeds)
图 6.15 鸭茅植物学特征（A. 鸭茅草地；B. 叶片和圆锥花序；C. 叶舌；D. 种子）

Biological characteristics

Orchardgrass prefers mild and humid climate. Optimal day time temperature is 22℃ , and night time temperature is 12 ℃ . When day time temperature exceeds 35 ℃ , dry matter accumulation decreases. Heat resistance is poorer than meadow fescue and slightly better than the ryegrass. Cold resistance is poor, and it is susceptible to frost damage. It is shade-enduring, so it is often cultivated in the orchard.

Orchardgrass has broad adaptation ability. It can grow in peat soils or sandy soils, while it grows best on clay and clay loam soils. It can grow on poor and dry soil. but not in sandy

生物学特性

鸭茅喜温和湿润气候，生长适宜温度为昼温 22℃、夜温 12℃。日温 35℃干物质产量减少。耐热性不如牛尾草，略强于黑麦草。耐寒性差，易受冻害。能耐阴，故常在果园栽种，因而又名“果园草”。

鸭茅对土壤要求不严格，在泥炭土及沙壤土上均能生长，而以黏土或黏壤土为适宜，在较瘠薄和干燥土壤也能生长，但在沙土则不甚相宜。需水，但不

soil. It needs water, but flood tolerance is poor. Drought resistance is better than timothy. Acid tolerance is moderate, lent alkali tolerance is poor. It is responsive to N fertilizer.

Growth of orchardgrass is slower than other grasses. It does not flower or set seed in the spring seeding year. Regrowth capacity is very good, producing more leaf and less stem, resulting in low yield, but very good quality. It is long lived and can continue to grow as long as 8 year.

Seed is small and light. 3~4 months postripeness are important for seed maturity. Sowing with the newly harvested seed can cause emergence irregularity and delay. Germination rate of seed is maintained for 2-3 years, losing germinability after 8-12 years of storage.

Cultivation and Utilization

Orchardgrass seedlings grow slowly and tiller late, compete poorly with weeds. Good seedbed preparation is necessary to benefit tiny seedlings. Sowing in spring and autumn is suitable; seeding earlier in autumn will avoid frost damage of seedlings. Autumn sowing should be earlier than late September in locations south of the Yangtze River. Winter wheat or winter oats can be useful a as nurse crop to avoid frost damage. Planting should be no later 45 d before a hard froze. Row spacing is 15~30 cm. Seeding rate is 11.25~15 kg/hm^2. There are some empty seeds. Seeding rate should calculated amount of based on pure live seed. Dense sowing in a row and covering lightly is good. A planting depth of 1~2 cm is better.

Orchardgrass can be mixed with alfalfa, red clover, alsike clover, white clover, ryegrass, or tall fescue. In locations where red clover is adapted, orchardgrass and red clover growing in mixture, does not prevent orchardgrass seed harvest. Yield and quality of forage improves following harvest seed. Orchardgrass grows thickly, and if seeded is mixture with white clover, white clover can creep into interval spaces and provide nitrogen to the grass ensuring good growth. Orchardgrass mixed with legumes, seeding rate is 3~3.75 kg/hm^2.

Requirements of Fertilizer for orchardgrass are high, especially N fertilizer. In addition to applying sufficient base

耐水淹。能耐旱，其耐旱性强于猫尾草。略能耐酸，不耐盐碱。对氮肥反应敏感。

鸭茅的发育较其他禾本科牧草慢，春播当年常不能开花结实。再生能力强，叶多茎少，产量较低，品质则较好。寿命较长，长成之后，颇能持久，可持续生长达 8 年之久，普通以利用 5~7 年为度。

鸭茅的种子小而轻。种子成熟后有 3 ～ 4 个月的后熟期，用新收获的种子播种出苗不齐，且拖延时间甚长；种子的发芽率可保留 2~3 年，贮存 8~12 年后，发芽力即丧失。

栽培利用技术

鸭茅苗期生长缓慢，分蘖迟，幼苗细弱，与杂草竞争能力差，整地要求精细，以利出苗整齐。春秋播种均可，秋播宜早，以免幼苗遭受冻害，长江以南各地，秋播不应迟于 9 月中下旬，可用冬小麦或冬燕麦作保护作物同时播种，严霜前 45 d 播种。条播行距 15~30 cm，播种量为 11.25~15 kg/hm^2。种子空粒多，应以实际播量计算。密行条播较好，覆土宜浅，以 1~2 cm 为宜。

鸭茅可与苜蓿、红三叶、杂三叶、白三叶、黑麦草、牛尾草等混种。在能生长红三叶的地区，鸭茅与红三叶混种时，红三叶并不妨碍鸭茅的种子收取问题，而收种后草料的产量及质地均已提高。鸭茅丛生，如与白三叶混种，白三叶可充分利用其空隙匍匐生长并供给禾本科草以氮素使其生长良好。鸭茅与豆科牧草混种时，种子用量为 3~3.75 kg/hm^2。

鸭茅需肥量较大，尤其对氮肥敏感。因此，除施足基肥外，生育过程中宜适当施追肥。在一定范围内其产量与施氮肥成正比关系。据试验，施氮量为 562.5 kg/hm^2 时，鸭茅干草产量最高，

fertilizer, topdressing during the growing season enhance yield. Yield is proportional to N application, within limit range. The highest hay yield of orchardgrass is 18 t/hm^2 when N application was 562.5 kg/hm^2 according to research. Lower yields were deserved when N application exceeded 562.5 kg/hm^2. Quantities and properties of N fertilizer can not only affect yield and composition of forage but also palatability. Also forage consumption and production performance of livestock is affected.

It is best to cut at the heading stage when stems and leaves are tender and quality is good. Later, fiber increases, quality declines and regrowth is reduced. According to research, cutting at early flowering decreased regrowth yield by 15%~26% compared with heading stage harvest. In addition, stubble can not be too low, or regrowth is seriously affected.

Seeding sparsely when we reserve seed for planting, and nitrogen fertilizer application should not be excessive. Common orchardgrass have a strong shattering. We should collect seed at the ripening stage in time. Seed yield is about 300~450 kg/hm^2.

Common varieties: The foreign registration varieties are **'kay'** **'currie'** **'rano'** **'ina'** **'juno'**. The registration of varieties in China is wild cultivar **'gulin'** dactylis glomerata, **'baoxing'** orchardgrass, **'chuandong'** cocksfoot, and introduced varieties of **'amba'** cocksfoot.

达 18 t/hm^2；但若施氮量超过 562.5 kg/hm^2 时不仅降低产量，而且会减少植株数量。氮肥种类和性质除影响牧草的产量和成分外，对适口性也有影响，进而影响牧草的消耗量和家畜的生产性能。

鸭茅以抽穗时刈割为宜，此时茎叶柔嫩，质量较好。收割过迟，纤维增多，品质下降，还会影响再生。据测定，初花期与抽穗期刈割相比，再生草产量下降 15%~26%。此外，割茬不能过低，否则将严重影响再生。

留种时宜稀播，氮肥不宜施用过多。种子落粒性强，应在种子蜡熟期及时采收，种子产量为 300~450 kg/hm^2。

常见品种：国外登记的品种主要有 'kay' 'currie' 'rano' 'ina' 'juno'。国内审定登记的品种主要有野生栽培品种'古蔺'鸭茅、'宝兴'鸭茅、'川东'鸭茅以及引进品种'安巴'鸭茅。

6.10 *Elymus* L. 披碱草属

The gensus *Elymus* L. includes about 20 species worldwide. It distributes in the temperate frigid zones of northern hemisphere. There are about 10 species in china, and distribute in the grassland and alpine steppe widely.

披碱草属植物有 20 余种，多分布于北半球的温带和寒带。我国有 10 余种，广泛分布在草原及高山草原地带。

6.10.1 Siberian wildryegrass

Scientific name: *Elymus sibiricus* L.

Origin and geographical distribution: Siberian wildryegrass began to cultivate in Russia and Britain hundreds

6.10.1 老芒麦

学名： *Elymus sibiricus* L.

英文名： Siberian wildryegrass

别名： 西伯利亚碱草、垂穗大麦草

years ago, and introduction and domestication began in Jilin province of China in the 1950s. In China it mainly distributes in the areas of northeast, northwest, Inner Mongolia, Hebei, Shanxi and Sichuan provinces. It is one of the main members of meadow steppe and meadow community. It also distributes in Mongolia, Korea and Japan.

起源与地理分布：老芒麦栽培始于俄罗斯和英国，栽培史达百余年，我国最早于 20 世纪 50 年代在吉林开始引种驯化，在我国分布于东北、内蒙古、西北、河北、山西、四川等地，是草甸草原和草甸群落中的主要成员之一。蒙古、朝鲜和日本等地也有分布。

6.10.1.1 Botanical characteristics

Siberian wildryegrass is a perennial and bunch grass. Root is fibrous with no underground stems. Plants height is 60~90 cm. Stem is tufted with smooth leave. Auricles are short. Leave blade is soft. Spikes is loose and drooping, and 15~20 cm long. Usually per node with 2 spikelets, and spikelet is gray-green or slightly purple. Spikelet length is 12~25 mm, contains 4~5 flowers. Glume is lanceolate with densely hairy. The first lemma length is 8~11mm, and awn length is 15~20 mm. Awn is curve at outward. Seed is long, circle and oblate. The thousand seed weight is 3.5~4.9 g. The botanical characteristics of siberian wildryegrass as shown in Fig.6.16.

6.10.1.1 植物学特征

老芒麦为疏丛型多年生禾草，根系呈须状，无地下茎；株高 60~90 cm。茎秆簇生。叶平滑，无叶耳。叶舌短，叶片柔软。穗状花序较疏松而下垂，长 15~20 cm，通常每节生 2 枚小穗。小穗灰绿色或稍带紫色，长 12~25 mm，含 4~5 朵小花。颖披针形，密生微毛。第一外稃长 8~11 mm，芒长 15~20 mm，稍向外反曲。种子长扁圆形，易于脱落，千粒重为 3.5~4.9g。老芒麦植物学特征如图 6.16。

6.10.1.2. Biological characteristics

Siberian wildryegrass like humid environment and fertile soil. It is poor with drought and barren resistance. Cold resistance is good. It survives in subalpine meadow in the altitude of 3 000 m. It was introduced successfully in the northwest alpine region of China. It is an important forage grass to transform abandoned land into artificial pasture in the humid and semi humid climate zone. It germinates after seeding 7~10 d when the water temperature is appropriate. It can head, blossom or seed in spring seeding year. Siberian wildryegrass turns green early usually in mid April, heading in the end of June, flowering in July, seed maturing in the end of August. It is taking about 134 d from turning green to maturing. The ability of tiller is strong and the number of tiller can reach 5~11 in spring seeding year. The tiller node is at 3~4 cm surface soil layer. Vegetative branches is dominant in seeding year, reproductive branches is absolute advantage in second year.

6.10.1.2 生物学特性

老芒麦喜湿润环境和肥沃土壤，不耐旱，不耐贫瘠，但有一定的耐寒能力，能在海拔 3 000 m 的亚高山草甸出现。在西北较高寒地区引种获得成功，是半湿润、半干旱气候地带改造撂荒地、人工种植的重要牧草。在水温较适宜时，老芒麦播后 7~10 d 即可出苗，春播当年可抽穗、开花，甚至结实。老芒麦返青较早，通常 4 月中旬返青，6 月底抽穗，7 月开花，8 月底种子成熟，由返青至种子成熟约 134 d。老芒麦分蘖能力强，春播当年可达 5~11 个，分蘖节位于表土层 3~4 cm 处。播种当年以营养枝占优势，第二年后以生殖枝占绝对优势。

6.10.1.3 Cultivation and utilization

The field of siberian wildryegrass need deep ploughing and enough base fertilizer in autumn. It needs suppress and cultivated land in harrow before sowing. The seed has a long

6.10.1.3 栽培利用技术

种植老芒麦的地块，秋季深耕，施足基肥。播前须耙耱和镇压。老芒麦种子具长芒，播前应去芒。有灌溉条件或春墒较好的地方，可春播；无灌溉条件

Fig. 6.16 The botanical characteristics of Siberian wildryegrass
图 6.16 老芒麦植物学特征

awn which need to be get rid of before sowing. Spring sowing in the place with good irrigation conditions and soil moisture is suitable, or it is sown in summer or autumn. It can be sown in ice crust between late autumn and early winter in the area of short growing season. Drill spacing is 20~30 cm, the unicast seed rate is 22.5~30 kg/hm^2 for hay, and 15.0~22.5 kg/hm^2 for harvesting seed. Sowing depth is 2~3 cm. It should be irrigated at elongation stage and the time after cutting if possible. Siberian wildryegrass is a top grass suit for cutting. It should be cut between heading stage to flowering period and processed into silage or hay. The seed is easy to fall off after mature. And should be harvested in time when seed of bottom spike is mature.

Common varieties include: '**jilin**' *Elymus sibiricus* and '**tongde**' *Elymus sibiricus*.

的干旱地方，以夏秋季播种为宜；在生长季较短的地方，可采用秋末冬初寄籽播种。条播行距 20~30 cm，单播播种量收草者为 22.5~30 kg/hm^2，收种者为 15.0~22.5 kg/hm^2，播深 2~3 cm。有灌溉条件时，应在拔节期和每次刈割后浇水。老芒麦属上繁草，适于刈割利用，宜在抽穗期至始花期进行，可青饲或调制成干草。老芒麦种子成熟后极易脱落，采种应在穗状花序下部种子成熟时及时进行。

常见品种： ‘川草 1 号’‘川草 2 号’‘吉林’老芒麦、‘农牧’老芒麦、‘青牧 1 号’‘同德’老芒麦。

6.10.2 Drooping wildryegrass

Scientific name: *Elymus nutans* Griseb.

Origin and geographical distribution: Drooping wildryegrass distributes in Inner Mongolia, Hebei, Shaanxi, Gansu, Ningxia, Qinghai, Xinjiang, Sichuan, Tibet and other provinces. The introduction and cultivation time for this grass is not long, it adapts to the wet alpine, but cannot grow well in arid region.

6.10.2.1 Botanical characteristics

Drooping wildryegrass is a perennial bunch herb, 50~70 cm high, the cultivated species are 80~120 cm high. Root system is developed and looks like whisker, and the rhizomes are sparse plexiform. Leaves are flat, 6~8 cm long, 3~5 mm wide, are puckering on both sides or smooth on the lower part. On upper part of leaves is pilose sparsely. Leaf sheaths are shorter than internodes except the base part. Ligules are short, 0.5 mm long. Spikes arrange are more tight, spikelets often incline to one side of the rachis, usually are curved, apex drooping, 5~12 cm long. There are two pieces of spikelet on one cob in each section generally. There is only one spikelet on each section near the top, with sterile spikelets on base part, spikelet pedicels are short or absent. Spikelets are green, with purplish after mature, 12~15 mm long, containing 3~4 florets per spikelet, but only 2~3 flowers among them are fertile. Glumes are oblong and have 3~4 veins, 4~5 mm long, with short awn 3~4 mm long. Lemma is long lanceolate and has 5 veins. The awn is 12~20 mm long, rough, outward retrorsely or unfolding slightly, palea and lemma are subequal with 5 veins, obtuse or truncate at the apex. The anther turns black after maturing. Caryopsis, seed is lanceolate, purple brown, and the thousand seed weight is 2.85~3.2 g. The botanical characteristics of drooping wildryegrass as shown in Fig.6.17.

6.10.2.2 Biological characteristics

Drooping wildryegrass has extensive habitat plasticity, which likes growing on plain, plateau and flat beach. It has strong cold resistance, as low as -38℃ . It has strong drought resistance because its root goes deep into the soil making full use of soil moisture. As poor resistant to waterlogging, it

6.10.2 垂穗披碱草

学名： *Elymus nutans* Griseb.

英文名： drooping wildryegrass

别名： 弯穗草、钩头草

起源与地理分布： 垂穗披碱草分布于内蒙古、河北、陕西、甘肃、宁夏、青海、新疆、四川、西藏等地。该草种引种栽培年限不长，适应高寒湿润地区，在干旱区生长不良。

6.10.2.1 植物学特征

垂穗披碱草为多年生疏丛型草本植物，高 50 ~ 70 cm，栽培种高 80 ~ 120 cm。根茎疏丛状，根系发达，呈须状。秆直立，3 ~ 4 节，基部节稍膝曲。叶片扁平，长 6 ~ 8 cm，宽 3 ~ 5 mm，两边微蹙或下部平滑，上面疏生柔毛，叶鞘除基部者外，其余均短于节间。叶舌极短，长约 0.5 mm。穗状花序排列较紧密，小穗多偏于穗轴的一侧，通常弯曲，先端下垂，长 5 ~ 12 cm，穗轴每节一般有两枚小穗；接近顶端处各节仅具一枚小穗，基部具不育小穗，小穗梗短或无；小穗绿色，成熟带紫色，长 12 ~ 15 mm，每小穗含 3 ~ 4 小花，其中仅 2 ~ 3 花可育。颖呈长圆形，具 3 ~ 4 脉，长 4 ~ 5 mm，具 3 ~ 4 mm 长的短芒；外稃长披针形，具 5 脉；芒长 12 ~ 20 mm，粗糙，向外反曲或稍展开，内稃与外稃具 5 脉，等长，先端钝圆或截平。花药成熟后变黑色。颖果，种子披针形，紫褐色，千粒重 2.85 ~ 3.2 g。垂穗披碱草植物学特征如图 6.17。

6.10.2.2 生物学特性

垂穗披碱草具有广泛的生境可塑性，喜生长在平原、高原平滩等地方。抗寒性强，能耐 –38℃低温。因根系入土深，可利用深层土壤水分，故具较强的抗旱性，但不耐水淹，时间过长则

Fig. 6.17 The botanical characteristics of drooping wildryegrass
图 6.17 垂穗披碱草植物学特征

may die in the long time suffering. It has tolerance to salt, can grow well in 7-8 pH of the soil. The tillering ability of root is very strong. Drooping wildryegrass doesn't bear seed in the seeding year, only tassels at the second year it turns green between late April and early May, tassels and blossoms between

受害死亡。有一定的耐盐碱性，能在pH 7 ～ 8 的土壤上良好生长。根茎分蘖能力很强。播种当年只抽穗不结实，第二年 4 月下旬至 5 月上旬返青，6 月中旬至 7 月下旬抽穗开花，8 月中、下

mid June and late July, and the mature of seed is in mid or late August, the growth period is 100~120 d.

旬种子成熟，生育期为 100 ~ 120 d。

6.10.2.3 Cultivation and utilization

The soil needs deep plowing and appropriate base fertilizer before sowing. Breaking awn treatment is also needed. Dill spacing is 20~30 cm. Seeding depth is 2~4 cm with seeding rate of 15~22.5 kg/ hm^2. It is generally sowed in spring in the Qinghai-Tibet Plateau, it can be earlier or sowed in summer and autumn in warmer regions. Level land and compacting are necessary to land at the sowing year. It should be sowed between late April and mid May in Qinghai-Tibet Plateau. Drooping wildryegrass should be sowed early in the irrigated area in order to increase the yield. We should pay attention to weeds at seedling stage, and 1~2 times irrigation at jointing stage can significantly increase the yield. It reaches the highest yield from the second to the 4th year, then decreases from the 5th year, therefore we should loosen soil, cut root and reseed grass seed since the 4th year to prolong the service life of grassland.

Common varieties: '**gannan**' *Elymus nutans* and '**kangba**' *Elymus nutans.*

6.10.2.3 栽培利用技术

播前应深翻，适当施入底肥，并作断芒处理。条播行距 20~30 cm，播深 2~4 cm，播种量为 15~22.5 kg/hm^2。在青藏高原一般春播，气候稍暖地区可以早播或夏秋播。当年播种时，应对土地进行耙耱、镇压。在青藏高原以 4 月下旬至 5 月中旬播种为宜。有灌溉条件的地方应早播，以提高当年产量。苗期注意防除杂草，拔节期灌水 1 ~ 2 次可显著增产。生长 2 ~ 4 年的产量较高，第 5 年后产量开始下降，因此从第 4 年开始要进行松土、切根和补播草籽，可延长草场使用年限。

常见品种：'甘南' 垂穗披碱草、'康巴' 垂穗披碱草。

6.11 *Lolium* L. 黑麦草属

Lolium L. is annual or perennial herbaceous plants. It is a genus of about 10 species, mainly distributes in the temperate humid regions. The perennial ryegrass and annual ryegrass are cultivated worldwide, and also widely planted in China.

黑麦草属系一年生或多年生草本植物。黑麦草属植物约有 10 种，主要分布在世界温带湿润地区。其中多年生黑麦草与多花黑麦草为世界性栽培牧草，在我国也有广泛种植。

6.11.1 Perennial ryegrass

Scientific name: *Lolium perenne* L.

Origin and geographical distribution: Perennial ryegrass is native to southwest Europe, northern Africa and southwest Asia temperate, but is widely cultivated around the world currently. It mainly distributes in the eastern, central and southwest of China, and grows best in mountain area of the Yangtze River Basin. It is a kind of good forage grass, with

6.11.1 多年生黑麦草

学名： *Lolium perenne* L.

英文名： perennial ryegrass

别名： 英国黑麦草、宿根黑麦草

起源与地理分布： 多年生黑麦草原产西南欧、北非和西南亚的温带。目前世界各国均有栽培。在我国主要分布于华东、华中和西南等地，以长江流域的

a wide range of feeding value.

6.11.1.1 Botanical characteristics

Perennial ryegrass is a perennial bunch herb. Stems grow up to 80~100 cm (Fig.6.18A). Developed fibrous roots mainly distribute within top 15 cm soil layer (Fig.6.18B). Stem is erect, smooth and hollow, with light green color. The number of tillers in a single plant is 60~100; some can reach up to 250~300. Ribbed upper surface to leaf blade (leaf lamina), emerging leaf folded; The leaves are dark green and glossy, Shiny under surface to leaf blade, 15~35 cm long and 3~6 mm width; most of them lie on the base stem (Fig.6.18C). Leaf sheaths are longer than or equal with the internode, tightly wrapping the stem (Fig.6.18D); ligule is membranous, about 1mm long. The spike is 20~30 cm long, with 15~25 spikelets per panicle (Fig.6.18E). Spikelets have no handle, alternate on both sides of cob closely, 10~14 mm long, and with 5~11 flowers and 3~5 grain seeds. The first glume is usually vestigial, and the second glume has hard texture, with 3~5 veins, 6~12 mm long. Lemma is 4~7 mm long, thin, tip blunt, without awns, palea is as long as lemma, sharp on top, transparent, with fine wool on the edge. Caryopsis is fusiform. The thousand seed weight is 1.5~2.0 g (Fig.6.18F).

6.11.1.2 Biological characteristics

Perennial ryegrass likes warm humid climate. It is not resistant to high temperature and cold. When the temperature is over 35℃ , its growth is blocked or even died. When the temperature is below -15 ℃ , it overwinters unsteadily, or couldn't live through the winter. It requires cool in summer, not cold in winter. Perennial ryegrass can be cultivated in place where there is annual precipitation of 500~1 500 mm, the most appropriate precipitation condition is 900~1 000 mm. The optimum temperature for growth and development is 20~25 ℃ , it can also grow in 10℃ . Perennial ryegrass like fertilizer, is suitable for planting in fertile, moist, well drained loam or clay, also can grows in acidic soils, the suitable pH is 6~7. But it should not be planted in sandy soil or wetlands. Perennial ryegrass is a short-lived top grass, generally can be used 3~4 years, but can also endure for many years when the condition is appropriate. Perennial ryegrass grows and develops rapidly,

高山地区生长最好。是一种良好的牧草，具有广泛的饲用价值。

6.11.1.1 植物学特征

多年生黑麦草是禾本科黑麦属多年生疏丛型草本植物。株高 80 ~ 100 cm（图 6.18A）。须根发达，主要分布于 15 cm 深的土层中（图 6.18B）。茎直立，光滑中空，色浅绿。单株分蘖一般 60 ~ 100 个，多者可达 250 ~ 300 个。叶片深绿有光泽，长 15 ~ 35 cm，宽 3 ~ 6 mm，（图 6.18C），幼叶折叠。叶鞘长于或等于节间，紧包茎（图 6.18 D）；叶舌膜质，长约 1 mm。穗状花序长 20 ~ 30 cm，每穗有小穗 15 ~ 25 个，小穗无柄，紧密互生于穗轴两侧，长 10 ~ 14 mm（图 6.18E）；有花 5 ~ 11 枚，结实 3 ~ 5 粒。第一颖常常退化，第二颖质地坚硬，有脉纹 3 ~ 5 条，长 6 ~ 12 mm。外稃长 4 ~ 7 mm，质薄，端钝，无芒；内稃和外稃等长，顶端尖锐，透明，边有细毛。颖果梭形。种子千粒重 1.5 ~ 2.0 g（图 6.18F）。

6.11.1.2 生物学特性

多年生黑麦草喜温暖湿润气候。不耐高温，不耐严寒。遇 35℃以上的高温生长受阻，甚至枯死，遇 –15℃以下低温越冬不稳，或不能越冬。要求夏季凉爽、冬无严寒。在年降水量为 500 ~ 1 500 mm 的地方都可种植，最宜 900 ~ 1 000 mm 的降水条件。其生长发育的最适温度为 20 ~ 25℃，在 10℃时亦能较好生长。性喜肥，适宜在肥沃、湿润、排水良好的壤土或黏土上种植，亦可在微酸性土壤上生长，适宜的 pH 为 6 ~ 7。但不宜在沙土或湿地上种植。多年生黑麦草是短命上繁牧草，一般可利用 3 ~ 4 年，但条件适宜时，也可多年不衰。多年生黑麦

Fig. 6.18 The botanical characteristics of perennial ryegrass
(A. stand; B. fibrous root system; C. leaf blade and stem; D. leaf sheath; E. spike; F. seeds)
图 6.18 多年生黑麦草植物学特征
（A. 多年生黑麦草草地；B. 须根系；C. 叶片和茎秆；D. 叶鞘；E. 穗状花序；F. 种子）

its life cycle is 100~110 d, growth for 250 d per year.

6.11.1.3 Cultivation and utilization

Seeds of perennial ryegrass are small, and seedlings are slender which are weak to push up the earth. Cultivated fields need deep plowing and harrowing, crushing clods, leveling

草生长发育迅速，生育期为 100 ~ 110 d，全年生长天数为 250 d 左右。

6.11.1.3 栽培利用技术

多年生黑麦草种子小，幼苗纤细，顶土力弱，种植地要深翻松耙，粉碎土

the ground, water conservation to make the soil lose in the upper and firm in the lower, creating a good soil conditions for seed germination. We should apply sufficient base fertilizer depends on its soil fertility combining tillage. The general application of organic fertilizer is 15-22.5 t/hm^2, and compound fertilizer is 375-450 kg/hm^2. Perennial ryegrass can be planted in spring, summer and autumn, planting season can be determined according to the specific local conditions. Generally to the seed collection, it can be sown in fall, easy for the planting of other crops after seed collecting in the following year. Seed field uses single sowing, while grass harvesting can use both single sowing and mix sowing. Row spacing of seed field should be wider as 35~40cm and the grass field 20~30cm. The seeding quantity for seed collection is 11.25~15 kg/hm^2 and 15~18.75 kg/hm^2 for grass harvesting. Perennial ryegrass can mix sowed with white clover, red clover and birdsfoot trefoil, the sowing depth is 2~3 cm. The rain before the seedling emergence make the soil surface form hardening layer, attention to get rid of the hardened layer in order to facilitate the emergence and preserve the seedling. The seedling needs timely weeding, and attention to pest control. Grass harvesting is better from heading stage to milk stage; Grazing should be performed in the height of 15~20 cm. It needs harvest, transport and sunning, threshing, and store in time when 70% of the whole field spikes turn to yellow white because consistent seed maturation and high self threshing.

Common varieties: **'toyah'** **'peak'** and **'excellence'**.

块，整平地面，蓄水保墒，使土壤上虚下实，为种子出苗创造良好的土壤条件。结合翻耕，视土壤肥力情况施足底肥。一般施有机肥料为 15 ~ 22.5 t/hm^2，或复合化肥 375 ~ 450 kg/hm^2。春、夏、秋季均可播种，可根据当地具体条件来确定，一般收种可在秋季播种，便于翌年收种后翻种其他作物。种子田用单播，收草可单播，可混播。条播行距，种子田宜宽，为 35 ~ 40 cm，收草田宜窄，为 20 ~ 30 cm。收种播种量为 11.25 ~ 15 kg/hm^2，收草播种量为 15 ~ 18.75 kg/hm^2。可与白三叶、红三叶、百脉根混播，播种深度 2 ~ 3 cm。播种后出苗前遇雨，土壤表层形成板结层，要注意及时破除板结层，以利出苗，保全苗。幼苗期要及时除草，并注意防治虫害。收草宜在抽穗至乳熟期进行；放牧宜在株高 15 ~ 20 cm 时进行。因种子成熟先后不一致、自行脱粒，应在全田 70% 的种穗变黄白时及时收割、运晒、脱粒、收藏。

常见品种：‘托亚’‘顶峰’‘卓越’。

6.11.2 Multiflorum ryegrass

Scientific name: *Lolium multiflorum* L.

Origin and geographical distribution: Multiflorum ryegrass, also called Italian ryegrass or annual ryegrass, is native to southern Europe, northern Africa, Asia Minor and other places, it cultured in the northern part of Italy in thirteenth century, later it spreads to other countries, widely distributes in UK, America, Denmark, New Zealand, Australia, Japan and other temperate countries with more rainfall. In China, it's suitable to grow in the area in the south of the

6.11.2 多花黑麦草

学名：*Lolium multiflorum* L.

英文名：annual ryegrass

别名：意大利黑麦草、一年生黑麦草

起源与地理分布：多花黑麦草原产于欧洲南部、非洲北部及小亚细亚等地，13 世纪已在意大利北部栽培，之后传播到其他国家，广泛分布于英国、美国、丹麦、新西兰、澳大利亚、日本等温带

Yangtze River, there are cultivars in Jiangxi, Hunan, Jiangsu, Zhejiang and other provinces. Multiflorum ryegrass is also introduced for spring sowing in the northeast, Inner Mongolia and other provinces.

降水量较多的国家。在我国适生于长江流域以南地区，在江西、湖南、江苏、浙江等地均有人工栽培种。东北、内蒙古等地亦引种春播。

6.11.2.1 Botanical characteristics

Multiflorum ryegrass is an annual grass, with dense fibrous root that are mainly distributes in the depth of 15 cm in the soil. Culms turn into bunch, erect, 80~120 cm high (Fig.6.19A), the leaf sheath is loose and the ligule is small or unapparent. Compared with perennial ryegrass, except emerging leaf rolled (Fig.6.19B), tiller is larger with wider leaves than perennial ryegrass. The blade is 20~30 cm long, 7~10 mm wide. The spikes are 15~25 cm long, 5~8 mm wide, the spikelet is back to the spike-stalk, 10~18 mm long, with 10~20 florets (Fig.6.19C); the glume quality is hard, with 5~7 veins, 5~8 mm long; the lemma quality is thin, with 5 veins, the first lemma is 6 mm long, with fine awns is about 5 mm long (Fig.6.19D), the glume is equal to lemma. The thousand seed weight is 1.5~2.0 g.

6.11.2.1 植物学特征

一年生禾草，须根密集，主要分布于 15 cm 以上的土层中。秆成疏丛，直立，高 80 ~ 120 cm（图 6.19A），叶鞘较疏松，叶舌较小或不明显，叶片长 20 ~ 30 cm，宽 7 ~ 10 mm。与多年生黑麦草相比，除了幼叶内卷的区别外（图 6.19B），小穗较大，叶片较宽。穗状花序长 15 ~ 25 cm，宽 5 ~ 8 mm，小穗以背面对向穗轴，长 10 ~ 18 mm，含 10 ~ 20 个小花（图 6.19C）；颖质较硬，具 5 ~ 7 脉，长 5 ~ 8 mm；外稃质较薄，具 5 脉，第一外稃长 6 mm，芒细弱，长约 5 mm（图 6.19D），内稃与外稃等长，千粒重 1.5~2.0 g。

6.11.2.2 The Biological characteristics

Multiflorum ryegrass likes warm and humid climate, when the day and night temperature is 27℃ /12℃ , its growth is the fastest, it grows faster than other gramineous grass in autumn and spring but poorly even die in hot summer. It is resistant to moisture, but avoids seeper. Multiflorum ryegrass like loam, is also suitable for clay loam. The most suitable soil pH ranged from 6 to 7, can also adapt in the pH 5 and 8. Multiflorum ryegrass is not resistant to cold; its winter survival rate in Beijing is only 50%. In the south of the Yangtze River Basin, autumn sowing seedlings can overwinter safely, and provide high quality green feed in early spring. If sowing in September in Wuhan, we can harvest the first stubble in March on the second year, and cut 2~3 times before midsummer, heading and blossom of multiflorum ryegrass comes from late April to early May, seeds mature in early June, the plants die after fructifying aboveground. The propagation ability of shattering seeds is strong. Multiflorum ryegrass has

6.11.2.2 生物学特性

多花黑麦草喜温热和湿润气候，在昼夜温度为 27℃ /12℃时生长最快，秋季和春季比其他禾本科草生长快，夏季炎热则生长不良，甚至枯死。耐潮湿，但忌积水；喜壤土，也适宜黏壤土。最适宜土壤 pH 为 6 ~ 7，在 pH 为 5 和 8 时仍可适应。多花黑麦草不耐严寒，在北京越冬率仅为 50%。在长江流域以南，秋播可安全越冬，并可在早春提供优质青饲料，如在武汉地区，9 月播种，第二年 3 月即可收割第一茬，盛夏前可刈割 2 ~ 3 次，4 月下旬到 5 月初抽穗开花，6 月上旬种子成熟，地上部结实后植株死亡。落粒的种子自繁能力强。多花黑麦草分蘖多，再生迅速，春季刈割后 6 周即可再次刈割。耐牧，即使重牧之后仍能迅速恢复生长。

Fig. 6.19 The botanical characteristics of multiflorum ryegrass
(A. stand; B. rolled emerging leaf; C. spike; D. spikelets with fine awn)
图 6.19 多花黑麦草植物学特征
（A. 多花黑麦草；B. 幼叶内卷；C. 穗状花序；D. 带细芒的小穗）

more tillers, regenerates rapidly, it can be cut after 6 weeks to spring harvest. Multiflorum ryegrass has grazing tolerance, it can still restore growth quickly even after heavy grazing.

6.11.2.3 Cultivation and utilization

The cultivation of multiflorum ryegrass is basically the same as perennial ryegrass; it needs plowing and soil preparation before sowing, providing enough base fertilizer and superphosphate 150~225 kg/hm^2. It's suitable for autumn sowing in the south of the Yangtze River, both drilling and broadcasting are ok, and the dill spacing is 15~30 cm. The seeding rate is 15 kg/hm^2, sowing depth is 1.5~2 cm, it can be broadcasted in the adequate rainfall area, and increase seed-

6.11.2.3 栽培利用技术

多花黑麦草的栽培技术与多年生黑麦草基本相同，播前需要耕翻整地、施足底肥，施过磷酸钙 150 ~ 225 kg/hm^2。在长江以南地区宜秋播，可以条播或撒播，条播行距 15 ~ 30 cm。播种量为 15 kg/hm^2，播深 1.5 ~ 2 cm，在雨水充足的地区也可以撒播，适当增加播种量，约为 22.5 kg/hm^2。多花黑麦草与多年生黑麦草、红三叶，白三叶等混播，可提高产草量，也可与水稻、玉米、高粱等轮作，成为牲畜冬

ing rate properly, about 22.5 kg/hm^2. Multiflorum ryegrass can be mixed seeding with perennial ryegrass, red clover, white clover and other grass to increase the yield of grass, it can also be rotated with rice, corn, sorghum and so on, which is the main forages for livestock in winter and spring. It is suitable for spring sowing and using in North China because of its dimate of drought and cold in winter, and not too hot in summer. Nitrogen fertilizer can not only improve the yield, but also increase the content of crude protein, so that available nitrogen fertilizer should be topdressed in the growth period. The seeds fall off easily, they should be collected in time when most of the seeds is mature.

春主要饲草。在我国北方冬季干旱、寒冷，而夏季又不太炎热的地区适于春播，当年利用。增施氮肥不仅能提高产量，亦可提高其粗蛋白质含量，故生长期间应追施速效氮肥。种子易脱落，当大部分种子成熟后应及时收获。

6.12 *Arrhenatherum* L. —tall oatgrass 高燕麦草属—高燕麦草

There are 6 species in this genus while only one is cultivated species, named tall oatgrass.

Tall oatgrass

Scientific name: *Arrhenatherum elatius*(L.) Presl.

Origin and geographical distribution: Tall Oatgrass is native to the central and southern Europe, the shores of the Mediterranean, Western Asia and Northern Africa. It is cultivated widely in Central Europe countries, especially in France, Switzerland and Germany, it is also planted in Britain, Sweden and Australia. Tall oatgrass is cultivated in America since 1800. Tall oatgrass was introduced and planted in North China, Northeast China, Beijing, Gansu and other places, but the area is small and not spread yet in production.

Botanical characteristics

Tall oatgrass is a perennial herb, the fibrous root system is about 60~100 cm deep in soil. Bunch, erect stems, is 110~135 cm high with 4~5 sections. There are many tillers, the leaves are flat and smooth; Blade is 16~24 cm long, 4~9 mm wide, leaf sheaths split and are shorter than internodes, the ligule is membranous, about 1.5 mm long, the panicles are divergent, grey green with slightly purple, lustrous, 20~

高燕麦草属共有 6 种，其栽培种仅一种，即高燕麦草。

高燕麦草

学名： *Arrhenatherum elatius*(L.) Presl.

英文名： tall oatgrass

别名： 大蟹钩草、长青草、燕麦草

起源与地理分布： 高燕麦草原产于欧洲中南部、地中海沿岸及亚洲西部和非洲北部。中欧各国栽培甚广，尤以法国、瑞士、德国栽种较多，英国、瑞典及澳大利亚亦有栽种。美国栽种始于 1800 年。我国华北、东北、北京、甘肃等地都引种种植过，均属小面积试验，生产中尚未推广应用。

植物学特征

多年生草本，须根系，入土 60 ~ 100 cm，疏丛型，茎直立，株高 110 ~ 135 cm，4 ~ 5 节。分蘖多，叶片扁平，叶面较光滑；叶长 16 ~ 24 cm，宽 4 ~ 9 mm，叶鞘分裂，短于节间；叶舌膜质，长 1.5 mm 左右；圆锥花序散开，灰绿色略带紫色，具光泽，

30 cm long, branches verticillate, carry out when flowering, the spikelets are 7~8 mm long, only with stamens in lower flower, rotating and curved awns at lemma, pistils and stamens in upper flower, the glumes are awnless or with short awn near the top. The thousand seed weight is 2.91~3.15 g.

Biological characteristics

Tall oatgrass is a medium-lived perennial bunch grass, it usually lives for 5 to 7 years, its production begin to decline after fourth years. Tall oatgrass likes warm and humid climate, it is resistant to heat and drought strongly, but poor to cold, it has strong branch formation ability, especially at the first and second year of life, the number of branches in a line is lower than perennial ryegrass but higher than smooth brome, Alopecurus pratensis, chicken feet and Timothy grass. Therefore, in the mixed grass, tall oatgrass should be mixed sowing with those top grass which develop fast in the seeding year such as chicken feet grass, meadow fescue, red clover, and so on, the seeding quantity should not be too high, 6~8 kg/hm^2 is appropriate.

Tall oatgrass root system is developed, the plant assumes a Pyramid type, and it likes fertile, well drained soil, it suits growing on clay loam and loam, but has poor growth on sandy loam, it is not resistant to waterlogging. The salt and alkali resistance of tall oatgrass is medium, and it can grow normally at pH 7~8 of the soil.

Cultivation and utilization

Tall oatgrass is suitable for spring sowing in north China and autumn sowing in south China, it can be sown in summer in Inner Mongolia, northeast and some other places. Seed liquidity is poor, the awns should be removed before seeding. The content of single drill is 45~75 kg/hm^2, row spacing is 20~30 cm, soil depth is 3~4 cm, because tall oatgrass is intolerant to shade, it doesn't need cover crops typically, in order to avoid inhibiting the growth of tall oatgrass. The regeneration rate of tall oatgrass is fast, the regrowth has more tillers, and its yield is high, so it is the most suitable species for cutting. Tall oatgrass can be mixed seeding with those grass which has the same growth period with it but doesn't grow as rapidly as it,

长 20 ~ 30 cm，分枝轮生，开花时开展；小穗长 7 ~ 8 mm，下部花仅具雄蕊，外稃生旋转而弯曲的芒，上部花有雌雄蕊，外颖无芒或近顶端处有短芒。千粒重为 2.91 ~ 3.15 g。

生物学特性

高燕麦草为中寿多年生疏丛型禾本科牧草，通常寿命 5 ~ 7 年，第 4 年后产量开始下降。喜温暖潮湿气候，耐热和耐旱能力较强，耐寒性较差，具有很强的枝条形成能力，特别是在生活第一、二年。一个株丛中的枝条数低于多年生黑麦草而高于无芒雀麦、大看麦娘、鸭茅及猫尾草。因此，在混播牧草中，适应与播种当年发育较快的上繁草如鸭茅、牛尾草、红三叶等混播，其播种量不宜过高，以 6 ~ 8 kg/hm^2 为宜。

高燕麦草根系发达，株丛呈金字塔形，性喜肥沃、排水良好的土壤，适于黏壤土及壤土，沙壤土上生长不良，不耐水淹。抗盐碱中等，在 pH7 ~ 8 的土壤上能正常生长。

栽培利用技术

北方宜春播，南方可秋播，内蒙古、东北等地可以夏播。种子流动性很差，播种前应做去芒处理。条播单播量为 45 ~ 75 kg/hm^2，行距 20 ~ 30 cm，覆土 3 ~ 4 cm，由于高燕麦草不耐阴，所以通常不需要保护播种，以免保护作物抑制其生长。高燕麦草再生速度快，再生草分蘖多、产量高，故最适于刈割。高燕麦草可选择生长期与它相同而生长速度不同者与它混播，如鸭茅、牛尾草、红三叶等。收草用高燕麦草应在抽穗或开花初期刈割。高燕麦草不耐家畜践踏，不宜做放牧用。种子生活力以第一年为最强，以后逐年降低，4 年后完全丧失生活力。

such as chicken feet grass, meadow fescue, red clover and so on. The tall oatgrass for grass using should be cut at heading or early blossom stage. Tall oatgrass is intolerant to livestock trampling, it is not suitable for grazing. Seed viability is the strongest in the first year, later it decrease year by year, and lost viability completely after four years.

6.13 *Phalaris* L. 虉草属

Phalaris (*Phalaris* L.) is an annual or perennial grass. This genus contains about 20 species, is native to Europe, Asia and North America, widely distributes in North temperate.

虉草属为一年生或多年生草本。本属约有20种，原产于欧、亚、北美三洲，广泛分布于北温带。

6.13.1 Reed canarygrass

Scientific name: *Phalaris arundinacea* L.

Origin and geographical distribution: Reed canarygrass occurs in Europe, Asia (temperate regions), North America and South Africa, and widely distributes in northeast, northwest, north and central of China. In low-lying swamp meadow and vulnerability to flood meadows, reed canarygrass grow best in tidal slag sandy soil that with high content of organic matter, and is also adapted to fertile loam and clay.

6.13.1.1 Botanical characteristics

Reed canarygrass is a perennial herb, with rhizomes. The culms are usually solitary or few, 60~140 cm tall, with 6~8 internodes (Fig.6.20A). Leaf sheaths are glabrous; the lower part is longer than internodes while the upper part is shorter than internodes (Fig.6.20B). Ligules are membranous, 2~3 mm long. Leaves are flat, the tender leaves are slightly rough, 6~30 cm long, 1~1.8 cm wide. Panicles are dense and erect, 8~15 cm long (Fig.6.20C). The spikelets are dense, and 4~5mm long, containing 1 bisexual floret and its attached two (sometimes one) lemmas which has degenerated into linear or scalelike. The caryopsis is enwrapped tightly in the lemma, the seeds are gray to black, about 3 mm long, and the thousand seed weight is 0.7~0.9 g (Fig.6.20D).

6.13.1 虉草

学名： *Phalaris arundinacea* L.

英文名： reed canarygrass

别名： 草芦、丝带草

起源与地理分布： 分布于欧洲、亚洲（温带地区）、北美和南非；中国的东北、西北、华北、华中地区广为分布。在低洼沼泽草甸和易受洪水淹没的草地，以及有机质含量高的潮渣沙质土上生长最好，也适应于肥沃的壤土和黏土。

6.13.1.1 植物学特征

虉草为多年生草本。有根茎。秆通常单生或少数生，高60～140 cm，有6～8节（图6.20A）。叶鞘无毛，下部者长于节间，而上部者短于节间（图6.20B）；叶舌薄膜质，长2～3 mm。叶片扁平，幼叶粗糙，长6～30 cm，宽1～1.8 cm。圆锥花序紧密，长8～15 cm（图6.20C），分枝上举，密生小穗，小穗长4～5 mm，含1枚两性小花及附于其下的2枚（有时为1枚）退化成线形或鳞片状的外稃。颖果包紧于稃内，种子淡灰至黑色，长约3 mm，千粒重0.7～0.9 g（图6.20D）。

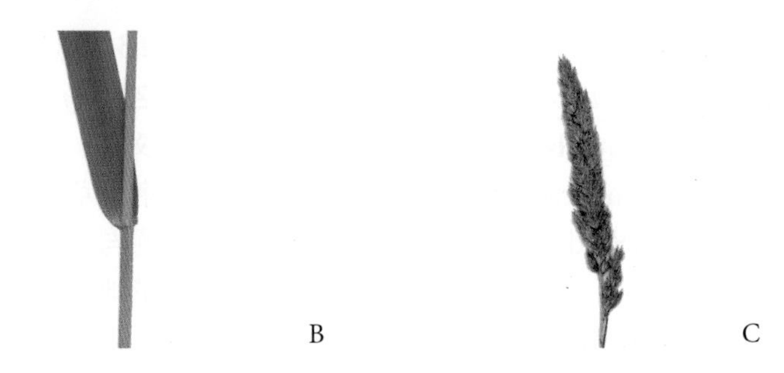

Fig. 6.20 The botanical characteristics of reed canarygrass
(A. stand; B. leaf sheath; C. panicle; D. seeds)
图 6.20 虉草植物学特征
（A. 虉草草地；B. 叶鞘；C. 圆锥花序；D. 虉草种子）

6.13.1.2 Biological characteristics

Reed canarygrass prefers warm, it often grows in the floodplain, lakes, bottomland and swamp, often mixed with reed, has submergence tolerance. The cold resistance, winter hardiness and adaptability of reed canarygrass are all strong, it can grow in the colder regions, its overwintering is better than perennial ryegrass but not as good as Timothy and smooth brome.

Reed canarygrass is not critical to soil, it can grow on all kinds of soil, but grow best on clay or clay loam, it can also grow on sand if only there is sufficient water, reed canarygrass is resistant to acid, it can grow well on pH 4~4.5 of soil.

6.13.1.3 Cultivation and utilization

Reed canarygrass can be propagated by seeds, it can also reproduce by rhizome asexually. The seeds can be sown in both autumn and spring, seeds can germinate normally when the ground temperature reach 5~6℃ . Row spacing is 30~45 cm with 3~4 cm depth of seeding, the sowing quantity is 22.5~30 kg/hm^2, the quantity can be halved if used for seed collection. Reed canarygrass is tall and leafy. It is not mixed sowing with other grass. Reed canarygrass reproduces by clonal propagation, row spacing is 40 cm, hole spacing is 30cm, it should be planted with 2~3 tillers each hole, be buried 5-6 cm depth, it can survive more easily if watered after planting in a dry weather.

Reed canarygrass likes water, it needs irrigation after tillering or each cutting, proper nitrogen fertilizer is also needed. When reed canarygrass is used for haymaking, it must be cut before heading; otherwise delayed mowing will make the quality change bad, and decrease the palatability and nutritive value. Reed canarygrass seed scatters easily, so it should be collected in time. The seed yield is 225~300 kg/hm^2.

6.13.2 Bulb canarygrass

Scientific name: *Phalaris tuberosa* L.

Origin and geographical distribution: Bulb canarygrass is native to southern Europe, the temperate regions at shore of the Mediterranean; it is cultivated in Europe, American, Australia, New Zealand and other countries. It was introduced

6.13.1.2 生物学特性

虉草喜温，常生长在河漫滩、湖边、低洼地、沼泽地，常与芦苇混生，耐水淹。抗寒性较强，越冬性好，适应性强，可在较寒冷的地区生长，越冬性较多年生黑麦草好，但不如猫尾草和无芒雀麦。

虉草对土壤要求不严，在各类土壤上都可生长，但在黏土或黏壤土上生长最好，在水分充足的情况下，沙土上也可生长，耐酸性，在 pH 4 ~ 4.5 土壤中生长亦良好。

6.13.1.3 栽培利用技术

虉草可用种子繁殖，也可用根茎无性繁殖。种子繁殖可秋播，也可春播，当地温达 5 ~ 6℃时，种子可以正常发芽。条播行距 30 ~ 45 cm，播深 3 ~ 4 cm，播量为 22.5 ~ 30.0 kg/hm^2，收种者可减半。虉草植株高大、叶茂，不宜与其他牧草混播。无性繁殖，行距 40 cm，穴距 30 cm，每穴栽植 2 ~ 3 个分蘖，埋土深 5 ~ 6 cm，如天气干旱，栽后浇水更易成活。

虉草喜水，在分蘖和每次刈割后应灌水，适当追施氮肥。调制干草时，必须在抽穗前进行刈割，迟刈将使草质变劣，适口性及营养价值均下降。虉草种子极易散落，应及时采收。种子产量为 225 ~ 300 kg/hm^2。

6.13.2 球茎虉草

学名： *Phalaris tuberosa* L.

英文名： blub canarygrass

别名： 球茎草芦

起源与地理分布： 球茎虉草原产于南欧、地中海沿岸的温带地区，欧洲、美国、澳大利亚、新西兰等国有栽培。我国 1974 年从澳大利亚引进，在广西、湖南等地生长好，西北地区生长中等。

to China from Australia in 1974, it grows well in Guangxi, Hunan provinces, and moderately in northwest area.

6.13.2.1 Botanical characteristics

Bulb canarygrass is a perennial tall herb. The fibrous root system is buried deep. The basal part of stem swells into red globose. There are buds on section, they could continue to develop into new stems, and extend to the surrounding, forming dense grass. Stems erect, it is 1~2 m high, the blade are flat, soft and smooth, 30~45 cm long, 15 mm wide. Leaf sheaths are red. It has no auricles, but the ligules are big. Panicles are dense, 8~15 cm long, pale purple or grayish green. Spikelet with 1 floret is as a rule. Caryopsis is tightly wrapped by vernicose palea and lemma. The seeds are yellowish to brown. The thousand seed weight is 1.4 g.

6.13.2.2 Biological characteristics

Bulb canarygrass prefers cool and humid climates, it should be planted in area where is drought in summer, wet in winter, and the rainfall is 380~760 mm. It is resistant to cold, drought and also submergence, it can still survive in the hot and drought summer, but stop growing, it will begin to grow in the early autumn, and it can provide a large number of forage grasses when winter comes. Bulb canarygrass doesn't have strict requirements to soil; it is not only resistant to acid soil, but also alkaline soil. It grows best in rich clay. When bulb canarygrass is planted in the northwest, it is sowed in early April, tassels and blossoms in July, the seeds mature in August. Its growth period is about 140 d.

6.13.2.3 Cultivation and utilization

The seeds of bulb canarygrass are tiny, and the seedlings grow slowly. Therefore it requires fine soil preparation and weeding. It is sensitive to fertilizer; ground fertilizer is needed before sowing. It can be sowed in both autumn and spring. The row spacing is 40~50 cm, sowing rate is 7.5~15 kg/hm^2, sowing depth is 2~3 cm. Bulb canarygrass should be mixed sowing with white clover in rainy areas, with red clover, alfalfa and annual alfalfa in more arid area. We should pay attention to intertillage and weeding at seedling stage, and apply nitrogen fertilizer properly. Hay yield is 15~22 t/hm^2.

6.13.2.1 植物学特征

球茎虉草为多年生高大草本。须根系入土深。茎基部膨大成球状，呈红色。节上有芽，可不断发育成新茎，并向四周扩展，形成稠密的草丛。茎直立，株高 1 ~ 2 m，叶片扁平，质软光滑，长 30 ~ 45 cm，宽 15 mm。叶鞘红色。无叶耳，叶舌大。圆锥花序紧密，长 8 ~ 15 cm，呈淡紫色或灰绿色。小穗有 1 小花。颖果被有光泽的内外稃紧紧包住；种子淡黄至棕色，千粒重为 1.4 g。

6.13.2.2 生物学特性

球茎虉草喜凉爽湿润气候，宜在夏季干旱、冬季湿润，年降水量 380 ~ 760 mm 的地区种植。较耐寒、耐旱，还耐水淹，夏季炎热干旱时仍能成活，但生长停滞，初秋又开始生长，在冬季来临时可提供大量饲草。对土壤要求不严，既耐酸性土壤，又适应碱性土壤，但以肥沃黏土上生长最好。在西北地区栽种，4 月初播种，7 月抽穗开花，8 月结籽成熟，生育期 140 d 左右。

6.13.2.3 栽培利用技术

球茎虉草种子细小，幼苗生长缓慢。因此要求精细整地、清除杂草。对肥料敏感，播前应施基肥。秋播或春播均可。条播行距 40 ~ 50 cm，播种量 7.5 ~ 15 kg/hm^2，播深 2 ~ 3 cm。多雨地区宜与白三叶混播，较干旱地区宜与红三叶、紫苜蓿及一年生苜蓿混播。苗期应注意中耕除草，适当追施氮肥。干草产量为 15 ~ 22 t/hm^2。

6.14 *Pennisetum* Schumach. 狼尾草属

6.14.1 Napiergrass

Scientific name: *Pennisetum purpureum* Schumach.

Origin and geographical distribution: Africa, China's Hainan, Guangdong, Guangxi, Fujian, Jiangxi, Sichuan, Yunnan and other provinces and autonomous regions have all cultivated.

6.14.1.1 Botanical characteristics

Napiergrass is a large cluster perennial herb. Sometimes it has rhizomes. Stem is erect, 2~4 m tall, Smooth or with hairs on the joint, densely villous on the base of inflorescence. The blade is linear, flat, 20~50 cm long, 1~4 cm wide, rough edges, hydrophobic raw fuzz above, glabrous below. Panicle is 10~30 cm long, 1~3cm wide. Spikelet is usually solitary or 2~3 clustered, lanceolate, 5~8 mm long, first glume is about 0.5 mm or degradation, obtuse or with unequal two-lobed at the apex, veins are obscure, second glume is lanceolate, one third of the spikelet length, dropped or obtuse at the apex, with a pulse or no pulses, the first lemma is about four-fifths of the spikelet, with 5~7 veins. The second lemma is as long as the spikelet, with five veins. The botanical characteristics of napiergrass as shown in Fig.6.21.

6.14.1.2 The Biological characteristics

Napiergrass likes the warm, humid climate, with good adaptability. It can tolerate short-term light frost. The tolerance of drought is relatively strong. It can grow on sand, clay and acidic soil, but the soil that are deep, loose and fertile is the best. Under the conditions of appropriate moisture and temperature, emergence after 7~10 d, start tillering 15~20 d later. In Guangdong, Guangxi, Fujian and other places, it can grow every year from February to December, it grows vigorously from April to September and then weakened after October. Heading and flowering varies in different varieties, the little stem heading at September and October generally, big stem heading and flowering from November to next

6.14.1 象草

学名: *Pennisetum purpureum* Schumach.

英文名: napiergrass

别名: 紫狼尾草、乌干达草

起源与地理分布: 原产于非洲。中国海南、江西、四川、广东、广西、福建、云南等地已引种栽培。

6.14.1.1 植物学特征

高大多年生丛生型草本。有时具地下茎。茎秆直立，高 2 ~ 4 m，节上光滑或具毛，在花序基部密生柔毛。叶片线形，扁平，长 20 ~ 50 cm，宽 1 ~ 4 cm，边缘粗糙，上面疏生细毛，下面无毛。圆锥花序长 10 ~ 30 cm，宽 1 ~ 3 cm。小穗通常单生或 2 ~ 3 簇生，披针形，长 5 ~ 8 mm；第一颖长约 0.5 mm 或退化，先端钝或具不等 2 裂，脉不明显；第二颖披针形，长为小穗的 1/3，先端锐减或钝，具一脉或无脉；第一外稃长约为小穗的 4/5，具 5 ~ 7 脉。第二外稃与小穗等长，具 5 脉。象草植物学特征如图 6.21。

6.14.1.2 生物学特性

喜温暖、湿润气候，适应性强。能耐短期轻霜，耐旱力较强。对土壤要求不严，沙土、黏土和微酸性土壤均能生长，但以土层深厚、肥沃、疏松的土壤最为适宜。在水分、温度适宜的条件下，种植 7 ~ 10 d 出苗，15 ~ 20 d 后开始分蘖。在广东、广西、福建等地，每年 2—12 月均能生长，4—9 月生长最盛，10 月以后生长逐渐减弱。抽穗开花因品种不同而异，小茎种一般在 9—10 月抽穗；大茎种则在 11 月至翌年 3 月抽穗开花。

Fig. 6.21 The botanical characteristics of napiergrass
图 6.21 象草植物学特征

March. Seed setting rate is low, seed mature is inconsistencies, easily scattered. Seed germination rate is low. Seedlings grow slowly, and traits are unstable.

结实率很低，种子成熟不一致，容易散落。种子的发芽率也很低，实生苗生长慢，性状不稳定。

6.14.1.3 Cultivation and utilization

6.14.1.3 栽培利用技术

Stem is commonly used to plant in production. Usually it is planted with applying 22.5~37.5 t/hm^2 organic fertilizer as base fertilizer when seedbed preparation. There is no strict requirements to the planting period generally, at an average temperature of 13~14℃, they can be planted in Guangdong,

生产上常用种茎繁殖。整地时通常施入 22.5 ~ 37.5 t/hm^2 的有机肥做底肥。对种植时期要求不严，在平均气温达 13 ~ 14℃时即可栽种；在广东、广西为 2 月；在湖南长沙为 3 月；在海

and Guangxi in February, Changsha in March, and in Hainan, as long as there is water it can be planted all the year. Selecting robust stems whose growth period is more than 100 d as seeding stem, cut them into sections every 3~4 nodes, two rows of each plot, row spacing is 50~60 cm, seed stems slant upward, buried under 5~7 cm, soil depth seed stems requirement is 1 500- 3 000 kg/hm^2.

6.14.2 Pearl millet

Scientific name: *Pennisetum typhoidum* Rich.

Origin and geographical distribution: Pearl millet originated in Africa, widely cultivated in Africa and Asia, resilient, and cultivated some north and south provinces in our country.

6.14.2.1 Botanical Characteristics

Pearl millet belongs to gramineous Pennisetum, annual herbaceous plants. The root system is developed, fibrous, and the stem base can produce adventitious roots. Stems are erect, cylindrical, the diameter is 1~2 cm, plant height is 1.25~3 m. Each plant has 5~20 tillers. Leaf sheath is smooth, the ligule is not obvious, the leaf is linear, it is about 80 cm long and 2~3 cm wide, each plant has 10~15 leaves. Densely cylindrical spica, which has 20~35 cm long, its diameter is about 2~2.5 cm , the spindle is hard and straight, and it has dense pilose, the panicle is 3.5~4.5 cm long, which is obovate and usually twins clusters, the second flowers is bisexual. The seed is 0.3~0.35 mm long, which is prominent from inner to outer glume, and is easy to fall off when it is mature, the thousand seed weight is about 4.5~5.1 g.

6.14.2.2 The Biological Characteristics

Pearl millet is a temperature-bias plant, but it has a wide range adaption to the warm conditions, in the country of origin, the annual average temperature reaches 23 to 26 ℃ . Since it was introduced to our country, the annual average temperature is 6~8 ℃ , the accumulated temperature(≥ 10 ℃) is 3 000- 3 200℃ ,which can grow in temperate semi-humid, semi-arid areas. The optimum temperature for seed germination is 20~25℃ , and the optimum growth temperature is 30~35℃ . Its drought resistance is stronger and it can grow in 400 mm of

南，只要有水分，常年可栽种。选择生长 100 d 以上健壮的茎秆做种茎，按 3 ~ 4 个节切成一段，每畦 2 行，株距 50 ~ 60 cm，种茎向上斜插，覆土 5 ~ 7 cm，种茎用量为 1 500 ~ 3 000 kg/hm^2。

6.14.2 御谷

学名： *Pennisetum typhoideum* Rich.

英文名： pearl millet

别名： 珍珠粟、蜡烛稗、非洲粟

起源与地理分布： 原产于非洲，广泛栽培于非洲和亚洲各地，适应性很强，在我国南北一些省区都有栽培。

6.14.2.1 植物学特征

御谷为禾本科狼尾草，属一年生草本植物。根系发达，须根，茎基部可生不定根。茎直立，圆柱形，直径 1 ~ 2 cm，株高 1.25 ~ 3 m。基部分枝，每株可分蘖 5 ~ 20 个，叶鞘平滑，叶舌不明显，叶线形，长 80 cm，宽 2 ~ 3 cm，每株有叶片 10 ~ 15 个。密生圆筒状穗状花序，长 20 ~ 35 cm，直径 2 ~ 2.5 cm，主轴硬直，密被柔毛，小穗长 3.5 ~ 4.5 cm，倒卵形，通常双生成簇，第二花两性。种子长 0.3 ~ 0.35 mm，成熟时自内向外颖突出而易脱落，千粒重 4.5 ~ 5.1 g。

6.14.2.2 生物学特性

御谷为喜温植物，但对温度条件适应幅度大，在原产地年平均温度达 23 ~ 26 ℃，引入我国后，年平均温度为 6 ~ 8 ℃，积温（≥ 10 ℃）3 000 ~ 3 200 ℃的温带半湿润、半干旱地区均能生长。种子发芽最适温度为 20 ~ 25℃，生长最适宜温度为 30 ~ 35℃。耐旱性较强，在年降水量 400 mm 的地区可以生长，但在干旱地区和贫瘠土壤上生长需灌溉，否则生长不良、产量低。

precipitation, but it needs irrigation in the arid and barren soil condition, otherwise has both poor growth and lower productivity. The cold resistance is poorer, and it is unfit for early spring sowing in the early spring areas where cannot resist the heavy frost. It can adapt the poor barren soil, and acid soil, and also can survive in alkaline soils. It likes water, especially sensitive to the nitrogen. Pearl millet is a short-day crop, it from south to north, extended growth period. In the northern areas of China pearl millet growth period is more than 130~140 d.

抗寒性较差，在早春霜冻严重地区，不宜早春播种。耐瘠薄，对土壤要求不严，可适应酸性土壤，亦能在碱性土壤上定居。喜水肥，尤以氮肥敏感。御谷为短日照作物，由南向北推移，生育期延长。北方地区御谷生育期在 130 ~ 140 d 以上。

6.14.2.3 Cultivation and Utilization

The cultivation method of pearl millet is similar to sorghum or corn. The seed should be sowed sparsely, the row spacing is about 50~60 cm, the plant distance is 30~40 cm. It should be planted closely if used for harvesting fresh grasses, the row spacing is 40~50 cm , the plant distance is 20~30 cm. Sowing depth is 3~4 cm, it should be compacted after seeding. The sowing rate for silage is 15~22.5 kg/hm^2, for seed harvest is 4~8 kg/hm^2.

6.14.2.3 栽培利用技术

御谷栽培方法与高粱或玉米相似。收种者播种宜稀，条播行距 50 ~ 60 cm，株距 30 ~ 40 cm。收刈鲜草者宜密，行距 40 ~ 50 cm，株距 20 ~ 30 cm。播深 3 ~ 4 cm，播后覆土镇压。青饲用播种量为 15 ~ 22.5 kg/hm^2，收种用为 4 ~ 8 kg/hm^2 。栽培管理与玉米、高粱相同。

6.15 *Sorghum* Moench—sudan grass 高粱属——苏丹草

Sudan grass

Scientific name: *Sorghum sudanense* (Piper) Stapf.

Origin and geographical distribution: Sudan grass is originally produced in the plateau of North Africa, it distributed in northeast of Africa, the Nile basin upstream and Egypt. There are about 30 species all over the world. At present, it is cultivated widely in Europe, North America and Asia continent. In the 1930s, it was introduced into China, and now it has been becoming the major annual grass that cultivated widely throughout the country.

苏丹草

学名：*Sorghum sudanense*(Piper) Stapf.

英文名： sudan grass

别名： 野高粱

起源与地理分布： 苏丹草原产于北非高原地区，非洲东北、尼罗河流域上游、埃及境内都有野生种分布。全世界约 30 种。目前，欧洲、北美洲和亚洲大陆均有栽培。我国于 20 世纪 30 年代开始引进，现已作为一种主要的一年生禾草在全国各地广泛栽培。

Botanical Characteristics

Sudan grass is an annual gramineous forage which has developed root system, the root depth is more than 2 m, almost 60%~70% root distributed in farming layer, and its

植物学特征

苏丹草为一年生禾本科牧草，根系发达，入土深达 2 m 以上，60% ~ 70%

horizontal distribution can reach to 75 cm, there is some adventitious root in close to the ground that have the absorption ability. The height of steam is 2~3 m; the number of tillers can reach to 20~100. The leaf is bar-like, 45~60cm long and 4~4.5 cm wide, there are 7~8 blades of per stem, its surface is smooth, the margin is slightly rough, the main leaf veins is obvious, the obverse side is white, and the back is green. There is no auricle on the leaf, the ligule is veins. Its flower is panicle, and its length is about 15~80 cm, inflorescence types vary between varieties, which can be divided into around-dispersed, precision vertical and lateral types. Each pedicel has two opposite spikelet, one is sessile and strong, mature, and the cob internode and another stalk infertility spikelet will fall off together when matured. There are often three spikelets on the top cob, its central has a stalk, and both sides are sessile. The inferior palea apex has geniculate awn which length is about 1~2 cm. The caryopsis is flat and obovate, which is precisely born in inside of the glume, its color is red and brown, the thousand seed weight is about 10~15 g. The botanical characteristics of sudangrass as shown in Fig.6.22.

Biological Characteristics

Sudan grass is thermophilic, but it is not cold-resistant, especially at seedling stage. It is easy to suffer from cold when the temperature is 2~3 ℃ . The lowest temperature of seed germination is 8~10 ℃ , and the optimum temperature is 20~30 ℃ . Because the root system of sudan grass is developed, which can absorb nutrients and moisture from the different depth of soil, it has strong drought - resistant. It can be dormant temporarily when suffered extreme drought in growth period, but it can recover quickly after raining. However, the production of sudan grass is closely related to the water supply conditions in its growth period, especially the flowering period should be irrigated reasonably. It should be noted that sudan grass is not resistant to wet, if there is too much water, they will be vulnerable to various diseases, especially the rust disease. Sudan grass is the short-day plant with 100-120d growth period, and the accumulated temperature(≥ 0 ℃) needs 2 200~3 300 ℃ . In the northern area,

的根分布在耕作层，水平分布 75 cm，近地面茎节常产生具有吸收能力的不定根。茎高 2 ~ 3 m，分蘖多达 20 ~ 100 个。叶条形，长 45 ~ 60 cm，宽 4 ~ 4.5 cm，每茎长有 7 ~ 8 枚叶片，表面光滑，边缘稍粗糙，主脉较明显，上面白色，背面绿色。无叶耳，叶舌脉质。圆锥花序，长 15 ~ 80 cm，花序类型因品种而不同，可分为周散型、精密型和侧垂型 3 种。每个梗节对生两个小穗，其中一个无柄，结实，成熟时连同穗轴节间和另一个有柄不孕小穗一齐脱落。穗轴顶端生小穗常 3 枚，中央的具柄，两侧的无柄。外稃先端具 1 ~ 2 cm 膝状弯曲的芒。颖果扁平，倒卵形，精密着生于颖内，红黄褐色，千粒重 10 ~ 15 g。苏丹草植物学特征如图 6.22。

生物学特性

苏丹草喜温不耐寒，尤其幼苗期更不耐低温，遇 2 ~ 3℃气温即受冻害。种子发芽最低温度为 8 ~ 10℃，最适温度为 20 ~ 30℃。由于苏丹草根系发达，且能从不同深度土层吸收养分和水分，所以抗旱力较强。生长期若遇极度干旱可暂时休眠，雨后即可迅速恢复生长。不过，苏丹草产量与生长期供水状况密切相关，尤其是抽穗开花期需水较多，应合理灌溉。应注意的是，苏丹草不耐湿，水分过多，易遭受各种病害，尤其易感染锈病。苏丹草为短日照作物，生育期 100 ~ 120 d，要求积温（≥ 0℃）2 200 ~ 3 300℃。在北方地区，4 月底播种，5 月初齐苗，6 月上旬分蘖，6 月末拔节，7 月中下旬开始抽穗和开花，9 月大部分种子才成熟。

栽培利用技术

玉米、麦类和豆类作物都是其良好的前作，但以多年生豆科牧草或混播牧草为最好。生产中，苏丹草可与毛苕子

Fig. 6.22 The botanical characteristics of Sudan grass
图 6.22 苏丹草植物学特征

sowing at the end of April, the seedling emergence in early May, tillering in early June, the jointing stage in late June, the heading and flowering period in mid to late July and most of the seeds are mature in September.

Cultivation and Utilization

Corn, wheat, barley and beans are all good preceding crop, but the perennial leguminous or mixed sowing forages are the best. Sudan grass can be mix-sowed with hairy vetch, peas and other annual legumes in production. Sudan grass is the thermophilic plant, it can be sowed when the

和豌豆等一年生豆科植物混种。苏丹草为喜温作物，需在地下 10 cm 处土温达 10 ~ 12℃时播种，北方多在 4 月下旬至 5 月上旬接种。多采用条播，干旱地区宜行宽行条播，行距 45 ~ 50 cm，播量为 22.5 kg/hm^2；水分条件好的地区可行窄行条播，行距 30 cm 左右，播量为 22.5 ~ 30.0 kg/hm^2。播种深度 4 ~ 6 cm。播后及时镇压以利于出苗。

underground 10 cm soil temperature reached 10~12 ℃, in the north area, it is sowed at the end of April to early May. It is often sown in line, the arid regions should be sown with a width line, the line spacing of seeding is 45~50cm with 22.5 kg/hm^2 of seeding rate. In good moisture conditions, it can be sowed with a narrow line, the row spacing is about 30cm, the seeding rate is 22.5~30.0 kg/hm^2. The depth of sowing is 4~6cm.Timely repression after sowing is beneficial to the seedling emergence. In addition, the mixed sowing can improve the quality and yield of forage. It should be sown 22.5 kg sudan grass seeds and 22.5~45.0 kg legume seeds per hectare. It also can be sowed in installments, about every 20~25 d, in order to extend the utilization time of green fodder. Sudan grass grows slow in seedling stage and is not resistant to weeds, it should be weeded during seedling which height is about 20 cm, it is not afraid of weed's suppression after sealing ridge, according to soil hardening situation, we can choose to whether intertill once again or not.

另外，混播可提高草的品质和产量，每公顷播种 22.5 kg 苏丹草及 22.5 ~ 45.0 kg 豆类种子。也可分期播种，每隔 20 ~ 25 d 播一次，以延长青饲料的利用时间。苏丹草苗期生长慢，不耐杂草，需在苗高 20 cm 时开始中耕除草，封垄后则不怕杂草抑制，可视土壤板结情况再中耕一次。

常见品种: ‘奇台’‘乌拉特 1 号’‘乌拉特 2 号’ ‘宁农’ ‘盐池’ ‘新苏 2 号’ ‘内农 1 号’。

6.16 *Paspalum* L.
雀稗属

Most of *Paspalum* L. are the perennial herb, the genus has about 300 species, they are distributed in tropical and temperate regions of the world, most of them come from the Western Hemisphere, especially in Brazil. China has about 8 species, mostly are fine herbage.

雀稗属多为多年生草本，本属有 300 多种，分布于世界的热带和温带地区，多产于西半球，尤以巴西为最多。我国约有 8 种，大多为优良牧草。

6.16.1 Dallis grass

Scientific name: *Paspalum dilatatum* Pior.

Origin and geographical distribution: Dallis grass originally produced in the southeastern of Brazil, northern of Argentina, Uruguay and its vicinity and the subtropical area. Now it has been introduced and cultivated to Australia and New Zealand, USA, South Africa, India, Japan, Philippines and other countries. In 1962, China introduced the seeds from Vietnam, and planted experimentally in Guangxi and Hunan province. Now it is widely planted in Yunnan, Fujian,

6.16.1 毛花雀稗

学名: *Paspalum dilatatum* Pior.

英文名: dallis grass

别名: 金冕草、宜安草

起源与地理分布: 毛花雀稗原产于巴西东南部、阿根廷北部、乌拉圭及其附近和亚热带地区。现已被澳大利亚、新西兰、美国、南非、印度、日本、菲律宾等许多国家引种栽培。1962 年我国从越南引进种子，首先在广西、湖南

Guangdong, Jiangxi, Hubei, Guizhou provinces etc. except for Guangxi, Hunan province.

6.16.1.1 Botanical characteristics

Dallis grass is a gramineous perennial herbs that belongs to *paspalum*. It has developed root system and thick stem. The stem is smooth and cespitose, erect or base inclination, and its height is 80~180cm. Leaf sheaths is smooth and flabby, the ligule is membranous, and its length is 2~5 cm, the leaf is bar-like, glabrous and dark green, its length is 30~45 cm, width is about 0.5~1.5 cm. The fringe is racemes, and has 12~18 branches, the spikelet is ovate, the length is about 3~4 cm, its apex is acute, and 4 lines are arranged on the spike-stalk, glume and lemma edge have filamentous pubescence, short-hair with both sides. Seed is ovoid, hairy, milky white, yellow to pale brown, the thousand seed weight is 2 g. The botanical characteristics of dallis grass as shown in Fig.6.23.

6.16.1.2 Biological characteristics

Dallis grass loves warm and humid climate, adapts to subtropical area, and has a certain degree of cold hardiness ability, it can resist the low temperature of -10℃ , is the strongest cold hardiness in subtropical grasses, also can keep green in non-frost areas in winter. It needs sufficient water, can resist drought and waterlogging, but if it undergo a long period of drought, it will be in dormancy and poor growth. It has a wide adaptability and soil requirement, can grow in various soils, particularly in fecund and wet black clay.

6.16.1.3 Cultivation and utilization

Dallis grass propagates by seed and branching. It can be sown both in spring and autumn, spring sowing is in February to April, and autumn in October. The land should be prepared before sowing; it needs enough organic fertilizer as base fertilizer. It is sown in line, the row spacing is about 40~50 cm, the depth of sowing is 1~2 cm, and the seeding rate is 15~22.5 kg/hm^2. It can be carried out throughout the year by vegetation propagation during the rainy season, the row spacing is usually 40~50 cm, the spacing distance is 20~30 cm, it is planted 3~4 tillers per hole, and should be covered soil about 5~6 cm, and watered after planting. Dallis grass also can sow with carpet grass, bermuda grass, red clo-

试种。现在云南、广东、福建、江西、湖北、贵州等地均有种植。

6.16.1.1 植物学特征

毛花雀稗是禾本科雀稗属多年生草本植物。根系发达，茎秆粗壮，光滑丛生，直立或基部倾斜，株高 80 ~ 180 cm。叶鞘光滑，松弛；叶舌膜质，长 2 ~ 5 cm；叶片条形，长 30 ~ 45 cm，宽 0.5 ~ 1.5 cm，无毛，深绿色。穗状总状花序，分枝 12 ~ 18 个，小穗卵形，长 3 ~ 4 cm，先端尖，成 4 行排列于穗轴一侧，颖和外稃边缘有长丝状柔毛，两面贴生短毛。种子卵圆形，有毛，乳白、乳黄至浅褐色，千粒重 2 g。毛花雀稗植物学特征如图 6.23。

6.16.1.2 生物学特性

毛花雀稗喜温热湿润气候，适于亚热带地区种植，亦有一定的耐寒能力，可耐 –10 ℃低温，是亚热带牧草中抗寒力强的牧草，冬季无霜冻地区能保持青绿。需水较多，耐水渍亦较耐干旱，但长期干旱会处于休眠和生长不良。适应性广，对土壤要求不严，各种土壤都能生长，尤喜肥沃而湿润的黑色黏重土壤。

6.16.1.3 栽培利用技术

毛花雀稗既可用种子繁殖，也可分枝繁殖。春播、秋播均可，春播宜于 2—4 月进行，秋播在 10 月进行。播前需整好地，施足有机肥作基肥。条播，行距 40 ~ 50 cm，播深 1 ~ 2 cm，播种量为 15 ~ 22.5 kg/hm^2。分株繁殖全年雨季均可进行，一般按行距 40 ~ 50 cm、株距 20 ~ 30 cm 栽植，每穴栽 3 ~ 4 个分蘖，覆土深 5 ~ 6 cm，栽后浇水。毛花雀稗也可与地毯草、狗牙根、红三叶、苜蓿、胡枝子、黑麦草等混播，混播时播种量为 6 ~ 15 kg/hm^2。

Fig. 6.23 The botanical characteristics of dallis grass
图 6.23 毛花雀稗植物学特征

ver, alfalfa, lespedeza, and ryegrass, the seeding rate of mixture sowing is 6~15 kg/hm^2.

Dallis grass should fertilize after mowing, and must be irrigated when drought. The seeds maturation of dallis grass are inconsistent and easy to fall off, and should be timely harvested, the seed yield is about 300~450 kg/hm^2.

毛花雀稗刈割后应及时追肥，在干旱时必须进行灌溉。种子成熟不一致，且易脱落，宜及时采收，种子产量为300 ~ 450 kg/hm^2。

6.16.2 Broadleaf paspalum

Scientific name: *Paspalum wettsteinii* Hackel.

Origin and geographical distribution: Broadleaf paspalum originates from South American Brazil, Paraguay, northern Argentina and other subtropical rainy area. In 1974, China imported it from Australia, now it has been become a main variety and used successfully in pasture improvement. The planting area is more than 266 hm^2 in Guangxi, in addition, it has been an widely introduced cultivation in Yunnan, Guizhou, Guangdong, Fujian, Hunan, Jiangxi provinces and so on.

6.16.2.1 Botanical characteristics

Broadleaf paspalum is the semi-procumbent perennial bunchgrasses. The plant height is 50~100 cm, with short rhizomes, the lower part of stem is creeping on the ground, close to the ground part can grow adventitious roots, and its fibrous root is developed. The blade length is 12~32 cm, the width is 1~3 cm, both surfaces and backs are covered with densely white pilose, leaf margin has sawtooth, leaf sheath is dark purple, the color of upper leaf sheath part is lighter. The length of raceme is 8~9 cm, there are usually 4~5 arranging on the main shaft, and spikelet is solitary, it is arranged in two rows on the one side of rachis. the seed is ovoid, one side is uplift, and the other side is flattening, the thousand seed weight is 1.35~1.4 g. The botanical characteristics of broadleaf paspalum as shown in Fig.6.24.

6.16.2.2 Biological characteristics

Broadleaf paspalum likes the high temperature, rainy climate and fertile soil with good drainage, it also can grow in the arid slope of red or yellow soil, but its leaves are narrow and aging easily. In the subtropical regions of China, it can keep evergreen, and most luxuriant in summer and autumn, the growth stops during the winter frost and its tip turns yellow, and restore growth after the frost. Seeds can germinate in 20℃ , it is sowed in March in Nanning, full stand of seeding is at the beginning of April, after two weeks it enter the tillering stage, its elongation stage is in late May, the heading is in late June, flowering is in mid-July, and fructification is in mid-August. The flowering and fruiting period is com-

6.16.2 宽叶雀稗

学名： *Paspalum wettsteinii* Hackel.

英文名： broadleaf paspalum

起源与地理分布： 宽叶雀稗原产南美巴西、巴拉圭、阿根廷北部等亚热带多雨地区。我国 1974 年从澳大利亚进口，现已作为一个当家品种在草地改良中推广应用。广西种植面积超过 266 hm^2，此外，云南、贵州、广东、福建、湖南、江西等地均有引种栽培。

6.16.2.1 植物学特征

宽叶雀稗为半匍匐丛生型多年生禾草。株高 50 ~ 100 cm，具短根状茎，茎下部贴地面呈匍匐状，着地部分节上可长出不定根，须根发达。叶片长 12 ~ 32 cm，宽 1 ~ 3 cm，两面密被白色柔毛，叶缘具小锯齿，叶鞘暗紫色，茎上部叶鞘色较浅。总状花序长 8 ~ 9 cm，小穗单生，通常 4 ~ 5 个排列于总轴上，呈两行排列于穗轴的一侧。种子卵形，一侧隆起，一侧压扁，千粒重 1.35 ~ 1.4 g。宽叶雀稗植物学特征如图 6.24。

6.16.2.2 生物学特性

宽叶雀稗性喜高温多雨的气候和肥沃排水良好的土壤，在干旱贫瘠的红、黄壤坡地亦能生长，唯叶子明显变窄，而且易老化。在我国南亚热带地区可四季常青，而以夏秋季节生长最茂盛，冬季下霜期间生长停止，叶尖发黄，霜期过后即恢复生长。种子在气温稳定在 20 ℃时即可萌发，在南宁 3 月份播种，4 月初全苗，出苗两周后即进入分蘖期，5 月下旬拔节，6 月下旬抽穗，7 月中旬开花，8 月中旬大量结实。花果期较长，一年可收种子两次，种子产量为 375 ~

Fig. 6.24 The botanical characteristics of broadleaf paspalum
图 6.24 宽叶雀稗植物学特征

paratively long , we can collect seeds twice annually, the seed yield is 375~450 kg/hm^2. The fresh forage yield is 45~75t/hm^2. It has strong tillering and renewable ability, it can resist grazing, and be mixture sowed with large wing beans, Stylo, Desmodium, wild soybeans, which can form a good stand in the seeding year.

6.16.2.3 Cultivation and utilization

The land should be plowed or raked repeatedly before the sowing of broadleaf paspalum in order to facilitate seed implantation. It should be sowed when spring rain season is coming. In southern of China, it is usually sowed in March

450 kg/hm^2。单播人工草地鲜草产量为 45 ~ 75 t/hm^2。分蘖力和再生力强，且耐牧，可与大翼豆、柱花草、山蚂蝗、野大豆等混播，当年即可形成良好的草群。

6.16.2.3 栽培利用技术

宽叶雀稗播种前应把土地进行翻耕或用重耙反复耙平耙碎，以利种子着床。待春暖下雨季节进行播种。华南地区通常在 3—4 月份播种，播种后不需要覆土。如作收种用，可行条播。

to April, it need not to cover earth after sowing. Driu seeding method is usually used in seed in seed production field. Seeding rate is 7.5 kg/hm^2, the seed should be mixed with calcium, magnesium, phosphorus or ash, and make into pill garment seeds while sowing . If used as grazing land, its can be mixture sowed with siratro, stylosanthes, desmodium and clover. Broadleaf paspalum is an excellent forage in grazing land, its leaf to stem ratio is 46%, 54% respectively and the drying rate is both 26%, water buffalo and cattle like it. When the seed is few, we can expand the planting area by transplantation, it can easily survived in the rainy season.

播种量为7.5 kg/hm^2,播种时可用钙、镁、磷肥或草木灰与种子拌匀或做成丸衣种子再播。如作放牧地用，可与大翼豆、柱花草、山蚂蝗、三叶草等混播。宽叶雀稗是放牧地的优等牧草，水牛、黄牛均喜食，茎叶比分别为46%和54%，风干率为26%。在种子缺乏时，亦可采用分株移植的办法来扩大种植面积，在雨季移植极易成活。

Chapter 7

Forbs

其他科牧草

Forbs are the broadleaf herbs of forage species other than grasses, legumes, and woody species, grown under cultivated conditions. As a group, forbs comprise a relatively small number of forage species, and they are grown on far fewer acres than are either grasses or legumes. Their primary importance arises from their use as specialty crops or to fill a niche within a forage system.

其他科牧草是指除了禾本科、豆科和木本饲用植物外的一类牧草，多为阔叶类，这类牧草数量相对较少，栽培面积远远少于禾本科和豆科牧草，它们的重要性体现在可以作为一种特殊植物用于牧草系统。

7.1 Forage chicory 饲用菊苣

Scientific name: *Cichorium intybus* L.

Family: Compositae

Chicory is a short-lived perennial forb native to Europe, western and central region of Asia, North Africa and South Africa. It can be commonly found in the mountains, field boundaries and wasteland in northwest and northeast of China. Chicory is a good nectar plant which flowering up to 2~3 months. In Europe, chicory is widely used as a kind of leafy vegetable (Fig.7.1).

Its root contains plentiful inulin and aromatic substanc-

学名： *Cichorium intybus* L.

英文名： chicory

别名： 欧洲菊苣、咖啡草、咖啡萝卜

菊苣是一种起源于欧洲、亚洲中西部、北非和南非的菊科多年生阔叶类牧草，在我国西北和东北的山区、平原和撂荒地上很常见。菊苣也是一种很好的蜜源植物，开花期长达2~3个月。在欧洲亦常被用作蔬菜（图 7.1），从

Fig. 7.1 Vegetable chicory
图 7.1 蔬菜菊苣

es, which can be extracted as coffee substitute. The extracted bitter substance can be used to improve the vitality of digestive organs.

菊苣根中提取的菊粉和香味剂可作为咖啡的替代品，促进消化。

7.1.1 Botanical characteristics

Forage chicory produces rosette growth with leaves emanating from a region of compacted internodes in the crown at the top of the taproot (Fig.7.2). Its taproot is long, thick and fleshy. Lateral roots are thick and well distributed horizontally or diagonally. The main stem is erect and hollow, sparsely covered with coarse hair. It is capitulum, cylindrical involucres, blue corolla and cuneiform achene. (Fig.7.3). The thousand seeds weight is 1.2~1.5 g.

7.1.1 植物学特征

饲用菊苣前期生长呈莲座状（图 7.2），即由主根顶端紧缩的节间组成的根冠部长出大量叶片。主根长，肉质，鲜嫩多汁，侧根发达呈水平分布或斜向下分布。主茎直立中空，具条棱，疏被粗毛，叶分为基生叶和茎生叶。头状花序，花冠蓝色，瘦果楔形（图 7.3）。种子千粒重 1.2~1.5 g。

7.1.2 Biological characteristics

Chicory tolerates a wide range of soil conditions, including acid soils, but performs best on moderately drained, deep soils with medium to high fertility and a pH above 5.5. It produces a deep taproot capable of extracting moisture from great depths in the soil during drought time. Waterlogged and heavy clay soils tend to limit persistence. The crop has performed well in both pure stands and mixtures with other cool-season forages.

7.1.2 生物学特性

菊苣喜温暖湿润气候，耐寒性较强，抗旱性能较好，较耐盐碱。喜肥喜水，对氮肥敏感，对土壤要求不严，旱地、水浇地均可种植。耐酸性土壤，但在排水良好、土壤肥力中等或高肥及 pH 在 5.5 以上的土壤上表现最好。主根深，可以吸收土壤深层水分，尤其当遇到长时间的干旱时。积水和黏重土壤对菊苣生长不利。菊苣无论是单作还是与其他冷季型牧草混播表现都很好。

Fig. 7.2 Rosette seedling stage of forage chicory
图 7.2 苗期莲座状饲用菊苣

Fig. 7.3 The stand of forage chicory at flowering stage
图 7.3 开花期的饲用菊苣

7.1.3 Cultivation technology

Spring planting is best for planting of forage chicory at mid-temperate latitudes. Drill seeding at depth of 0.6~1.3 cm is recommended, using a cultipacker-type seeder is preferred when planting into clean-tilled seedbeds. Sod-seeding into existing pasture stands also has been successful if herbicide is applied prior to seeding as a sod-suppression technique. Seeding rate for pure stands of chicory is 4.5~5.6 kg/ hm^2. For mixtures, 2.3~3.7 kg/ hm^2 of chicory seed and two-thirds of the usual seeding rate for the other component are recommended.

Nitrogen from fertilizer or a complementary legume is essential for adequate productivity and quality. Weed control should precede sowing. If waterlogged it should be drained immediately to avoid root rot disease and plant mortality.

7.1.4 Utilization and management

The most common system of livestock utilization for forage chicory is grazing. Careful management grazing is essential, to obtain the most satisfactory combination of yield, quality, and persistence. Continuous grazing or prolonged grazing periods can severely damage existing stands because of damage to crowns caused by either overgrazing (more likely with sheep) or excessive trampling (more likely with cattle).

New stands seeded in spring should not be grazed for at least 80~100 d after sowing. Rotational grazing management produces the best combination of chicory productivity and persistence. Grazing should be managed to provide a stubble height of at least 5 cm at the end of each grazing cycle.

7.1.3 栽培技术

在中纬度地区新建菊苣草地，春季播种最好。条播深度 0.6~1.3 cm，可使用镇压播种机在耕翻好的清洁苗床上播种。在现有草地上补播菊苣时，如果播种前使用除草剂对已有植被进行抑制并进行条播也很容易成功。单播的播种量为 4.5~5.6 kg/ hm^2，与其他牧草进行混播时，推荐菊苣播种量为 2.3~3.7 kg/ hm^2，其他混播组分的播种量为常规播种量的 2/3 左右。

从化肥或互补的豆科牧草中获取氮素对提高产量和改善饲草品质至关重要。播种前应进行除草，生长期间若遇积水，应立即进行排水，以防根腐致植物死亡。

7.1.4 利用和管理

菊苣最常见的利用方式是放牧，科学的放牧管理对草地的高产、优质和持久十分重要。然而持续放牧或延长放牧时间会严重破坏草地，因为过度放牧（如绵羊）或严重踩踏（如牛群）会破坏菊苣的根冠部分。

春季新建草地的首次放牧时间应在播种后 80~100 d，轮牧是草地发挥高产和提高持久性的最好方式。生长最快速阶段的轮牧周期为 25~30 d，在面积受限的草地上推荐缩短放牧时间。放牧留茬高度不低于 5 cm，并且做到均匀利用。

7.2 India lettuce 苦荬菜

Science name: *Lactuca indica* L.

Family: Compositae

India lettuce was a kind of wild plant in China that is widely distributed in all parts of the country. It is also distributed in Korea, Japan, India and other countries. After years of domestication and breeding, India lettuce has become a popular, high yielding and high quality forage crop. It has been planted successfully in the south, north and northeast regions. It is an excellent succulent feed for all kinds of livestock. In addition, India lettuce can stimulate appetite and decrease the blood pressure. Through further processing it can also be made into frozen food and beverage.

7.2.1 Botanical characteristics

India lettuce is annual or biennial herbage belongs to compositae family. It has a taproot system; its axial roots are stout and fusiform. Roots may extend to a depth of over 2 m. The root contagiously distribute in the 0~30 cm soil layer. The stem is erect, with many smooth branches, about 1.5~3.0 m in height. Leaves are fascicular, numbering 15~25 with no obvious petiole. Leaf shape is different, generally oval or lanceolate. Leaves are 30~50 cm long, 2~8 cm wide. The leaf margin is cleft toothed to pinnatifid. Stem leaves are smaller, 10~25 cm long, alternate and sessile, base amplexicaul. All parts of plant contain white milky juice that tastes bitter. The flower is capitulum, ligulate and canary. The fruit is achene, long ovate, purple black when mature, with a white crested in top. The thousand seeds weight is 1.0~1.5 g (Fig. 7.4).

7.2.2 Biological characteristics

India lettuce prefers a warm climate. It has resistance to cold and heat. It will bloom blossom and bear fruit in area where the frost free period is over 150 d, or cumulative temperature (≥10℃) accumulated more than 2 800℃. Seed will

学名： *lactuca indica* L.

英文名： India lettuce

别名： 苦麻菜、鹅菜、凉麻、山莴苣、八月老。

苦荬菜原为我国野生植物，几乎广布全国各地。朝鲜、日本、印度等国也有分布。经过多年的驯化和选育，苦荬菜已成为深受欢迎的高产优质饲料作物，并在南方和华北、东北地区大面积种植，是各种禽畜的优良多汁饲料。此外，苦荬菜具有开胃和降血压的作用，经过深加工还可制成冷冻食品和饮料。

7.2.1 植物学特征

菊科莴苣属一年生或越生草本植物。直根系，主根粗大，纺锤形，入土深 2 m 以上，根系集中分布在 0~30 cm 的土层中。茎直立，上部多分枝，光滑，株高 1.5~3.0 m。叶丛生，15~25 片，无明显叶柄，叶形不一，披针形或卵形，长 30~50 cm，宽 2~8 cm，全缘或齿裂至羽裂；茎生叶较小，长 10~25 cm，互生，无柄，基部抱茎。全株含白色乳汁，味苦。头状花序，舌状花，淡黄色。瘦果，长卵形，成熟时为紫黑色，顶端有白色冠毛，千粒重 1.0~1.5 g（图 7.4）。

7.2.2 生物学特性

苦荬菜喜温耐寒且抗热。无霜期 150 d 以上，积温（≥10 ℃）2 800 ℃以上的地区均可开花结实。土壤温度达 5~6 ℃时苦荬菜种子即能发芽，在 25~30 ℃时生长最快。苦荬菜抗热能力较强，在 35~40 ℃的高温条件下也能良好生长。需水量大，适宜在年降水量

Fig. 7.4 India lettuce
图 7.4 苦荬菜

germinate when the soil temperature is between 5~6℃ and the growth is fastest when the soil temperature is at 25~30℃. Good heat resistance and it grows well even at 35~40℃. More water needed during growth period, it is suitable for planting in regions where annual precipitation reaching 600~800 mm. Excessive water may cause root rot and death. It has some drought resistance. It has no strict edaphic requirement and can be planted in various kinds of soil. Fertile soil with good drainage is the most ideal. Because of its shade tolerance, it can be planted in orchards. India lettuce grows slowly in seedling stage. The growth period is about 120~130 d of early maturing varieties.

7.2.3 Cultivation technology

India lettuce is not suitable for sustained cropping. Seedling is very weak, so a fine soil preparation is essential. The seed need to be cleaned before sowing, which can be sowed by broadcasting, drill seeding, and also can be used sprigging cultivated by setting seedling in sowing. India lettuce is suitable for dense planting and usually does not require thinning

600~800 mm 的地区种植。不耐涝，积水数天可使根部腐烂死亡。具有一定的抗旱能力。对土壤要求不严，各种土壤均可种植，但以排水良好、肥沃的土壤最为适宜。较耐阴，可在果林行间种植。苦荬菜苗期生长缓慢。经培育而成的早熟苦荬菜品种，生育期 120~130 d。

7.2.3 栽培技术

苦荬菜不宜连作。幼苗出土力弱，要求精细整地。播前需要对种子进行清选，撒播、条播、穴播均可，也可育苗移栽。苦荬菜宜于密植，通常不间苗。出苗后要及时中耕除草。苦买菜生长迅速，需及时刈割。采种地要多施磷、钾肥，为防止倒伏，适量施用氮肥，不可过多。苦荬菜花期长，种子成熟不一致，而且落粒严重。在大部分果实的冠毛露出时收种为宜。为避免落粒损失，也可采取分期采收的方法。一般每公

out seedlings. Cultivating and weeding after budding is critical. India lettuce grows fast, so mowing must be timely. Seed field need sufficient phosphorus and potassium, and appropriate nitrogen fertilizer in order to reduce lodging. It has long flowering period, seed maturity is not very concentrated and the seed shatter losses is very serious, so harvest when most seeds maturing is essential. Seed yield is about 375~750 kg/hm^2.

7.2.4 Feeding value

India lettuce produced many crisp, juicy and nutritious leaves, especially the higher crude protein content is similar to that of alfalfa, which contains 0.49% lysine, tryptophan 0.25%, methionine 0.16%, is a high quality protein feed. In addition, crude fat, nitrogen free extract and vitamin contents are high. India lettuce leaves are leafy, bitter and palatable; swine and poultry prefer to eat. India lettuce can promote appetite and digestion, reduce internal heat and reduce disease, also don't need to replenish vitamin. India lettuce can be used as fresh forges or processed into silage and hay.

顷产种子 375~750 kg。

7.2.4 饲用价值

苦荬菜叶量大，脆嫩多汁，营养丰富，特别是粗蛋白质含量较高，与苜蓿相似，所含蛋白质中氨基酸种类齐全，其中含赖氨酸 0.49%，色氨酸 0.25%，甲硫氨酸 0.16%，是一种优质的蛋白质饲料。另外，它的粗脂肪、无氮浸出物、维生素含量也很丰富。苦荬菜叶量丰富，略带苦味，适口性特别好，猪、禽类最喜食。苦荬菜还有促进食欲和消化、祛火去病的功能。饲喂苦荬菜，可节省精料、减少疾病，也不必补饲维生素。苦荬菜可青饲利用，也可调制成青贮饲料和干草。

7.3 Cup plant 串叶松香草

Scientific name: *Silphium perfoliatum* L.

Family: Compositae

Cup plant is native to the highlands of northern America, mainly distributed in the southeastern mountains of the United States. It was introduced to Europe at the end of 18th century. The Soviet Union and a number of European countries introduced it as a cultivated silage crop in the 1950s, and began to expand to a large area since the 1960s. Our country introduced it from North Korea in the 1970s. It is cultivated in most of the provinces at present. Cup plant is not only a high yield and quality forage, it has a good ornamental and nectar plants because of a long flowering period, beautiful color, a fragrant smell. The roots has medicinal value, is a kind of traditional Native American herbal medicine.

学名： *Silphium perfoliatum* L.

英文名： cup plant

别名： 松香草、菊花草、串叶菊花草

串叶松香草原产于北美中部的高原地带，主要分布在美国东部、中西部和南部山区。18 世纪末引入欧洲，20 世纪 50 年代苏联及一些欧美国家将其作为青贮作物引入进行栽培，60 年代开始大面积推广利用。1979 年从朝鲜引入我国，目前大部分省区均有栽培。串叶松香草花期长，花色艳丽，有清香气味，既是一种高产优质的牧草，也是良好的观赏植物和蜜源植物；其根还有药用价值，是印第安人的传统草药。

7.3.1 Botanical characteristics

Cup plant is a tall perennial herbage plant, native to central North America that grows up to 2~3 m tall. This species has square stems and leaves that are mostly opposite, egg-shaped, toothed, with cuplike bases that hold water. The flower heads are rich, golden yellow, 2.5 cm in diameter, and closely grouped at the tips of the stems. The small, tubular disk flowers are in the middle of the flower and are sterile and do not produce fruits (Fig.7.5). Its fruit is brown heart-shaped achene. The thousand seed weight is 20~25 g.

7.3.2 Biological characteristics

Cup plant prefers a warm climate and is resistant to

7.3.1 植物学特征

串叶松香草为菊科松香草属多年生草本植物。根系发达粗壮，多集中在 5~40 cm 的土层中；根茎节上着生有由紫红色鳞片包被的根茎芽。茎直立，四棱，高 2~3 m。叶分基生叶和茎生叶两种，长椭圆形，叶面粗糙，叶缘有缺刻。头状花序，黄色花冠（图 7.5）。瘦果心脏形，褐色，每个头状花有种子 5~21 粒。种子千粒重为 20~25 g。

7.3.2 生物学特性

串叶松香草为喜温耐寒抗热植物。需水较多，特别是在现蕾期和开花期。

Fig. 7.5 The stem and flower of cup plant
图 7.5 串叶松香草

cold and heat. Water demands are high, especially squaring stage and flowering stage. It has moderate drought tolerance, strong resistance to water logging and poor resistance to salt and barren. It prefers a warm and humid climate, survives to cold -38℃ and heat of one week with high temperatures of 40℃, but it's not tolerant to dry-hot wind. It is a long-day and light loving plant. It prefers slightly acidic to neutral soil (pH 6.5~7.5), isn't tolerant to saline soil and barren.

7.3.3 Cultivation technology

Cup plant prefers good moisture and high fertility soil, seeds can be sowed directly, but seedling transplant should be the better way to establish successfully. Generally, it can be sowed from spring to autumn, good quality seed should be soaked in warm water (30℃) for 12 h before sowing. Seeding rate is 6~9 kg/hm^2, the row space is 40~50 cm for forage production and 100~120 cm for seed collecting. Timely inter-tillage and weeding are necessary in seedling stage.

Harvest should be done from bud stage to early bloom stage when the plant height reaches 60~70 cm, 2~4 times annually and the fresh yield is about 150~225 t/hm^2. It is critical to irrigate and fertilize after each harvest. Seed field need more sufficient phosphorus and potassium, it is not recommended to harvest forage in seed production field. Less fertilization and irrigation during the late growth stage can prevent lodging. Seed maturity is inconsistent and prone to loss, so seed harvest must be timely when 2/3 achenes turned yellow. Seed yield usually reaches 750~1 500 kg/hm^2.

7.3.4 Feeding value

Cup plant produced high yields, of good quality, rich in crude protein, amino acids and carbohydrates. Calcium, phosphorus and carotene content are also high. Nutritional value is good for cattle, sheep, pigs, rabbits, poultry and fish. According to reports, the forage contains pine vanilloid diterpene and polysaccharides called rosin glycosides. Long-term feeding will lead to cause animal poisoning; consequently, it

具有一定的抗旱能力，耐涝性较强，抗盐性及耐瘠薄能力差。喜温暖湿润气候，抗寒（最低 -38℃），耐热（连续一周的 40℃高温），但不耐干热风；喜光，为长日照作物。喜欢微酸性及中性土壤（pH 6.5~7.5），不耐盐渍土，不耐瘠薄。

7.3.3 栽培技术

串叶松香草不耐贫瘠，要选择水肥充足、便于管理的田块种植。可以用种子直播，也可以育苗移栽，这样更容易成功。春夏秋皆可播种，一般播前将种子用 30℃温水浸种 12 h 有利于及时出苗，播种量为 6~9 kg/hm^2，收草地和收种地行距分别为 40~50 cm、100~120 cm；苗期要及时中耕除草。

一般在孕蕾至初花期，高度为 60~70 cm 时刈割为宜。年刈 2~4 次，鲜草产量 150~225 t/hm^2。刈割后及时灌水施肥是关键。采种田一般不刈割，并多施磷、钾肥。为减少倒伏，生育后期要减少施肥和灌水。种子成熟不一致，而且容易落粒，因而在 2/3 的瘦果变黄时即可采收，一般种子产量 750~1 500 kg/hm^2。

7.3.4 饲用价值

串叶松香草不仅产量高，而且品质好，粗蛋白质和氨基酸含量丰富，富含糖类。另外，钙、磷和胡萝卜素的含量也极为丰富，是牛、羊、猪、兔等家畜和家禽以及鱼类的优质饲料。据资料显示，串叶松香草含有的松香草素、二萜和多糖，称松香苷，长期饲喂会导致动物中毒，应掌握好配比。鲜叶可直接喂牛、羊、兔。与燕麦、苏丹草、青饲玉米等混合青贮可提高奶牛的产奶量；切细拌精料发

must be in a reasonable ratio. Fresh leaves can be fed directly to cows, sheep and rabbits. Making mixed silage with oats, sudan grass, corn that can increase milk production of dairy cattle. Chopped forages fermented for 12~24 h may be fed to pigs. Hay meal is very nutritious, 5%~10% of it as formula feed is good.

酵 12~24 h 喂猪；干草粉营养丰富，加 5% ~10%作配合饲料。

7.4 Common comfrey 聚合草

Scientific name: *Symphytum pezegrinum*

Family: Boraginaceae

Comfrey is native to Former Soviet Union, North Caucasus and Siberia and grows on the shore of rivers and lakes, edge of forest and mountain grasslands. China introduced comfrey from Japan, Australia and Korea in 1964 and 1972 successively, and promoted as a kind of forage in the northeast, north and northwest. Since 1977, with government vigorous promotion provinces have planted it all. The largest cultivated areas are concentrated in mainly north of the Yangtze River and south of that region, including Jiangsu, Shandong, Shanxi and Sichuan.

学名： *symphytum pezegrinum*

英文名： common comfrey

别名： 爱国草、友谊草、肥羊草、紫草

聚合草原产于北高加索和西伯利亚等地，生长在河岸边、湖畔、林缘和山地草原。1964 年和 1972 年我国先后从日本、澳大利亚和朝鲜引进，分别在东北、华北和西北地区推广试种。1977 年以来，在国家的大力推广下，各省区均有种植。栽培面积最大的地区集中在长江以北和以南地区，其中以江苏、山东、山西、四川等地栽培最多。

7.4.1 Botanical characteristics

Comfrey is a perennial herb and clusters. The root is stout and fleshy. The taproot is about 3 cm in diameter and up to 80 cm long. The difference between taproot and lateral roots is not obvious. Primary lateral roots are obvious with most roots distributed in the 30~40 cm soil layer. Plant may reach 80~150 cm high, with short white bristles. There are two kinds of leaves, stem leaf and basal leaf. Basal leaves are clustered showed as rosette; stem leaves have a short stalk or no petiole. Inflorescence is cyme, shaped as scorpioid and seed-setting is very low. The stem is cylindrical and upright, ridge is thinner than bottom. There are latent buds and branches in leaf sheath. The regenerative capacity of the stem is very strong, since it can produce new shoot, root

7.4.1 植物学特征

聚合草为紫草科聚合草属多年生草本植物，丛生。根粗壮发达，肉质，主根直径 3 cm 左右，主根长达 80 cm，侧根发达，主侧根不明显，主要根系分布在 30~40 cm 的土层中。株高 80~150 cm，全身密被白色短刚毛。叶有茎生叶和基生叶两种。基生叶簇生呈现莲座状，具有长柄；茎生叶有短柄或无柄。蝎尾状聚伞花序，结实率极低。茎秆为圆柱形，直立，向上渐细。在叶片、叶鞘处有潜伏的芽和分枝。茎的再生能力很强，能产生新芽和根，可育成

and even be bred to new strains (Fig. 7.6).

新株（图 7.6）。

7.4.2 Biological characteristics

As a mesophyte, comfrey has a high water demand and should be planted in which annual precipitation is 600~800 mmn, less than 500 mm or more than 1000 mm lead to poor growth. Life span may be as long as 20 years. It prefers humid conditions, and resists cold to -40℃, tolerates heat to 39℃, grows fastest within 22~28℃. Roots extended to depths of 30~40 cm. Comfrey has good drought resistance, but does not tolerate water logging. Soil requirements are not strict. It is not suitable for low-lying, heavy saline land but suitable for drainage, soil deep, fertile loam or sandy soil. It is fast-growing, requiring high fertility, organic matter and nitrogen.

7.4.2 生物学特性

中生植物，对水分要求较高，宜在年降水量 600~800 mm 的地区种植，年降水量小于 500 mm 或大于 1 000 mm 的地区生长差。寿命大约 20 年。喜温湿气候，抗寒（最低 -40℃），耐热（最高 39℃），生长最快温度为 22~28℃。根系发达，入土深（30~40 cm）。抗旱性较强，不耐涝；对土壤要求不严，不宜在低洼、重盐碱地种植，最适宜排水良好、土层深厚、肥沃的壤土或沙质土。高产速生，要求高肥，对有机质和氮肥要求多。

Fig. 7.6 Common comfrey
图 7.6 聚合草

7.4.3 Cultivation technology

Comfrey is commonly propagated by vegetative and propagation coefficient is 100~200. As there are many residual roots in the soil, easily lead to weed problem. Since comfrey is shade tolerance, it can be intercropped with corn, cabbage or rape, but not suitable for field rotation. Plant spacing bases on the degree of soil fertility, fertilizering levels, irrigation and farm management level. Interplanting and weeding is necessary after survived and before closing of crop. After mowing, soil needs to irrigate, apply fertilizer and exclude water logging. Harvest for 4~6 times per year in the south and in north for 3~4 times. Residual stubble height is generally about 5 cm.

7.4.4 Feeding value

Comfrey provides a juicy feed for pigs, cattle, sheep, poultry and fish with abundant leaves, high quality and good palatability. Comfrey is rich in protein and vitamins, especially lysine, methionine and arginine, and these amino acids are essential for animal growth and development. In addition, it contains large amounts of allantoin and vitamin B_{12}, which is effective against the enteritis after eating. Because of dense stiff bristles, domestic animal don't like to eat, but it is soft and juicy and be ate by pigs, cattle, sheep and camels after crushing and beating. Comfrey is nutritious and has high digestibility, protein digestibility is 61.20%, and crude fiber digestibility is 60.44%. However, comfrey is not suitable to be long-term and sole feed, as it contains pyrrolizidine alkaloids (comfrey hormone), which can do harm to the central nervous and liver.

7.4.3 栽培技术

聚合草一般采用无性繁殖，繁殖系数为 100~200。聚合草栽植时，因其在土壤中残留的根系多，易造成草害；耐阴，可与玉米、白菜、油菜等进行套作，但不宜大田轮作。株行距大小根据土壤肥沃程度、施肥水平、灌溉及田间管理水平等而定。成活后、封行前中耕除草，刈割后灌水施肥、排除积水。南方一年收获 4~6 次，北方 3~4 次。留茬高度一般约为 5 cm。

7.4.4 饲用价值

聚合草是一种叶量丰富、优质、利用期长的牧草，是猪、牛、羊、禽、鱼等的多汁饲料。聚合草含有丰富的蛋白质和各种维生素，特别是富含赖氨酸、甲硫氨酸和精氨酸等动物生长发育不可缺少的氨基酸。此外，还含有多量的尿囊素和维生素 B_{12}，可治疗肠炎，牲畜食后不拉稀。全身密被粗硬的刚毛，家畜不喜采食，粉碎打浆后，柔软多汁，为猪、牛、羊、骆驼等多种家畜采食。聚合草不仅营养丰富，而且消化率高，如蛋白质消化率为 61.20%，粗纤维消化率 60.44%。但聚合草含双稠吡咯生物碱（聚合草素），对中枢神经和肝脏有害，不宜长期单喂。

7.5 Grain amaranth 籽粒苋

Scientific name: *Amaranthus hypochondriacus* L.
Family: Amaranthaceae

Amaranthus hypochondriacus L., a native of tropical and

学名： *Amaranthus hypochondriacus* L.

别名： 千穗谷、蛋白草。

subtropical regions in Central America and Southeast Asia, is a crop for food and feed, and a major food for the Indians in South and Central America. *Amaranthus hypochondriacus* L. can be used for ornamental flowers and plants, and can also be used as raw materials for foods such as bread, biscuits, pastries and car. China has developed dozens of health foods containing *Amaranthus hypochondriacus* L. ingredients.

7.5.1 Botanical characteristics

Amaranthus hypochondriacus L. is an annual herb with a tap root. Taproot is about 1.5~3.0 m. Lateral root distribution in soil of 20~30 cm. Stems are erect, growing to a height of 2~4 m and a diameter up to 3~5 cm. Stems are green or purple red, with many branches. Leaves are alternate, leaf margin is ovate-orbicular and green or purple red. It is spic state panicle, terminal or axillary, erect and branching. Flowers are small, unisexual and monoecious. Utricles are ovoid (Fig. 7.7). Seeds are spherical; color is purple black, brown or yellow color and has a glossy surface. Seed weight is 0.5~1.0 g per thousand.

7.5.2 Biological characteristics

Amaranth has a short growth period of about 120 days. It is a C_4 short-day plant, and prefers a warm and humid climate. Cold resistance is poor. The most suitable temperature: for germination is 12~24℃ and for growth is 24~26℃. The root system is buried deeply and well-developed to resist drought; Water logging may lead to root rot and death. Amaranth has board edaphic requirements, has tolerance of barren and salt. Loose and fertile soil is the best. Amaranth has high assimilation rate. Amaranth grows quickly under land favorable conditions.

7.5.3 Cultivation

Planting amaranth depletes soil fertility, and cannot be continuously cultivated. Rotating grain amaranth with wheat and beans is feasible. Amaranth seed is small and emerging is poor, so the seed bed requires careful cultivate, deep plowing and frequent raking to make plow layer scattered. *Amaran-*

籽粒苋原产于中美洲和东南亚热带及亚热带地区，为粮食、饲料、蔬菜兼用作物，在中南美洲为印第安人的主要粮食之一。籽粒苋也可作为观赏花卉，还可作为面包、饼干、糕点、饴糖等食品工业的原料，目前国内已研制出含有籽粒苋成分的保健品数十种。

7.5.1 植物学特征

籽粒苋属苋科苋属一年生草本植物。直根系，主根入土深达 1.5~3.0 m，侧根主要分布在 20~30 cm 的土层中。茎直立，高 2~4 m，最粗直径可达 3~5 cm，绿或紫红色，多分枝。叶互生，全缘，卵圆形，绿或紫红色。穗状圆锥花序，顶生或腋生，直立，分枝多。花小，单性，雌雄同株。胞果卵圆形（图 7.7）。种子球形，紫黑、棕黄、淡黄色等，有光泽，千粒重 0.5~1.0 g。

7.5.2 生物学特性

生育期短，约 120 d；喜欢温暖湿润气候，耐寒性差。最适温度为：发芽 12~24 ℃，生长 24~26 ℃。短日照 C_4 植物；根系发达，入土深，耐干旱，不耐涝，积水地极易烂根死亡。籽粒苋对土壤要求不严，耐瘠薄，抗盐碱，以疏松、肥沃的土壤最好；同化率高，生长速度快。

7.5.3 栽培技术

籽粒苋耗地，忌连作，可与麦类、豆类轮作、间种。种子小而顶土力弱，所以要求精细整地、深耕多耙，耕作层疏松。籽粒苋为高产作物，需肥量较多。4 叶期后及时间苗、定苗和中耕除草。刈割后要及时灌水施肥（主要以氮肥尿素为主）。鲜草产量 750~150 t/hm^2。

Fig. 7.7 Grain amaranth
图 7.7 籽粒苋

thus hypochondriacus L. is a high-yield crop, and requires a lot of fertilizer. When the plants grow to four leaf stage, timely thinning, final singling, intertillage and weeding is necessary. After mowing, timely irrigation and fertilization (mainly urea) is required. The fresh forage yield is 750~150 t/hm^2.

7.5.4 Feeding value

Stems and leaves of grain amaranth are nutritious tender with a delicate fragrance, and high in essential amino acids, especially lysine. It is good feed for cattle, sheep, horses, rabbits, pigs, poultry and fish. Seeds can be used as high-quality concentrates feed. Stems and leaves of Amaranthus contain protein quality suitable for supplemental feed; Nutritional value is equivalent to alfalfa and corn grain. *Amaranthus hypochondriacus* L. can be used both as green feed and be processed into silage, hay and hay powder, very palatable for livestock.

7.5.4 饲用价值

籽粒苋茎叶柔嫩，清香可口，营养丰富，必需氨基酸含量高，特别是赖氨酸含量极为丰富，是牛、羊、马、兔、猪、禽、鱼的好饲料。其籽实可作为优质精饲料利用，茎叶的营养价值与苜蓿和玉米籽实相近，属于优质的蛋白质补充饲料。籽粒苋无论青饲还是调制青贮、干草和干草粉均为各种家畜家禽所喜食。

7.6 Kochia prostrata 木地肤

Scientific name: *Kochia prostrate* (L.) Schrad.

Family: Chenopodiaceae

Kochia occurs mainly in desert and steppe; include the Northeast Region, Inner Mongolia, Shaanxi, Gansu and Ningxia in China and in the south of the Mediterranean and Siberia.

7.6.1 Botanical characteristics

Kochia prostrata belongs to Chenopodiaceae, it is a perennial shrub. Roots are stout and well-developed, buried to a depth of 2~2.5 m. Stems are highly branched, erect or inclined, reaching a height of 10~90 cm. Leaves are alternate, narrow, 8~25 mm long, and 1~2 mm wide. It is compound spike, and has 5 perianthes, and pubescence is dense (Fig. 7.8). Seed is ovate or suborbicular. Seed weight is about 1 g per thousand.

7.6.2 Biological characteristics

Kochia initiates spring growth early and has long growing season and strong drought resistance. Its taproot is thick and lateral roots are well-developed; roots grow rapidly at

学名： *Kochia prostrate* (L.) Schrad.

英文名： kochia prostrata

别名： 伏地肤

木地肤主要分布在荒漠地带及草原地区，我国东北、内蒙古、陕西、甘肃、宁夏等地区以及地中海南岸和西伯利亚等都有分布。

7.6.1 植物学特征

木地肤是藜科木地肤属多年生小灌木。根系粗壮发达，入土深达 2~2.5 m。茎多分枝，直立或斜生，高 10~90 cm。叶互生，狭条形，长 8~25 mm，宽 1~2 mm。复穗状花序，花被 5 片，密被柔毛（图 7.8）。种子卵形或近圆形。千粒重约为 1 g。

7.6.2 生物学特性

返青早，生长期长，具有很强的抗旱能力。主根粗大，侧根发达，根系生长速率很快，可超过地上部分 1~1.5 倍。

Fig. 7.8 Kochia prostrata
图 7.8 木地肤

a rate equivalent to 1~1.5 times the aerial parts. Due to its well-developed root system, it can absorb deep soil moisture. Stems and leaves have short and dense pubescence, which can limit transpiration. Kochia prostrata is fast growth, early maturity, high regeneration ability combine with winter hardiness, heat resistance, strong alkali resistance and wide ecological plasticity. It usually grows on sandy steppe and desert areas, loamy sand or gravel soil, and is occasionally associated with needle grass. Individuals may grow together, forming a synusia in the desert steppe.

由于发达的根系，能吸收土壤深处的水分；它的茎秆和叶片上密生短而密的柔毛，能节制蒸腾。木地肤生长快、成熟早，再生能力强，抗寒、抗热、耐碱性较强，具有广泛的生态可塑性，常生长在草原和荒漠区的沙质、沙壤质或砾石土壤上，偶与针茅草伴生，多单株生长，在荒漠草原形成层片。

7.6.3 Cultivation and utilization

Kochia has small seeds with weak emergence characteristics. Germination requires good soil moisture content. Optimal and timely sowing is critical.

Kochia grows early and retains winter branches well. It has high crude protein content, which can provide long-term use for livestock. Cattle, sheep and camels prefer to eat green and fresh stems, leaves and inflorescences. Sheep and goats can be fattened in autumn. Hay harvested before flowering can be used to feed poultry. Crude protein content of leaves is higher than stems. Kochia has been used as feeding for a long time and has high yield. In arid regions it is an excellent feed.

7.6.3 栽培和利用

种子较小，顶土能力弱，发芽要求较好的墒情，选地和适时播种是关键。

木地肤返青早，冬季残枝保存完好，粗蛋白含量较高，能被牲畜长期利用。青鲜茎叶、花序为牛、羊、骆驼喜食，秋季对山羊、绵羊有催肥的作用，开花期前收割制成的干草各种家畜都喜食。叶量丰富，叶中的粗蛋白质含量远较茎秆多。饲用时间长，产量高，是干旱地区的优良饲料。

7.7 Ceratoides lanata 驼绒藜

Scientific name: *Ceratoides lateens* (J.T.Gmel.)

Family: Chenopodiaceae

In China, Ceratoides lanata is mainly distributed in Inner Mongolia, Gansu province and other areas; More widely distributed in abroad. In Eurasia, from Spain to Siberia, extended to Iran and the arid areas of Pakistan.

学名: *Ceratoides latens*（J.T.Gmel.）

英文名: ceratoides lanata

别名: 优若藜

主要分布在我国内蒙古、甘肃等地区；国外分布较广，在欧亚大陆，西起西班牙，东至西伯利亚，南至伊朗和巴基斯坦的干旱地区都有分布。

7.7.1 Botanical characteristics

Ceratoides is subshrub, 30~100 cm tall, with many branches and stellate hairs. Leaves are alternate, and oval-lan-

7.7.1 植物学特征

半灌木，高 30~100 cm，多分枝，

ceolate, 1~2 cm long and 2~5mm wide. The leaf is acute or obtuse; the base is cuneate and entire. Flowers are unisexual and monoecious. Male flowers integrated spikes occur at the end of the branches; female flowers are axillary, without perianth. Two bracts form a tube, out of which there are 4 branches of hair whose length is equal to the tube. Utricles are elliptic or obovate. The seed has the same shape with utricle (Fig. 7.9).

有星状毛。叶互生，条形，长圆披针形，长 1~2 cm，宽 2~5 mm，先端尖或钝，基部楔形，全缘。花单性，雌雄同株，雄花在枝端集成穗状花序；雌花腋生，无花被；两苞片，全生成管，果期管外具 4 束与管长相等的长毛（图 7.9）。胞果椭圆形或倒卵形，种子与胞果同形。

Fig. 7.9 Ceratoides lanata
图 7.9 驼绒藜

7.7.2 Biological characteristics

Ceratoides is tolerant to drought, cold and barren soil. The crown is stout, often exposed at the surface, buried 60 cm around the taproot with poor lateral root development. Exposed roots above the soil outside are easily withered. Ceratoides has short-lived seeds. Germination capacity is generally retained for only 8~10 months. After one year, germination is poor. Ceratoides is a temperate xerophytic semi-shrub and its growth is suitable for arid and semi-arid climatic conditions. Suitable soils are brown soil, gray soil, and gray-brown or brown desert soil. It is mainly distributed in the desert and the desert steppe zone. In mountainous region, piedmont, mountain valley, river bank and sand dune of a foremen-

7.7.2 生物学特性

驼绒藜是一种抗旱、耐寒、耐瘠薄的半灌木。根颈较粗壮，常裸露于地表，主根入土 60 cm 左右，侧根发育较差，根系暴露在土外较多，容易枯死。驼绒藜种子的寿命较短，发芽能力一般只能保持 8 ~ 10 个月，超过一年则发芽较差。驼绒藜是一种温带旱生半灌木，适宜在干旱与半干旱的气候条件下生长，土壤为棕钙土、灰钙土、灰棕荒漠土或棕色荒漠土。主要分布于荒漠地带，也可进入荒漠草原地带。在上述地带的山地、山麓、山间谷地

tioned places, ceratoides forms a simple dominant species community or different types of ecological communities with other shrubs and perennial grasses.

以及河岸沙丘等处，形成单纯优势种群落，或与其他小半灌木及多年生禾草等共同组成不同类型的驼绒藜群落。

7.7.3 Cultivation technology

Ceratoides is difficult to establish successfully in desert areas. Transplanting is preferable. Soils with adequate water and fertilizer for spring sowing should be better. After seedling emergence, seedlings can moved to a cultivated field after in spring of the following year. In arid regions, sowing must take the advantage of rainy weather, when soil moisture content is suitable. Seeds should be mixed with wet sand, and then sown. Covering soil should not be too thick so as not to affect emergence.

7.7.3 栽培技术

驼绒藜在荒漠地区直播很难成功，以采取育苗移栽的方法为宜，即选择水肥充足的土壤，进行早春播种，待培育出幼苗，翌年春暖后移入大田栽培。若在干旱地区直播，必须趁阴雨天气，抢墒播种。通常先将种子与湿沙混合拌匀，然后播下，覆土不宜过厚，以免影响出苗。

7.7.4 Feeding value

Ceratoides is a subshrub with better forage quality. Camels, goats and sheep like to eat it anytime, especially in autumn and winter. Sheep and goats select young shoots, as well inflorescence and fruit. Horses feed on it during all seasons, but it has poor palatability for cattle. Ceratoides is nutritious, containing high crude protein, calcium and nitrogen-free extract. During the winter, it retains protein and the aboveground stems are well preserved in winter, which is valuable for livestock. Ceratoides's branches and leaves are very dense, good quality and palatability and the yield can reach 750~2250 kg of forage per hectare. It has drought resistance, so it has tame value in arid desert region. It can be used as feed and it is also valuable as a windbreak and sand for stabilization, and soil and water conservation. It's one of the most promising plants for improving natural grasslands.

7.7.4 饲用价值

驼绒藜为中上等饲用半灌木。骆驼与山羊、绵羊四季均喜食，秋冬最喜食，绵羊与山羊除喜食其嫩枝外，亦喜采食其花序及果实，马四季均喜采食，对牛的适口性较差。驼绒藜为富有营养价值的植物，它含有较多的粗蛋白质、钙及无氮浸出物，尤其在越冬期间，尚含有较多的蛋白质，且冬季地上部分茎保存良好，这对家畜冬季采食有一定的意义。驼绒藜的枝叶繁茂，每亩产草量可达50~150 kg，它的品质和适口性均较好，并具有耐旱能力，在干旱的荒漠地区有引种驯化的价值。驼绒藜除饲用外，还可用以防风固沙、保持水土，是改良天然草场最有前途的植物之一。

Chapter 8

Cereal Forage Crops

禾谷类饲料作物

8.1 Corn 玉米

Scientific name: *Zea Mays* L.

Most historians believe corn was domesticated in the Tehuacan Valley of Mexico. Corn is the most widely grown grain crop in the world, second only to wheat in cultivated area, and America has the largest area. Its position in livestock production far exceeds that in food production; approximately 75% of the crop is used for animal production. Countries with the highest yields per unit area are Austria, Italy, America and Canada. It is concentrated in the northeast, north and southwest mountainous areas in China, roughly forming a long narrow strip from northeast to the southwest.

8.1.1 Botanical characteristics

Corn is an annual crop, the fibrous root system are mostly concentrated in 30 cm soil layer, some reaching depths up to 150~200 cm(Fig. 8.1). The stem is commonly composed of 8~20 internodes of 18 cm length with alternating leaves. A leaf grows from each node, which is generally 6~10 cm in width and 70~100 cm in length. The plant height has much to do with varietal characteristics, normally 1~4 m in height (Fig. 8.2). It is monoecious. The apex of the stem ends in the tassel, an inflorescence of male flowers. When the tassel is mature and conditions are suitably warm and dry, anthers on the tassel dehisce and release pollen. Elongated stigmas, called silks, emerge from the whorl of husk leaves at the end of the ear (Fig. 8.3). They are often pale yellow and 18 cm in length, like tufts of hair in appearance. At the end of each is a carpel, which may develop into a "kernel" if fertilized by a pollen grain (Fig. 8.4). Grains can be flint, dent, intermediate or others (Fig. 8.5). Grain's color is mainly yellow and white.

学名： *Zea mays* L.

英文名： maize

别名： 玉蜀黍、包谷、包米、玉茭、玉麦、棒子

大多数历史学家认为玉米从墨西哥特瓦坎山谷流域驯化而成，是世界分布最广的作物之一，种植面积仅次于小麦，美国栽培最广。玉米是重要的粮饲兼用作物，它在牧畜业生产上的地位远远超过它在粮食生产上的地位，大约75%的玉米被用作饲料。玉米单产最高的国家是奥地利、意大利、美国和加拿大。我国集中分布于东北、华北和西南山区，大致形成一个从东北向西南的狭长形地带。

8.1.1 植物学特征

玉米为禾本科玉米属一年生草本植物。须根系，根多集中分布在30 cm的土层中，最深可达150~200 cm（图8.1）。茎由8~20个节间组成，每个节间长约18 cm。每个茎节上长出一个叶片，互生，叶片长度70~100 cm，宽6~10 cm。株高一般1~4 m，与品种有关（图8.2）。雌雄同株异花，雄花着生在植株顶部，为圆锥花序，成熟后雄穗上的花药张开，花粉散落；雌性花序着生在植株中部，为肉穗花序。从雌穗苞叶延伸出的柱头呈胡须状，也叫花丝，淡黄色，长约18 cm。授粉结实后发育成果穗（图8.3，图8.4）。籽粒有硬粒型、马齿型、中间型等多种（图8.5），籽粒主要为黄色和白色。

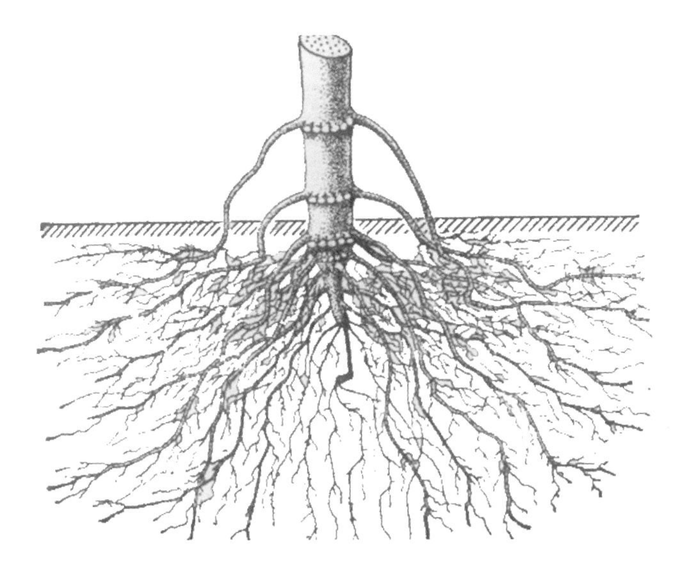

Fig. 8.1 The fibrous root system of corn
图 8.1 玉米的须根系

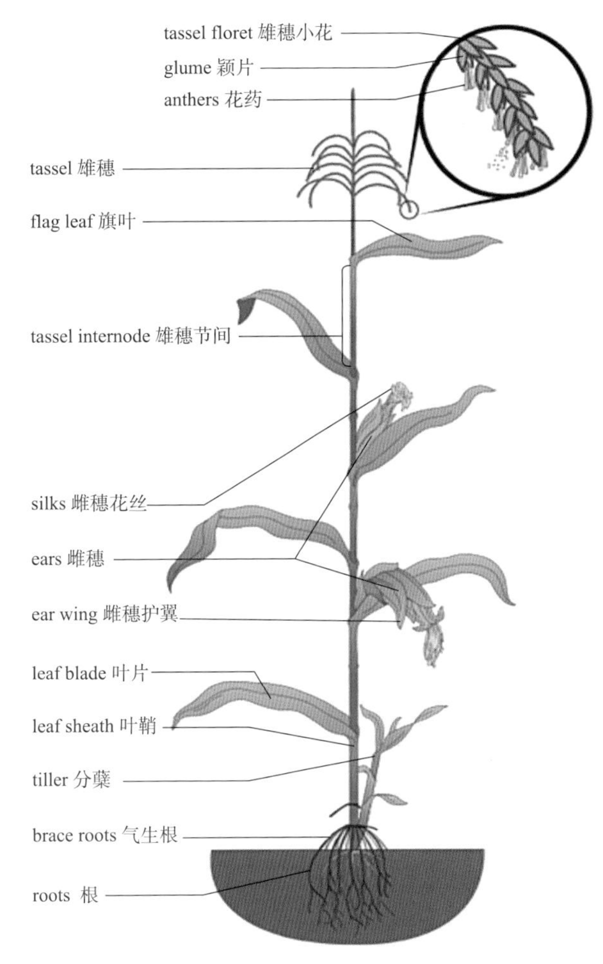

Fig. 8.2 Corn plant diagram and full-grown corn plants
图 8.2 玉米植株示意图和发育完全的植株

Fig. 8.3 Male (left) and female (right) inflorescence, with young silk
图 8.3 玉米的雌（右）雄（左）花序

Fig. 8.4 Male flowers and mature kernels
图 8.4 雄花和成熟籽粒

Fig. 8.5 Three various subspecies (from left to right: flint, dent and intermediate) related to the amount of starch
图 8.5 硬粒型、马齿型、中间型三个玉米变种

8.1.2 Biological characteristics

Corn is a thermophilic crop. It is divided into three types according to the growth period, which are early maturing (90~100 d), medium maturing (100~120 d) and late maturing (120~150 d). For early and medium type, effective accumulative temperature is 1 800~3 000℃, late maturing type is 3 200~3 300℃. Minimum temperature for seed germination is 6~7℃. Spring planting cannot be too early to avoid seedling frost damage. It is suitable for sowing when the temperature of 5~10 cm soil depth stabilizes at 10~12℃.

Corn requires plentiful water and fertilizer during the whole growth process. It mainly focuses on the root growth during the seedling stage and less water is needed. Soil water content of 60% of field capacity is adequate. Leaf growth at the jointing stage is very rapid, and ear start to differentiate. Soil moisture should be maintained at 70% ~ 80% of the field capacity. Heading and flowering stages are critical periods for adequate moisture for corn. Soil moisture should be maintained at 80% of field capacity, otherwise the pollen will dehydrate and silk will wither because of the high temperature (32~35℃)and low air humidity(less than 30%). During grain filling stage and maturation, both photosynthesis and nutrient transport require large amount of water. Soil water content should be kept at 70%~80% of the field capacity.

The nitrogen fertilizer requirement of corn is much higher than other cereal crops. Corn is short-day C_4 plant. It prefers to get more sun light, not shade tolerance. Planting should not be too dense to avoid barren stalk. Production can be successful under a wide range of soil conditions. Deep fertile soil with good structure, loosen, good permeability and rich in organic matter is preferable. The optimum pH is 5~8.

8.1.3 Cultivation techncIogy

Continuous cropping of corn should be avoided, but preceding cropping rotation is flexible. Corn should be planted in flat, well drained, deep and fertile soil. It should be plowed deeply and cultivated carefully. Varieties selection depends

8.1.2 生物学特性

玉米为喜温作物，根据生育期长短分为早熟品种（生育期 90~100 d）、中熟品种（生育期 100~120 d）和晚熟品种（生育期 120~150 d）3 种类型，早、中熟品种需有效积温为 1 800~3 000℃，晚熟品种则需 3 200~3 300℃。种子发芽最低温度 6~7℃，春玉米播种不能过早，防止幼苗霜冻，一般 5~10 cm 地温稳定在 10~12℃为适宜播种期。

玉米一生需水肥较多。苗期主要长根，需水较少，土壤含水量达田间持水量的 60% 即可；拔节期茎叶旺长，穗分化，土壤需维持田间持水量的 70%~80%；抽穗开花期，也是玉米的水分临界期，需维持田间持水量的 80%，否则出现高温“晒花”（温度达到 32~35℃，空气湿度小于 30% 时，花粉失水，花丝干枯）；灌浆成熟期，光合及养分运输均需大量水分，需要维持田间持水量的 70%~80%。

玉米对氮肥的需要量远高于其他禾本科作物。玉米属短日照 C_4 植物，喜光不耐阴，种植不宜过密，否则空秆多，降低产量。对土壤要求不严，以土层深厚肥沃，结构良好、疏松，通透性好，富含有机质的土壤为好。适宜 pH 5~8。

8.1.3 栽培技术

玉米对前作要求不严，忌连作，要选平坦、排水良好、土层深厚、肥力较高的地块种植，要深耕细耙。选用品种因栽培目的、气候、土壤条件而定。生产籽粒需选用成熟良好、高产稳产的中晚熟品种；青饲、青贮宜选用植株高大、茎叶繁茂、分枝多、叶直立的中晚熟品种。若做种用则选用穗大粒饱、排列整齐、符合特性的种。

播前晒种 2~3 d 可提高出苗率，

on the purpose of cultivation, climate and soil conditions. Grain production should choose high yield late-maturing varieties. Forage and silage production should use tall plants with lush stems stalk, plenty of branches, upright leaves in late-maturing varieties. If using as a seed, choose the seeds with big spike and grain bread, neatly arranged in line and match characteristics.

Before seeding, drying seeds in the sun for 2~3 d can increase germination rate and early emergence 1~2 d and reduce head smut. Seeding date depend on region and production purpose. In the north spring, corn can be seeded when soil temperature stabilize at 10~12℃ (in mid to late April). In northeast of China, the soil temperature should reach 8~10℃ or above (in mid May) when seeding. Corn can be monoculture or intercropping with shade tolerant legumes and potatoes. If it is cut for forage, the seeding rate is 75~90 kg/hm^2, 60~75 kg/hm^2 as silage and 45~60 kg/hm^2 as grain. The seeding depth depends on soil texture, soil moisture, climatic and seed size, generally for 4~6 cm, deeper to 6~8 cm in sandy soil or bad soil moisture, but no more than 10 cm. To get the highest emergence rate and maintain high yield, sometimes reseeding and re-planting is essential. It is necessary thinning the seedling when the corn plant has 3~4 leaves and final singling at 5~6 leaves of seedling stage. Corn should be intertilled and weeded 2~3 times. Prevention and controlling of plant diseases and insect pests (such as cutworms, mole crickets and grubs, etc.) is necessary. Control irrigation can promote the root growth and restrain the shoot development at emerging to jointing stage to strong the seedling, this period will last for one month for spring corn and 20 d for summer corn. For forage maize, a heavy application of base fertilizer followed with an appropriate application at jointing stage. The fertilizer scheme for silage corn: more fertilizer at jointing stage and less at tassel stage.

Harvest time varies based on cultivation purpose. For grain production, seeds should be harvested at dough stage, the bracts are dry and loose, and seeds are hard and shiny. If using for pigs, forage maize should be cut from jointing stage, feeding just after cutting; whereas for cattle: the moderate

并提早出苗 1~2 d，减轻丝黑穗病。播期因地域及栽培目的而异，华北春玉米在地温稳定在 10~12℃（4 月中、下旬）适宜播种，而在东北地区地温在 8~10℃以上最佳（5 月上、中旬）；可单播或与耐阴株矮蔓生豆科植物、马铃薯等间作套种；作青刈玉米时播种量为 75~90 kg/hm^2；青贮为 60~75 kg/hm^2；收获籽粒时 45 ~60 kg/hm^2；根据土质、墒情、气候条件、种子大小而决定播种深度，一般 4~6 cm，沙土或干旱时，应深至 6~8 cm，但不宜超过 10 cm。为了保证出苗率和高产，播种后根据情况可采取补种、移栽、适时间苗（3~4 真叶时）、定苗（5~6 真叶时），中耕除草 2~3 次，防治病虫害（地老虎、蝼蛄、蛴螬）等措施。此外，出苗至拔节期应该减少灌溉蹲苗促壮（春玉米为 1 月，夏玉米为 20 d）。青刈玉米应该重施苗肥、拔节肥；青贮玉米应该轻苗肥，重拔节肥，轻穗肥。

收获时间因栽培目的而异，收获籽粒的玉米应该在在蜡熟末期，苞叶干枯松散，子粒变硬发亮时收获；青刈玉米拔节后边割边喂（喂猪）或吐丝到蜡熟期收获（喂牛）；青贮玉米一般在乳熟至蜡熟期收获。

8.1.4 饲用价值

玉米籽粒是重要的能量精料，营养丰富，籽粒富含淀粉，还有胡萝卜素、核黄素、维生素 B_1、B_2 等。100 kg 玉米饲用价值相当于 135 kg 燕麦、125 kg 高粱或 130 kg 大麦，因此玉米被称为“饲料之王”。但玉米三酸不足（赖氨酸、甲硫氨酸、色氨酸），猪和禽不宜单独饲用，饲用应补充豆科饲料。青刈玉米味甜多汁，适口性好，是马、牛、羊、猪的良好青饲料。青

harvest time is silk to dough stage; harvest at the milky dough stage for silage use.

贮玉米品质优良，可大量贮备供冬春喂用。带穗青贮供冬春用，兼具干、青草的特点。青贮果穗喂母猪产仔率多，成活率高。日饲喂量青年牛为5~8 kg，冬季种母牛15~20 kg，母羊1.5~2 kg，占去势牛、羊日粮的70%。

8.1.4 Feeding value

Corn is an important energy feed which providing rich nutrition, such as starch, carotene, riboflavin, vitamin B_1 and B_2. The feed value of 100 kg corn kernal is equivalent to that of 135 kg oats, 125 kg sorghum or 130 kg barley. Corn is known as "the king of feeds", but three typical amino acids (lysine, methionine, tryptophan) are deficient. It would be better to mix feed with legumes to complement nutrition when feeding pig and poultry. Forage corn is sweet and juicy with good palatability and is good forage for horses, cattle, sheep, and pigs. Excellent quality corn silage can be stockpiled for winter and spring feeding. Silage with ear is good for winter and spring use; because it has both the feature of hay and green grass. Corn silage can increase farrowing rate and support a high survival rate. Typical daily feed quantifies include: young cattle 5 ~ 8 kg; wintering cows 15 ~ 20 kg; ewe 1.5~2.0 kg; up to 70% of diet DM for castrated bulls and sheep.

8.2 Oats 燕麦

Scientific name: *Avena sativa* L.

Oats are one of the most important cereal crop. Majority are planted in the Russia, followed by the United States, Canada, France, Australia, Poland, and Germany. It distributes mainly in the North of China where average annual temperature is 2~6℃ in the alpine zone. Oats are an important forage crop for Inner Mongolia, Qinghai, Gansu, Xinjiang and other major pastoral areas. Oats are divided into two types: with lemma and without lemma. Oats with a lemma are used as forage. The type without lemma is mainly used as food grain. There are two seasonal types, namely spring oats and winter oats, spring oats are mainly used as a feed grain.

学名： *Avena sativa* L.

英文名： oats

别名： 铃铛麦、草燕麦、莜麦

燕麦是重要的谷类作物之一。俄罗斯栽培最多，其次为美国、加拿大、法国、澳大利亚、波兰、德国等。在我国，燕麦主要分布于北方年均温2~6℃的高寒地带，是内蒙古、青海、甘肃、新疆等各大牧区的主要饲料作物。燕麦分带稃和裸粒两大类型。带稃燕麦为饲用，裸燕麦以食用为主。栽培上燕麦又分春燕麦和冬燕麦两种生态类型，饲用以春燕麦为主。

8.2.1 Botanical characteristics

Oats is an annual crop. Plant height is 80~120 cm, with tufted and erect stems. The fibrous root system is well developed; depths of 90~150 cm (Fig. 8.6). Oats tillers are prolifically. Stem is hollow and round with 4~9 internodes. An auxiliary bud grows on one side of node. Leaves are wide, long and flat; the surface of leaf has white powder. Panicles are open casual; spike-stalk is upright or drooping (Fig. 8.7). Spikelet grows in upper branches. Caryopses are fusiform. A short awn on the tip of lemma, some species has no awn.

8.2.1 植物学特征

燕麦为禾本科燕麦属一年生草本植物。株高 80~120 cm，丛生，茎秆直立。须根系发达，入土深达 90~150 cm（图 8.6）。分蘖较多。茎 4~9 节，圆形，中空，节部一侧着生有腋芽。叶片宽而长，平展，幼时表面被白粉。圆锥花序开散，穗轴直立或下垂（图 8.7），小穗着生枝顶端，颖果纺锤形，外稃具短芒或无芒。

Fig. 8.6 The full-grown oats
图 8.6 完全发育的燕麦

Fig. 8.7 Spikelet of oats
图 8.7 燕麦麦穗

8.2.2 Biological characteristics

Growth period of spring oats is 80~125 d (as short as 70 d); winter oats need more than 250 d. Oats are suitable for cold and humid climate but not high temperatures and arid conditions. Minimum temperature for seed germination is 3~4℃, optimum is 15~25℃. Mature plants are affected by frost damage at -5~-6℃. Flowering and seed development will be interrupted when temperatures reach 3~6℃. Oats are suitable for planting north of 38℃ isotherms. Growth period needs 1 300~2 100℃ accumulated temperature (≥5℃). Oats

8.2.2 生物学特性

春燕麦生育期为 80~125 d（短的只有 70 d），冬燕麦 250 d 以上；燕麦喜冷凉湿润气候，不耐高温干燥，种子发芽最低温度为 3~4℃，最适温度为 15~25℃，成株 -5~-6℃受冻害，3~6℃开花结实受阻，适宜在 38℃等温线以北种植。生育期需活动积温（≥ 5℃）1 300~2 100℃。燕麦属长日照植物，较大麦耐阴，可与豆科牧草混播；燕麦需

are long day plants. Oat has better shade-tolerant than wheat, can be mixed sowing with legumes forage. Oats need reliable moisture and are suitable for planting in regions of 400~600 mm of annual rainfall. Oats have broad soil requirements, but do well in low-lying land with high moisture and organics rich loam. Optimum pH is 5.5 ~ 8. 0.

8.2.3 Cultivation technology

Oats should not be used in continuous cropping pattern. They are sensitive to nitrogen fertilizer. Beans, cotton, corn, potato or sugar beets are good preceding crops. Oats have a short growth period. Soybeans, corn, sorghum or root crops may be grown following oats. The way of sowing is drilling. Deep and fertile soil and careful seedbed preparation are needed. Carefully weed control and timely fertilization and irrigation are required. At tillering stage, it should be weeding control once, and pay more attention to irrigation at tillering, jointing, and booting stages. Nitrogen-based fertilizer is also used as top application.

8.2.4 Feeding value

Oats grains with 12%~18% of protein content, some up to 21%, high content of amino acids, is an excellent fodder. Forage oat stalks are soft, palatable with high protein digestibility; the feeding value is higher than that of rice, wheat, and millet straw. Forage oat can be used to make silage, hay and for grazing. Silage has fragrant smell and is high quality forage for animals during the season of lack for green feed.

Study showed that feeding the oat silage which is processed by full maturity plant for cows and beef cattle, 50% concentrate can be saved. It's a low cost and high value way. Oat bars and lemma is soft, the crude protein is higher than that of the other wheat crops. Oat has low crude fiber; it can be used by to cows and horses.

水较多，适宜在年降水量 400~600 mm 的地区种植。对土壤要求不严，在黏重潮湿的低洼地上表现好，以富含有机质的壤土最为适宜。最适 pH 5.5~8.0。

8.2.3 栽培技术

燕麦忌连作，对氮肥敏感，以豆类、棉花、玉米、马铃薯、甜菜等为前作最好。燕麦生育期短，之后可复种大豆、玉米、高粱和块根作物。播种时应选纯净粒大的种子，一般采用条播，要求土层深厚、土壤肥沃、整地精细。苗期注意中耕除草、及时追肥和灌水。分蘖前后应中耕除草一次。在分蘖、拔节、孕穗期应注意浇水，追肥以氮肥为主。

8.2.4 饲用价值

燕麦籽粒含蛋白质 12%~18%，高者达 21% 以上，氨基酸含量高而全，是优良的精饲料。秸秆的饲用价值高于稻、麦、谷等秸秆。青刈燕麦茎秆柔软，适口性好，蛋白质可消化率高、营养丰富，可青饲、青贮、制干草和放牧。青贮饲料质地柔软、气味芬芳，是畜禽冬春季节青饲料缺乏时期的优质青饲料。据报道，用成熟期燕麦调制的全株青贮饲料喂奶牛和肉牛，可节省 50% 精料，成本低，价值高。燕麦秆和稃壳质地软，较其他麦类的秆壳蛋白质含量高，粗纤维低，可喂牛、马。

8.3 Barley 大麦

Scientific name: *Hordeum vulgare* L.

Barley, as one of the most important cereal crops in the world, originated in central Asia and China and has a long history of cultivation. The cultivated area ranks sixth in the world among cereal crops. Barley is the general name of bran barley and naked barley. Traditionally barley is referred to the bran barley. Naked barley is generally regarded as highland one, known as the mainly food crop in Qinghai and Tibet rangeland area. Winter barley and spring barley represent the two types by the planting area. The main areas of winter barley are in the provinces that along the Yangtze River, such as Jiangsu, Hubei, Sichuan and Zhejiang. Spring barley is mainly grown in the part of the northeast, Inner Mongolia, Shanxi, northern Xinjiang and Qinghai-Tibet Plateau. Barley is widely adapted, can endure infertile soil, with a short growth period, very nutritious and palatable. Barely is one of the important grain-forage crops.

学名： *Hordeum vulgare* L.

英文名： barley

别名： 有稃大麦、草大麦

大麦起源于中亚细亚及中国，栽培历史悠久，为世界上最主要的谷类作物之一，栽培面积居谷类作物第 6 位。大麦是有稃大麦和裸大麦的总称，习惯上所称的大麦为有稃大麦。裸大麦一般是指青稞，是我国青海、西藏牧区的主要粮食作物。根据其栽培地区的不同，我国的大麦可分为冬大麦和春大麦，冬大麦的主要产区为长江流域各省，如江苏、湖北、四川、浙江等地；春大麦则分布在东北、内蒙古、青藏高原和山西及新疆北部。大麦适应性强、耐瘠薄，生育期较短，营养丰富，饲用价值高，是世界上主要的粮饲兼用作物之一。

8.3.1 Botanical characteristics

Barley is an annual crop. Plant height is about 1m (Fig. 8.8).It can produce lots of tiller, the fibrous roots reach a depth up to 1 m, mainly distributed in 30~50 cm soil layer. The stem consists of 5~8 nodes, there are some lateral buds growing on the nodes. Leaves are lanceolate, wide and light green which covered with white powder when immature, with big size auricles and ligules. Barely have spike inflorescence. According to characters of development and seed setting of the barley spikelet, barley can be divided into two subspecies: six-rowed and two-rowed barley. Six rowed barley has a short spike internodes and a high seed setting density. Multi-rowed has two types of with or without awn (Fig. 8.9). Only two arrays of the spikelet can develop into seeds, so it is called two-rowed barley (or beer barley).

8.3.1 植物学特征

大麦为禾本科大麦属一年生草本植物。株高 1 m 左右（图 8.8），能产生多级分蘖。须根入土深可达 1 m，主要分布在 30~50 cm 的土层中。茎由 5~8 节组成，节具潜伏腋芽。叶为披针形，宽厚，淡绿色，幼嫩时具白粉。叶耳、叶舌较大。穗状花序。根据大麦小穗发育特性和结实性，可分为六棱大麦和二棱大麦。六棱大麦穗轴节片短，着粒密，每个穗轴节上对生的两个三联小穗均能发育结实(六棱大麦中的裸大麦是青稞）（图 8.9）；二棱大麦每穗轴节上 3 个小穗，仅中间一个小穗正常发育，两行结实，故称二棱大麦（啤酒大麦）。

Fig. 8.8 Barley stand and seeds
图 8.8 大麦田和种子

Fig. 8.9 Subspecies of barley[left: two-row type; middle to right: multi-row type (with and without awn)]
图 8.9 大麦亚种（二棱大麦、六棱大麦（有芒和无芒））

8.3.2 Biological characteristics

Growth period is 75~110 d for spring barley, 160~250 d for winter barley. It prefers cold climate. Naked barley has better cold tolerance than hulled barley. The lowest germinate temperature of spring barley is 3~4℃. The optimum germination temperature is about 20℃. Seedlings can tolerant low temperatures to some extent, but flowering stage is cold sensitive. Temperature for barley mature period should be above 18℃. Barley is drought tolerant, suitable for areas with 400~500 mm annual rainfall. Less water is needed at seedling

8.3.2 生物学特性

春大麦的生育期为 75~110 d，冬大麦 160~250 d。大麦喜冷凉气候，裸大麦耐寒力强于有稃大麦，春大麦最低发芽温度 3~4℃，最适温度为 20℃左右，幼苗耐低温，但花期不耐寒，成熟期温度须高于 18℃。大麦耐旱，适于在年降水量 400~500 mm 的地方种植，苗期需水少，分蘖后逐渐增加，抽穗开花期最多。大麦为长日照 C_3 植物，喜光。

stage and gradually increase after tillering stage, heading and flowering stage need most water. Barley is long-day C_3 plant. It prefers full sunlight. Barley is adapted to a range of soil conditions. It can tolerant saline-alkaline but not acid. Barley does well in well-drained, fertile, neutral sandy loam or clay loam soil. The optimum pH is 6~8.

8.3.3 Cultivation technology

Barley has a short growth period, and is well suited to both crop rotation and continuous cropping. Land for barley production should be flat, fertile soil and with good drainage. Seeds to be planted should be large plump grain with high purity and high germination rate. Drill seeding is used most frequently. Row space is generally 15~30 cm. When used as forage, it should be slightly narrow with seeding rate for 150~225 kg/hm^2 and sowing depth at 3~4 cm. After sowing, soil should be cultipacked. Barley is a fast-growing and narrow-planting crop which does not need thinning. Fertilization and irrigation should be timely. In late growth stages, weed control is important. Pests and diseases should be controlled and remove diseased plants.

8.3.4 Feeding value

Barley grain is concentrated feed for livestock and poultry, the nutritional value is slightly lower than corn. Sprouted barley grain can be feed to breeder stock and young animals as a vitamin supplement. Barley straw is better than wheat straw, or corn straw as coarse fodder. Green forage barley is an excellent, succulent feeds livestock. Silage barley stems are tender, juicy and rich nutritious. It is high quality coarse fodder for cattle, horses, pigs, sheep, rabbits and fish.

大麦对土壤要求不严，耐盐碱不耐酸，以排水良好、肥沃、中性沙壤土或黏壤土为好。最适 pH 为 6~8。

8.3.3 栽培技术

大麦生育期短，耗地轻，在轮作中可灵活安排，可连作。应选地势平坦、土质肥沃、排水良好的地块。选粒大饱满、纯净度高、发芽率高的种子播种，多用条播。行距 15~30 cm，做青刈则稍窄，播种量 150~225 kg/hm^2，播深 3~4 cm，播后镇压。大麦为速生密植作物，无需间苗，应及时追肥和灌水，生育后期注意防除杂草。对病虫害应及时防治，并拔除病株。

8.3.4 饲用价值

大麦籽粒是畜禽的精饲料，营养价值低于玉米。大麦发芽饲料是种畜和幼畜的维生素补充饲料。大麦秸秆也是优于小麦秸秆、玉米秸秆的粗饲料。青刈大麦是畜禽优良的青绿多汁饲料。青贮大麦茎叶柔嫩多汁，营养丰富，是牛、马、猪、羊、兔和鱼的优质粗饲料。

8.4 Sorghum 高粱

Scientific name: *Sorghum bicolor* (L.) Moench

Sorghum originated in the tropics and is one of the most traditional crops with a cultivation history more than 4 000 years. It is an important food and fodder crop in northern

学名： *Sorghum bicolor* (L.) Moench

英文名： sorghum

别名： 蜀黍、茭子、芦粟、秫秫

高粱原产于热带，是最古老的作物

China. Sorghum is one of the four major cereal crops in the world, only exceeded by wheat, rice and corn. Grown mainly in Asia, India has the largest cultivated area, followed by United States of America, Nigeria and China. It is also an important raw material providing for starch, brewing and alcohol industries. Green stems and leaves can be used for silage; straw for papermaking, building and weaving. It tolerates drought, water logging and salinity. Sorghum produces high yields, widely adapted, versatile, easily cultivated and is an excellent grain and forage crop.

8.4.1 Botanical characteristics

Sorghum is an annual cereal crop. Short varieties are 1.0~1.5 m tall, standard varieties are 1.5~2.0 m tall and tall varieties are 2.0~3.5 m (Fig.8.10). Stems are erect, solitary or tufted. The fibrous root system extends as deep as 1.5~2.0 m. External stem consists of thick and film cells, so it's relatively hard and rough. Stems consist of many nodes and some lateral buds grow on the nodes. Large leave have a central midrib. Flowers are panicles.

8.4.2 Biological characteristics

Sorghum is thermophilic and heat-resistant but does not tolerate cold. The minimum germination temperature is 8~10℃. Optimal growth temperature is 20~30℃. It is more drought resistant than corn. Sorghum has high water use efficiency under drought conditions. Sorghum is a short day C_4 plant has a high fertilizer demand, not critical to soil conditions, but grows well in soil which has rich organic matter. Good water logging and salt tolerance. The optimum soil pH is 6.5~8.0. Sorghum is often regarded as a pioneer in the saline-alkali soil crops.

8.4.3 Cultivation technology

Sorghum is not suitable for continuous cropping. Requirement for the preceding crop is flexible. A good seedbed preparation is very important as seedlings cannot resist weeds competition. Fertilizer requirements are high. Seeds should be well filled and with a high germination rate, drill seeding by a wide row. Before sowing, seeds should be placed

之一，迄今已有 4 000 年的栽培历史。它是我国北方重要的粮食作物和饲料作物。高粱是世界四大谷类作物之一，其地位仅次于小麦、水稻和玉米。主要分布在亚洲，印度的栽培面积最大，其次是美国、尼日利亚和我国。它也是淀粉、酿造和酒精工业的重要原料。它的青绿茎叶可作青贮饲料，秸秆是造纸、建房和编织的原料。它具有抗旱、耐涝、耐盐碱能力，产量高，适应性强，用途广，栽培容易，是一种优良的粮饲兼用作物。

8.4.1 植物学特征

高粱为禾本科高粱属一年生草本植物。矮茎种 1.0~1.5 m，中茎种 1.5~2.0 m，高茎种 2.0~3.5 m。直立、单生或丛生。须根系，入土深达 1.5~2.0 m（图 8.10）。茎外部由厚膜细胞组成，较坚硬，品质粗糙。茎多节，节上具潜伏腋芽。叶片肥厚宽大，中央有中脉。圆锥花序。

8.4.2 生物学特性

高粱属喜温植物，抗热不耐寒（发芽最低温度为 8~10℃，生长最适温度为 20~30℃）。抗旱性远比玉米强，在干旱条件下能有效利用水分。高粱为短日照 C_4 植物，需肥多，对土壤要求不严，但以富含有机质的土壤为宜。耐涝和抗盐碱能力很强，适宜的 pH 为 6.5~8.0，常作为盐碱地先锋作物。

8.4.3 栽培技术

忌连作，对前作要求不高。苗期不耐杂草，需精细整地。需肥较多。选粒大饱满、发芽率高的种子进行宽行条播。播种前晒种 3~4 d，必要时药剂拌种，及时中耕除草、追肥灌水。

Fig. 8.10 Sorghum crop
图 8.10 高粱

in the sun and dried for 3~4 d, dressing seeds with special drug agent when necessary. Weed control, fertilization and irrigation must be timely.

8.4.4 Feeding value

Generally speaking, feeding value and palatability of sorghum are inferior to corn. Sorghum grain is very nutritious and benefit to the piglets weaned earlier, grow rapidly and healthy; it also can improve the labor ability of horse. It contains tannin anti-nutritional factors, with antidiarrheal effect; too much sorghum feed can cause constipation. Sorghum's palatability is not so good as corn, and short of digestible protein, lysine and tryptophan, should be compatible with leguminous fodder while feeding. Young stems and leaves contain large amounts of hydrocyanic acid (HCN), which can lead to poisoning. The HCN content of common varieties is higher than that of sweet stem varieties. HCN levels are higher in young than mature plants, in leaves than in stems, upper leaves are higher than that in lower leaves; branches are higher than the main stem, sunny days higher than cloudy wet days. Fresh sorghum forage has higher HCN than that of hay or other stored sorghum forages. Selection has been made for

8.4.4 饲用价值

总体来说，高粱的食用、饲用价值及适口性均次于玉米。籽粒营养丰富，喂仔猪使其断乳早、增重快，令猪体健壮，且能提高马的使役力。但高粱含有单宁抗营养因子，具有止泻作用，多喂可引起便秘。此外，缺乏可消化蛋白质及赖氨酸、色氨酸，饲喂时应与豆类饲料配合饲喂。高粱幼嫩茎叶含有大量氢氰酸（HCN），多食易中毒。一般HCN含量普通品种高于甜茎种，幼嫩时期高于成株期，叶高于茎，上部叶高于下部叶，分枝高于主茎，晴天高于阴湿天，新鲜高于干草及贮藏。青贮选用甜茎种，茎叶青绿，可鲜喂或青贮，调制干草不易（茎干粗水分不易蒸发）。调制青贮饲料，茎皮软化，适口性好，消化率高，是家畜的优良贮备饲料。

sweet stalks as silage variety, because green stems and leaves can be fresh feed or silage. Hay processing is difficult, because much water is kept in the strong stems, which reduce the drying process. Silage is more workable for sorghum, softens stem, good palatability, high digestibility, and is an excellent reserve feed for livestock.

8.5 Foxtail millet 谷子

Scientific name: *Setaria italica* (L.) Beauv.

Millet originated in China and has a long history of cultivation. It is mainly produced in north of the Huai River. North and northeast account for almost 60% of the total area in China. It is also grown in India, Pakistan and Russia.

8.5.1 Botanical characteristics

Millet is an annual cereal crop. Plant height is 100~150 cm and clumped, green or purple-green. Well developed, numerous fibrous roots extend deeply in the soil. Aerial root grows out from 1~2 nodes above ground. It is round stems and spicate, thick and short heads for forage millet, but thick and long for edible one (Fig.8.11). Seed is a caryopsis. Seeds are round and small. Thousand seeds weight is 1.7~4.5 g.

8.5.2 Biological characteristics

Millet prefers warm temperatures and is not clod resistant. Average temperature is about 20℃ during growth period. It has good drought resistance, which can be planted in the areas with 400 mm annual precipitation. Millet prefers full sunlight and is a short-day plant. It cannot tolerant shade well, especially in the heading stage. Soil requirements for foxtail millet are not strict, but deep fertile soil is preferable. The optimum pH is 5~8. When the soil pH is more than 8.5, it must be acidified before planting.

8.5.3 Cultivation technology

Millet is not suited to continuous cropping. Beans, wheat, corn and other crops are acceptable preceding crops in rota-

学名： *Setaria italica* (L.) Beauv.

英文名： foxtail millet

别名： 粟、小米

谷子原产于中国，栽培历史悠久。现主要产于我国淮河以北地区，华北和东北的种植面积将近占全国总面积的60%。另外，在印度、巴基斯坦、俄罗斯等地均有种植。

8.5.1 植物学特性

禾本科狗尾草属一年生草本。株高100~150 cm，丛生，全株呈绿色或者绿紫色。须根发达，根量多且入土较深，地面上有 1~2 节产生气生根。茎圆形，穗状圆锥花序。饲用谷子穗大而短，食用谷子穗大且长(图 8.11)。果实颖果，种子圆形细小，千粒重 1.7~4.5 g。

8.5.2 生物学特性

谷子喜温不耐寒，生育期平均气温在 20℃左右。耐旱性比较强，素有旱谷子之称，在年降水量 400 mm 的地方都能旱作。谷子为喜光短日照植物，耐阴性较差，尤在穗期更为敏感。谷子对土壤要求不严，但以土层深厚、土壤有机质含量较高的壤土和沙壤土为宜。适宜的土壤 pH 为 5~8，超过 8.5 时需改良后才能种植。

Fig. 8.11 Foxtail millet
图 8.11 谷子

tion. Nitrogen and phosphorus needs are highest during boot stage. From jointing stage to boot stage, potassium are mostly needed. Therefore, irrigation combined with fertilization should be satisfied from the jointing stage until heading. Dry land fertilization should be done after rain.

8.5.4 Feeding value

Grain millet seed should be harvested immediately when completely ripe. At this time, the straw is still valuable; mature seeds loss are limited. For forage use, it is appropriate to cut millet from heading to flowering to achieve high yield and high quality forages. In addition to green feed, foxtail millet can also be made into hay or as a raw material for silage. Foxtail millet may have straw chestnut virus. It is palatable for horse, but it may result in kidney damage or arthroncus. Mix-feeding with other feed can reduce the risk effectively.

8.5.3 栽培技术

谷子最忌连作，前茬以豆类、麦类、玉米等作物为宜。孕穗期是谷子一生中需要氮磷最多的时期，拔节期到孕穗期是谷子需要钾最多的时期，为此，拔节和抽穗阶段应该结合灌溉进行施肥，旱作地应该借助降雨施肥。

8.5.4 饲用价值

收籽谷子应在完熟初期收割为宜，此时谷草尚有价值，且可避免成熟种子落粒损失。青饲用以抽穗至开花期进行刈割为宜，不仅产量高，而且品质好；除青饲外，还可调制成干草或作青贮料。需注意的是，谷子秸秆含栗毒，马喜食，但大量采食对肾脏有害，会使关节肿胀，若与其他饲草料搭配则可有效降低危害。

8.6 Rye 黑麦

Scientific Name: *Secale cereale* L.

Rye (*Secale cereale*) is a grass grown extensively as a grain, a cover crop and as a forage crop, mainly distributed in the temperate zone and the frigid area of Eurasia. Germany, Poland, Russia, Belarus and China represent the top five big producers in the world at current. China's cultivated area is small, mainly located in Heilongjiang, Inner Mongolia and Qinghai, Tibet and other Alpine areas and high altitude mountains.

8.6.1 Botanical characteristics

Rye is an annual herbaceous plant with developed fibrous roots, root depth up to 1.0~1.5 m. Erect and stout stems, about 1.1~1.3 m high, high tillering ability, it can produce 30~50 branches per plant. Flat blade, length 10~ 20 cm, width 5~10 mm.The spikes inflorescence at the stem terminal is 5~15 cm length (Fig. 8.12), generally 2 florets in each spikelets , alternate, 15 mm in length and lower florets seed setting but top floret sterile. The caryopsis is slender and ovate, reddish brown or dark brown (Fig. 8.13).

8.6.2 Biological characteristics

Rye is suitable for growing in cold areas. There are two types, winter type and spring type. In alpine regions only spring rye can be grown, but in warm areas either can be cultivated. Rye has excellent cold tolerance capability; it can survive at -25℃. It can overwinter successfully as low as -37℃ under snow. It is intolerant of high temperatures and water logging. Soil requirements are not strict, but it grows well in sandy loam soil rather than in saline soils. Rye is barren tolerant; however, adequate soil nutrients can increase yield, improve quality, and enhance regeneration.

8.6.3 Cultivation technology

In North China and other relatively warm regions, typical silage crops include corn, rye, sorghum, millet, and

学名： *Secale cereale* L.

英文名： Rye

别名： 粗麦、洋麦

黑麦是一个广泛种植的谷类作物、保护作物或饲料作物，主要分布在欧亚大陆的温带和寒带。目前世界上栽培面积最大、产量最高的国家有德国、波兰、俄罗斯和白俄罗斯等。中国栽培面积较小，主要分布在黑龙江、内蒙古和青海、西藏等高寒地区与高海拔山地。

8.6.1 植物学特征

黑麦为禾本科黑麦属一年生草本植物，须根发达，入土深度 1.0~1.5 m。茎秆直立粗壮，高 1.1~1.3 m，分蘖力强，30~50 个分枝。叶片扁平，长 10~20 cm，宽 5~10 mm。穗状花序顶生，长 5~15 cm（图 8.12），小穗含 2 朵小花，互生，长 15 mm，下部结实，顶部小花不育。颖果细长呈卵形，红褐色或暗褐色（图 8.13）。

8.6.2 生物学特性

黑麦喜冷凉气候，有冬性和春性两种，在高寒地区只能种春黑麦，温暖地区两种都可以种植。黑麦的抗寒性强，能忍受 -25℃的低温，有雪时能在 -37℃低温下越冬，它不耐高温和湿涝，对土壤要求不严格，但以沙壤土生长良好，不耐盐碱。黑麦耐贫瘠，但土壤养分充足时产量高、质量好、再生快。黑麦再生能力较强，在孕穗期刈割，再生草仍可抽穗结实。

8.6.3 栽培技术

在华北及其他较温暖的地区，黑

Fig. 8.12 Ear of rye
图 8.12 黑麦麦穗

Fig. 8.13 Rye grains
图 8.13 黑麦籽粒

soybeans. After the previous crop harvest, seedbed for rye should be prepared by ploughing the stubble with a round disk, followed by organic fertilizer application, plowing, and packing.

麦一般为玉米、高粱、谷子、大豆的后作。前茬作物收割后，用圆盘耙灭茬，然后施有机肥，耕翻，镇压，播种。

8.6.4 Feeding value

Rye is high in protein and minerals with overall high nutritional value. In particular, rye bran contains extremely rich fructan which can regulate the intestinal flora balance, lower blood cholesterol and reduce postprandial blood sugar. At the same time, it can increase gastrointestinal motility, promote the excretion of harmful substances, and has a good effect on the prevention of intestinal diseases. Rye has a large number of leaves; stems are soft and good palatability. It is high quality forage for cattle, sheep, and horse. Rye grains are good concentrate feed for pigs, chickens, cows and horses. The dairy industry has developed rapidly in recent years, as forage and silage in north of China, rye plays an important role and has good effect.

8.6.4 饲用价值

黑麦富含蛋白质、矿质元素及多种微量元素等，具有很高的营养价值和饲喂价值。尤其黑麦麸，含有极为丰富的果聚糖等，具有调节肠道菌群平衡、降低血中胆固醇、降低餐后血糖的作用，同时还能增加胃肠道的蠕动，促进有害物质排出，对预防肠道疾病具有良好的作用。黑麦叶量大，茎秆柔软，营养丰富，适口性好，是牛、羊、马的优质饲草。黑麦籽粒是猪、鸡、牛、马的精料。近几年城市奶牛业发展较快，北方广泛用黑麦做青饲、青贮，效果很好。

Chapter 9

Forages of Tuberous Root, Tuber and Melon Vegetables 根茎瓜类饲料作物

9.1 Sweet potato 甘薯

Scientific name: *Ipomoea batatas* L.
Family: Convolvulaceae

The origin and domestication of sweet potato is thought to be in tropical regions of Mexico and northwestern region of South America. In the 15th century it was introduced to Europe, Asia and Africa and introduced into China in 1594. Sweet potato is a worldwide crop; China is the largest sweet potato production country that contributes nearly 80% of the total output.

Sweet potato is an important food, feed, industrial raw materials and raw materials of new energy. Sweet potato consumption in China has gone through 3 stages, in 20th century it was mainly food-oriented in 1950s—1960s; in 1980s mainly sweet potato was used as forage, edible and processing material, in 1990s it was process-oriented, food and forage. About half of the sweet potato is used for livestock feed.

9.1.1 Botanical characteristics

The sweet potato is a dicotyledonous plant, perennial in tropical zones, annual in temperate areas. Seed propagation is primarily used in breeding. Vegetative reproductive by cuttings of stems are used in the field production. Sweet potato roots consist by tuberous roots, poorly developed root and undeveloped fibrous roots according to the development status, among which tuberous root are the main harvest organ (Fig. 9.1). The stems of sweet potato are overgrown and slender, and internodes can produce root. The varieties are divided into three types according to the stem length: long vine type (spring vine is more than 3 m and summer vine is more than 2 m), short vine type (spring vine less than 1.5 m, summer vine less than 1 m) and intermediate vine type. For the long and intermediate vine types, stems and leaves can be harvested as green feeds and are the first choice for forage production (Fig. 9.2). Tuberous roots are mainly harvested from the short vine type. Leaves are heart shape and have

学名： *Ipomoea batatas* L.
英文名： sweet potato
别名： 红薯、地瓜、红苕

甘薯原产于南美洲西北部和墨西哥的热带地区，15 世纪传入欧洲，后传入亚洲和非洲，1594 年传入我国。甘薯在世界范围内分布很广，但主要产区在亚洲，占世界的 90%，中国是世界上最大的甘薯生产国，占世界总产出的 80% 以上。

甘薯是重要的粮食、饲料、工业原料及新型能源原料。我国甘薯的消费经历了 3 个过程，20 世纪 50—60 年代主要以食用为主，80 年代饲用、食用和加工并重，90 年代以后以加工为主，食、饲兼用，约 50% 的甘薯用作饲料。

9.1.1 植物学特征

甘薯属于旋花科甘薯属蔓生草本植物。在热带为多年生植物，温带为一年生植物。种子繁殖常在育种中采用。大田生产一般采用茎秆扦插，进行无性繁殖，甘薯的根均为不定根，根据发育状况分为块根、柴根和须根，其中块根是主要的收获器官(图 9.1)。甘薯的茎蔓生、细长，茎节处可生根。根据其茎蔓的长短将其分为长蔓型（春蔓 3 m 以上，夏蔓 2 m 以上）、短蔓型(春蔓小于 1.5 m，夏蔓小于 1 m) 和中间型 3 种。其中长蔓型、中间型以收获茎叶做青饲料为主，为饲用首选类型（图 9.2）；短蔓型以收获块根为主。单叶心形具长柄，无托叶，叶缘分全缘、深裂和浅裂。花单生或若干朵聚成聚伞花序。花冠漏斗状，花紫色、淡红色、白色（图 9.3）。果

long petiole and estipulate. Leaf margin is entire or incision. Flowers are solitary or several flowers cluster into cymes. Its corollas are funnel-shaped, white, and purple or pink (Fig. 9.3). Fruits are spherical or flat spherical.

实球形或扁球形。

Fig. 9.1 The tuberous root of sweet potato
图 9.1 甘薯的块根

Fig. 9.2 The vine of sweet potato
图 9.2 甘薯的茎蔓

Fig. 9.3 The flower of sweet potato
图 9.3 甘薯的花

9.1.2 Biological characteristics

Sweet potato prefers warm climate and abundant sunshine, 130 ~ 150 frost-free days are required to reach maturity. Sweet potato is drought tolerance. Its transpiration coefficient is about 300. Soil moisture should maintain at 65% ~ 70%. If sweet potato suffers from drought stress in early growth stage, leaves will have grown poorly, and tuberous root formation will be hindered, and easily grow into thin roots; if the drought stress happened in the middle growth stage, it will influence the stems and leaves growth, tuberous root enlargement will be slowly, and finally result in a low yields. If drought stress appears at the later growth stage, leaves will premature senescence and yields will be very low. Conversely, with excessive soil moisture, stems and leaves are overgrowth and the sweet potatoes are abnormity and not suitable for storage. As the saying goes, "drought makes long and thin roots, wetness makes fibrous roots, while the middle makes tubers." Adaptability of sweet potato to soil conditions is widely, and it is suitable to growth in sandy soils with deep, loosens and moist soil which has good ventilation and organic matter.

9.1.3 Cultivation technology

Cultivation technique of sweet potato is seedling transplantation. Seedlings are usually grown about one month before field planting. Variety selection: great disease resistance, vigrous growth, high-yielding and good quality. The seedbed should be selected in leeward, sunny, well-drained, close to the water, and no rodent pests, manageable. Based on available heat sources, the seedbed can be divided into different heating types: artificial heat, bio heat, stuffed hot and solar heat. Nursery temperatures should be "high-medium-low" model. High temperature (38℃ or so) is usual in the early period of germination, and then decrease to 31℃, after emergence the temperature should be decreased to 25 ~ 28℃. The seedlings should be fully watered 5 to 6 days before cutting and then stop watering to restrain the aboveground growth of leaves and stems and promote the growth of roots, then reducing temperature to 20℃ to hardening of seedlings three

9.1.2 生物学特性

甘薯为喜温喜光作物，无霜期130~150d的地区才能种植。甘薯耐旱，蒸腾系数约为300。土壤湿度在65%~70%为宜。生长前期遇干旱，茎叶生长差，不利于块根形成，容易长成柴根；中期茎叶生长不良，块根膨大慢，产量低；后期若遇干旱，茎叶早衰，产量低。反之，土壤含水量过多，茎叶徒长，薯形细长不耐储藏。群众有谚语："干长柴根，湿长须根，不干不湿长块根"。甘薯对土壤适应性较广，但最适宜生长在土层深厚、土质疏松、含水量适中、通气性良好、含有机质多的沙质土壤。

9.1.3 栽培技术

甘薯的栽培采用育苗移栽技术，育苗一般于大田栽植前一个月左右进行，要选择抗病强、生长快、品质好的高产品种进行育苗。

苗床应选择在背风向阳、排水良好、靠近水源、无鼠害、便于管理的地方。苗床的类型依据热源不同分为人工加热、生物热源、酿热温床、太阳辐射热源。东北、华北育苗时以火炕育苗为主，南方以后3种为主。采用"前高中平后低"的先催后炼法。前期高温催芽，然后降为31℃，出苗后降为25~28℃，采苗前5~6d浇透水一次，以后停止浇水，进行"蹲苗"，采苗前3d降温至20℃"炼苗"。苗高15~18 cm时剪苗，气温稳定在15℃以上，地温17~18℃，无晚霜时开始剪苗。

移栽一般采用直插、平插和斜插3种方式。直插法适宜于干旱沙土或丘陵坡地及灌溉差的地方，茎长17~20 cm，3~5节。它的特点是结薯少、产量低、抗旱。平插法适宜于高水肥地，茎长

days before cutting the transplanting materials. When the seedling reach 15 ~ 18 cm long, the temperature keeps above 15℃ steadily and the soil temperature keeps at 17 ~ 18℃, it is the right time for cutting.

Transplanting methods: erect plug, flat plug and inclined plug. The in-line plug is suitable for arid sandy soil, hills and slope with poor irrigation, the length of stems is 17 ~ 20 cm and 3 ~ 5 nodes are required. Characteristics: low yields and good drought tolerance. The flat plug is suitable for cultivation under optimum water and fertilizer conditions, the length of stems is 20 ~ 30 cm and 5 ~ 8 nodes. Characteristics: big and even size of tuberous roots and poor drought tolerance. The inclined plug method involves burying 3 ~ 4 nodes in soil at 45 ° angle, suitable for hills, uplands and plains. The length of stems is about 23 cm and it requires 4 ~ 7 nodes. Characteristics: uneven size of the tuberous roots (upper larger and smaller lower depth), yields are higher than the in-lined plug and the flag plug, and seedlings survive well.

Field management at earlier stage includes seedlings checking, filling the gaps with seedlings, weed control, fertilization, irrigation and pest control. The goals of field management at middle stage are manage for stable growth of stems and leaves, promoting root enlargement. The specific measure is vines turning, but it is not advocated due to it will reduce the production by 10% to 20%; lifting vines could limit overgrowth; foliage top-dressing and draining flooded fields could control pests and diseases. The late stage is the period of tuber enlargement. The management goals are preventing premature of leaves and stems and promoting tuber enlargement. The main measures are top-dressing to prevent senescence and with appropriate irrigation.

Continuous cropping of sweet potato will lead to imbalance of soil nutrients, soil degradation, increasing pest pressure, and production will decrease gradually. However, crop rotation can restore soil fertility and reduce pests and weeds.

9.1.4 Feeding value

Yield of leaves should be considered when sweet potatoes

20~30 cm，5~8 节。它的特点是结薯大而匀，抗旱性差。斜插法是 45° 角入土 3~4 节，适宜于丘陵、山冈、平原旱地，茎长 23 cm 左右，4~7 节。它的特点是薯不均匀（上大下小），产量高于直插低于平插，苗容易成活。

田间管理前期的主要措施是查苗补苗、中耕除草、追肥、灌溉、防病虫，以促为主防徒长。中期田间管理的主要目标是控制茎叶平稳生长，促使块根膨大，主要措施是翻蔓，但会减产 10%~20%，因此不提倡；提蔓可抑制徒长；其他措施包括根外追肥、排涝和防病虫。后期是薯块膨大期，主要目标是防止茎叶早衰，促使块根膨大，主要措施是追肥（防早衰）和灌溉。甘薯连作会造成土壤养分失调、地力衰退、病虫害加重，产量逐年降低。轮作倒茬可充分培养地力，减轻病虫和杂草的危害。

9.1.4 饲用价值

饲用甘薯的收获要兼顾叶片的产量，也就是在藤叶和块根产量均高的时期收获。甘薯块根及茎蔓都是优良的饲草。块根中含有大量淀粉、维生素 B、维生素 C 等。新鲜甘薯营养价值为玉米的 25%~35%，因富含淀粉，其热能总值接近于玉米。甘薯生喂或熟喂，干物质和能量的消化率均相同，但以熟喂为好，消化吸收更容易。甘薯茎蔓中无氮浸出物含量较块根为低，但蛋白质含量较高，也含有较高的维生素。甘薯加工后的淀粉渣富含蛋白质和糖类，是猪和奶牛的好饲料。

are harvested for feeding. Harvest should be completed when vines, leaves and tubers have high yield. Vines and roots are excellent forage, and tubers contain high levels of starch, vitamins B and C. Nutritional value is 25% ~ 35% of that corn in fresh state and is similar to the total energy of corn because of high starch content. Sweet potatoes can be fed raw or cooked. Dry matter intake and the digestibility are the same either way, but feeding after cooking can improve the digestion and assimilation. The content of nitrogen free extract in vines is lower than roots, but vines are richer in protein and vitamin. Starch residue are rich in protein and carbohydrates, it is good feed for pigs and cows.

9.2 Potato 马铃薯

Scientific name: *Solanum tuberosum* L.

Family: Solanaceae

Potato is native to Andes in South America. Spain introduced potato in 1565 and then to Italy. Approximately in the 1620s—1650s, potato is introduced from Taiwan to Fujian then spread over China.

According to the reports of FAO, the world production of potatoes in 2013 was about 368 million tonnes. Just over two thirds of the global production is eaten directly by humans with the rest being fed to animals or used to produce starch. However, the local importance of potato is extremely variable and rapidly changing. China led the world in potato production, and nearly a third of the world's potatoes were harvested in China and India.

9.2.1 Botanical characteristics

Potato plants are annual herbaceous that grow about 60 cm high, it varies by variety(Fig. 9.4).Underground tubers are round, oval, elliptical and other form with red skin, yellow, white or purple(Fig. 9.5). Stems are prismatic, hairy, erect, half erect or prostrate. Primary leaf is simple, with the

学名： *Solanum tuberosum* L.

英文： potato

别名： 土豆、洋芋、山药蛋

马铃薯原产于南美洲的安第斯山脉，大约在 1565 年被西班牙引进栽培，后经过西班牙传到意大利。马铃薯在 17 世纪 20—50 年代，从台湾传入福建沿海地区，后传播到全国。

世界马铃薯年产量约 36 800 万 t（FAO，2013），其中 2/3 以上被人直接食用，其余 1/3 用于饲用或加工淀粉。然而，马铃薯在各国的重要性随着时间变化很大，目前中国处于世界马铃薯生产领先地位，世界 1/3 的马铃薯产自中国和印度。

9.2.1 植物学特征

马铃薯是茄科茄属一年生草本植物，株高约 60 cm，因品种而异，茎棱形，被绒毛，直立、半直立或匍匐，生育后期易倒伏（图 9.4）。地下块茎呈圆、

growth stage prolonged, compound leaves appears. There are white, pink, red, blue, or purple flowers with yellow stamens (Fig. 9.6). Potatoes are mostly cross-pollinated by insects. The potato mainly is propagated vegetatively by planting tubers, pieces of tubers cut to include at least one or two germ pores.

卵、椭圆等形，有芽眼，皮红、黄、白或紫色（图 9.5）。初生叶为单叶，随着植物的生长逐渐形成奇数羽状复叶，开花后叶片枯萎，花色艳丽，有白色、粉红色、蓝色或紫色，雄蕊黄色(图 9.6)。马铃薯属于虫媒花异花授粉植物，主要靠切分块茎无性繁殖，每个用于繁殖的块茎上至少保留两个芽眼。

Fig. 9.4 Potato plants
图 9.4 马铃薯植株

Fig. 9.5 Potato tuber
图 9.5 马铃薯块茎

Fig. 9.6 Potato flower
图 9.6 马铃薯的花

9.2.2 Biological characteristics

Potato has high yield and good adaptability. Like a cool climate, but cannot tolerate high temperature, and colds tress. Tubers begin to germinate at 7 ~ 8℃ and the optimal soil temperature is 15~18℃ at seedling stage; The ideal temperature for shoot growth is 21℃, tuber growth rate slow down when the soil temperature reaches 25℃ and stop growth at 30℃. Potato is drought tolerance, but keep the suitable soil moisture is vital to obtain high yield. Potato is not very strict with soil conditions, but a deep, good permeability, loose and fertile sandy loam soil is most suitable.

9.2.2 生物学特性

马铃薯产量高，对环境的适应性较强。喜冷凉气候，但不耐高温，也不耐寒。块茎7~8℃开始萌芽，苗期最适土壤温度15~18℃；地上茎叶生长适宜温度为21℃。土温达到25℃，块茎生长缓慢，30℃块茎生长停止。马铃薯较抗旱，但保持适宜的土壤持水量是取得高产的重要条件。马铃薯对土壤条件要求不十分严格，但最适宜在土层深厚、通透性好，疏松肥沃的沙壤土上生长。

9.2.3 Cultivation technology

Tubers are used for potato propagation by cutting disease free potatoes into pieces containing 2 ~ 3 buds and later sowing on ridge (Fig. 9.7). Seeding time of potato varies from north to south of in China, generally it is planted in autumn and winter in the south and spring in the north while soil temperature reaches 7 ~ 8 ℃ at 10 cm soil depth. In the northwest region, potato should be planted in early April to late May, early March in north of China and in mid-April to mid-May in the northeast and Inner Mongolia region.

The potato is suitable to ridge tillage as a row crop, sowing depth 6 ~ 12 cm, shallow seeding when soil moisture is good. The potato need proper fertilization and irrigation during growing stage, nitrogen is preferred at the seedling stage, and phosphate, potash at the tuber enlargement stage. Soil moisture content in more than 60% of field capacity should be maintained.

9.2.3 栽培技术

用块茎繁殖，把脱毒种薯切成带有2~3个芽眼的块状，起垄播种（图9.7）。我国马铃薯的播种时间南北各异，南方多在秋冬季节播种，北方主要是春播，以土壤表层10 cm温度达到7~8℃时作为马铃薯开始播种的标志。西北地区在4月上旬到5月下旬，华北地区在3月上旬，东北和内蒙古地区在4月中旬至5月中旬。

马铃薯是适合垄作的中耕作物，播种深度6~12 cm，墒情好时浅播。马铃薯生长期间要适当施肥和灌溉，前期以氮肥为主，后期以磷、钾肥为主，土壤含水量应保持在田间持水量的60%以上。

Fig. 9.7 Potato field
图9.7 马铃薯大田生产

9.2.4 Feeding value

Potato has high nutritional value. It is best known for its carbohydrate content and rich in vitamins A and C and some other minerals. Its high-quality starch content is about 16.5%. It also contains large amounts of lignin. Research showed that potato has good effects on the indigestion; it is also a health care medicine for stomach and heart disease.

Potatoes contain toxic compounds known as glycoalkaloids, of which the most prevalent are solanine. The green leaves and green skins of tubers exposed to the light are toxic.

9.2.4 饲用价值

马铃薯的营养价值很高，富含糖类、维生素 A、维生素 C 以及其他矿物质，优质淀粉含量约为 16.5%，还含有大量木质素等。研究表明，马铃薯对调解消化不良有特效，是胃病和心脏病患者的良药及优质保健品。

值得注意的是，马铃薯含有生物碱有毒化合物，最主要的是龙葵素，主要存在于绿色叶片和被光照射后发绿的块茎表皮中。

9.3 Forage pumpkin 饲用南瓜

Scientific name: *Cucurbita moschata* Duch.

Family: Cucurbitaceae

Forage pumpkin, also called winter crookneck squash, is native to Southeast Asia, and is mainly cultivated in India, Malaysia, Japan and all over in China. It is an important vegetable as well as a high yield forage crop. Forage pumpkin is rich in juice and sugar, which can replace some concentrated feeds. Its vines are also good forage.

9.3.1 Botanical characteristics

Forage pumpkin is an annual herb belongs to cucurbita of cucurbitaceae, it has developed root system which consist of taproot, lateral root and adventitious roots. Vines are sprawling, hollow and with obvious edge and white hairs. Vines are branching; the length is 5 ~ 10 cm and stem nodes are easily to form adventitious roots. Leaves are large and thick. Flowers are unisexual, yellow and monoecious. Corollas are large and flat, but not sagging (Fig.9.8). Pulp is yellow or deep yellow and generally is 1 ~ 3 cm thick (Fig.9.9). Fruits have aroma and contain lots of starch and sugars. Seeds are flat, white or light yellow, and thousand seed weight is 125 ~ 300 g, and its life span is 5 ~ 6 years.

学名： *Cucurbita moschata* Duch.

英文名： forage pumpkin

别名： 中国南瓜、倭瓜、番瓜、饭瓜

南瓜原产于亚洲东南部，印度、马来西亚、日本及中国栽培较多。我国各地均有栽培，是重要蔬菜，也是优质高产饲料作物。饲用南瓜不但多汁，而且含糖多，可代替部分精饲料，藤蔓也是良好青饲料。

9.3.1 植物学特征

南瓜为葫芦科南瓜属一年生草本植物。根系发达，具主根、侧根和不定根。茎蔓生，中空，具不明显的棱，上有白色茸毛。蔓分枝性能较强，长达 5~10 cm，茎节易产生不定根。叶片大而肥厚，花单性，黄色，雌雄同株异花。花冠裂片大，展开不下垂（图 9.8）。果肉黄色或深黄色，肉厚一般 1~3 cm（图 9.9），成熟时有香气，含淀粉和糖较多。种子扁平，白色或淡黄色，千粒重 125~300 g，寿命 5~6 年。

Fig. 9.8 Leaves and flower of forage pumpkin
图 9.8 饲用南瓜的叶和花

Fig. 9.9 Forage pumpkin
图 9.9 饲用南瓜

9.3.2 Biological characteristics

Forage pumpkin is native to tropical regions, prefers a warm and humid climate, and adapts to a wide range of temperature. The root is developed and very drought tolerant. Forage pumpkin is a photophilous and short-day plant, short-day light can stimulate early differentiation of female flowers and increase number of flowers. Forage pumpkin is not strict to soil condition, but well drained, fertile, loose, neutral or slightly acidic (PH 5.5-6.7) sandy loam is preferable.

9.3.3 Cultivation technology

Except for arable land, the marginal land is especially good for forage pumpkin planting. It also can be planted in breeding houses, slopes and courtyards with pergola. Roots of forage pumpkin are deep and the plow depth should be more than 20 cm.

9.3.4 Feeding value

Forage pumpkin pulp is compact, palatable, high yield and has good nutrition. It is easy to store and transport and a good feed for pigs, cattle, sheep, and chickens as well. But forage pumpkin should be smashed when feed for cattle to avoid being choked.

9.3.2 生物学特性

南瓜原产于热带地区，较喜欢温暖湿润气候，但对温度有较强的适应性。南瓜根系发达，吸水、抗旱能力强，直播时尤为突出。南瓜为喜光的短日照植物，短日照能促使雌花提早分化，使其数量增多。南瓜对土壤要求不严格，但以排水良好，肥沃疏松的中性或微酸性（pH 5.5~6.7）沙壤土为宜。

9.3.3 栽培技术

除耕地外，南瓜特别适合在边角地种植，可在圈舍、渠坡面、庭院搭架种植。南瓜根系较深，翻耕深度不得少于 20 cm。

9.3.4 饲用价值

南瓜肉质致密，适口性好，产量高，营养好，便于贮藏和运输，是猪、牛、羊、鸡的好饲料。喂牛时必须粉碎饲喂，以防噎食。

9.4 Beet
甜菜

Scientific name: *Beta vulgaris* L.

Family: Chenopodiaceae

Beet is native to Southern European and has been cultivated throughout the world, and mainly distribute in Europe, Asia, the United States. In China, it is mainly distribution in the northern area including Liaoning, Jilin, Ningxia, Inner Mongolia and Gansu. Fodder beet is palatable, nutritious, wide distribution plant with high yield (roots 45 ~ 75 t/hm^2, roots and leaves 75 ~ 300 t/hm^2) and good quality (sugar content is 4% ~ 20%). It is a multifunctional crop of food, feed and industrial material. For livestock, it is an excellent and juicy storage feed in cold northern regions in winter and spring.

学名： *Beta vulgaris* L.

英文名： beet

别名： 饲料萝卜、甜萝卜、糖菜。

甜菜原产于欧洲南部，适应性强，世界各地均有栽培，主要分布在欧洲、亚洲和美洲。我国主要分布在北方地区，辽宁、吉林、宁夏、内蒙古、甘肃较多。甜菜的适应性强，产量高（根 45~75 t/hm^2；根叶 75~300 t/hm^2），品质好（含糖量 4%~20%），适口性强，营养丰富，是粮、饲、工三料兼用作物。对畜牧业而言，是北方寒冷地区冬春季节优良多汁的贮备饲料。

9.4.1 Botanical characteristics

Fodder beet is a biennial plant belongs to *beta* of *chenopodiaceae*; it has four variants: sugar beet, forage beet, edible beet and leaf beet which is an ornamental plant as well (Fig.9.10, Fig.9.11, Fig.9.12). The taproot is the main organ harvested for forage use, it includes four parts: root apex,

9.4.1 植物学特征

甜菜为藜科甜菜属二年生草本植物，有糖用、饲用、食用和叶用（厚皮菜）4 个变种，其中叶用甜菜也可以作为观赏植物（图 9.10，图 9.11，图 9.12）。直根是饲用甜菜的主要利用部分，分根

Fig. 9.10 Sugar beet (left) and edible beet (right)
图 9.10 糖用甜菜（左）和食用甜菜（右）

To obtain high yield, turnip harvest should not be too early, but should be finished before -3~-4℃. Even though the shoot quality decreased, the primary product of fleshy root is high yield and good quality.

部分质量变差，但可以确保主产品肉质根高产优质。

9.7.4 Feeding value

As livestock feed, turnips provide fodder of high nutritional value and are very digestible. Turnip leaves are wide and thick, tender and juicy, but spicy flavors may require mixing with other feed to ensure adequate consumption.

9.7.4 饲用价值

芜菁甘蓝为营养价值高的多汁饲料，易消化。芜菁甘蓝叶片宽厚，柔嫩多汁，是家畜的优质饲料，但其有辛辣味，宜与其他饲料搭配饲喂。

Chapter 10

Forages Distribution and Regional Planning

牧草的分布和区划

10.1 Distribution of forages 牧草的分布

10.1.1 Origin of forages

According to research, about 10000 years ago, the demand for forages increased with people settling down and domesticated wild animals such as horses, sheep and cattle. However, forages cultivation has only several hundred years of history.

About the forages origin centre, according to research of American scientist J.R.Harlan(1981), there are four forages origin centers as follow: European (except for Mediterranean climate zones) Center, Mediterranean Basin and Near East (winter frost) Center, African Savannah (Tropical Steppe) Center and Tropical Americas Center.

10.1.2 Adaptation of cultivated forages

Long-term natural selection and artificial selection lead to a specific adaptation of forages. Compared with wild species, the cultivated forages have characteristics such as: Larger size of plants and vegetative organs, higher yield and good quality; More nutritious; Growth stages and maturity become more consistent and concentrate; Dormant are slightly weak and dormant periods becomes short; Self-protection mechanism and the ability to spread and reproduction decreased; Extent of environmental adaptation is narrowed.

10.1.3 Forages distribution

Basically, the cultivated forages distribute around the origin places of wild species and extend by the axis. Extent of the radiation range is determined by the following factors: The environmental and soil adaptation of the species, introduction and cultivation history, socio-economic conditions, production technology and human planting customs and social demands. For example, alfalfa has broad adaptation, but it distributed mainly in the Yellow River Basin and 14 provinces in north of China. Generally, the lowest temperature in winter and annual precipitation are the primary influence factors.

10.1.1 牧草起源概述

据考证，大约1万年前，人类由于对马、牛、羊等野生动物的驯化和围栏定居，增加了对饲草的需求。但实际上，人类栽培牧草的历史只有数百年。

关于牧草的起源，目前公认的是美国学者J. R. Harlan的4个起源中心论：即欧洲（不包括地中海气候带）中心、地中海盆地和近东（冬霜）中心、非洲萨瓦纳（热带干草原）中心和热带美洲中心。

10.1.2 栽培牧草的适应性

长期的自然选择和人工选择，使牧草具有某种特定的适应性。与野生种相比较，栽培牧草具有如下特点：植株体及营养器官变大，高产、优质；营养价值更高；生育期、成熟期整齐集中；休眠现象减弱，休眠时间缩短；自我保护功能和自行传播繁衍功能减退；对环境适应范围变窄。

10.1.3 牧草分布

牧草基本上以野生种原产地为轴心向周围辐射，辐射范围大小与下列因素有关：物种本身对环境和土壤的适应能力、引种栽培历史、社会经济条件、生产技术水平、人们的栽培习惯和社会需求。如苜蓿适应性很广，但在我国主要在黄河流域及以北的14个省区分布。一般来讲，牧草分布受冬季最低温度和年降水量影响最大。

10.2 Forages regional planning 牧草的区划

Forage regional planning is the premise of scientific cultivation of forages and it is particular important for pasture establishment and plays great important role in healthy animal husbandry.

10.2.1 Principles and basis for regional planning

(1) Principles for Regional Planning

① Principles for regional planning are mainly dominated by nature rules, combined with the agricultural economic rules meanwhile.

② The natural conditions, the development direction of agriculture and animal husbandry, overall plan and production measures should be consistent within the same region.

③ It is regardless of administrative boundaries of county or province (keep consistent as far as possible).

④ Also should be contiguous geographical area within the same region.

(2) Basis for Regional Planning

① The natural geography, topography and climate zones are basis for regional planning.

② The intended production purpose should be consistent with the ecological conditions.

③ The sub divisional designation is based on the ecological and biological characteristics, production conditions and use patterns of dominant forage species.

10.2.2 Characteristics of dominant forage species

High yield and good quality, good palatability, rapid growth and regrowth, good persistence, good stress resistance, with a long cultivation history in local place and been widely used in current production; Or, native or introduced forage species that do have a big potentiality in practice in the near future. The species should be easy for establishment and reproduction, multiple utilized such as feed, fertilizer, fuel, conservation of soil and water.

牧草的区划是科学种植牧草的前提，对人工草地的合理建植和畜牧业的健康发展具有重要的指导意义和现实意义。

10.2.1 区划的原则和依据

（1）区划的原则

① 以自然规律为主，与农业经济规律结合。

② 同一栽培区内，其自然条件、农牧业发展方向、布局、措施基本一致。

③ 区划时基本不考虑行政界线（尽可能一致）。

④ 同一区划范围保证地域连片。

（2）区划的依据

① 以自然地理位置和地貌、气候带为主。

② 生产发展方向和生态条件基本一致。

③ 以“当家”草种的生态生物学特性、生产条件和利用方式作为亚区分区条件。

10.2.2 “当家”草种应具备的条件

高产优质、适口性好、再生快、生长迅速、生存期长、抗逆性强，栽培历史悠久，当前生产上大面积应用；近期内确有发展前途的本地或引进草种。栽培简便，容易繁殖，兼具饲料、肥料、燃料、水保等多用途的牧草。

10.3 Binominal for regional planning 区划的方法和命名

Binominal: Geographic location, topography + Dominant forage species (Two kinds of legumes and two kinds of grasses respectively).

According to the "Perennial forages regional planning in China" and "Chinese perennial forages cultivation techniques", the regional planning districts of cultivated perennial forages can be divided into nine regions and 40 sub-regions.

区划名称使用双重命名法：按地理方位、地形地貌 + "当家"草种（2 种豆科、2 种禾本科）进行命名，依据"中国多年生栽培牧草区划"和"中国多年生草种栽培技术"，我国多年生栽培草种区域可划分为 9 个栽培区和 40 个亚区（图 10.1）。

10.4 Overview of the nine regional planning districts for forages in China 九大分区概述

10.4.1 *Leymus chinensis*, alfalfa, adsurgens, lespedeza cultivation area in northeast of China

This region includes Heilongjiang, Jilin and Liaoning provinces and eastern Inner Mongolia. Main climate features are continental climate (inland areas which are distant from oceans, with dry air, fewer clouds, sunshine, hot in summer and cold in winter, large temperature differences between day and night, lower rainfall and more variable seasonal distribution.). It is cold and less snow in winter, dry and windy in spring, humid in eastern areas and arid in western, low temperatures in northern and northwestern areas, short frost-free period. In the eastern areas of the region (Greater Khingan Mountains and Lesser Khingan Mountains), agriculture and forestry are the main industries. In the central (Songliao Plain and Sanjiang Plain) grain production is dominant, and is called "Northern Granary". In western hilly areas, a combination of crop and livestock production (Sanjiang horse, Northeast fine woo sheep, Chinese Merino sheep) occurs. The main forage species include: *Leymus chinensis*, alfalfa, adsurgens, smooth brome grass and lespedeza bicolor. Suitable varieties include 'jisheng' No.1, No.2, No.3 and No.4 *Leymus*

10.4.1 东北羊草、苜蓿、沙打旺、胡枝子栽培区

本区包括黑龙江、吉林、辽宁三省及内蒙古自治区东部。主要特点为：大陆性气候（指距离海洋较远的内陆地区，具有空气干燥、云雾少、日照充足、冬冷夏热、昼夜温差大、雨量少且季节分配不均等特点）冬季寒冷少雪，春季干旱多风，东部湿润，西部干旱，北部、西北部气温低，无霜期短。东部（大小兴安岭等山区）以林、农为主；中部（松辽、三江平原）是以粮（玉米、大豆）为主的"北大仓"；西部低山丘陵地区以农牧（三江马、东北细毛羊、中国美利奴羊）结合为主。主要栽培草种有羊草、苜蓿、沙打旺、无芒雀麦、二色胡枝子等。适宜的品种有：'吉生 1 号''吉生 2 号''吉生 3 号'和'吉生 4 号'羊草；'公农 1 号''龙牧 801''龙牧 803'和'肇东'苜蓿；早熟沙打旺等。亚区有：

chinensis; 'gongnong No.1' 'longmu 801' 'longmu 803' 'zhaodong' alfalfa; precocious adsurgens.

There are 6 sub-regions, they are:

(1) Sub-region of *Leymus chinensis*, alfalfa, adsurgens in Greater Khingan Mountains.

(2) Sub-region of alfalfa, smooth bromegrass, wild peas in Sanjiang Plain.

(3) Sub-region of *Leymus chinensis*, alfalfa, adsurgens in Songnen Plain.

(4) Sub-region of alfalfa, smooth bromegrass in Songliao Plain.

(5) Sub-region of alfalfa, lespedeza, smooth bromegrass in East of Changbai Mountain.

(6) Sub-region of adsurgens, alfalfa, *Leymus chinensis* in western Liaoning hills.

10.4.2 Alfalfa, adsurgens, *E.sibiricus*, Mongolia Hedysarum cultivation area in Inner Mongolia Plateau

This area is located in the Inner Mongolia Plateau, including Bashang area. Its main climate features are short frost-free period, limiting temperatures, often cold and blizzards in winter, severe sandstorms in spring, drought and scarce rainfall. Annual precipitation is about 50 ~ 450 mm. However, precipitation and warm temperatures occur during the same period in this region. About 70% of annual precipitation is concentrated in July, August and September. Major business in the region is nomadic livestock herding, hunting and fishing. Hetao Plain and Ningxia Yellow River Irrigation Area produce much of the commodity grain in China. The area is suitable for planting 'grassland No.1' 'grassland No.2' 'rambler' alfalfa and some other cold resistant varieties. Better adapted to the sand and high yield cultivation are *Hedysarum laeve* Maxim and the highly drought tolerant varieties such as Mongolian wheatgrass, 'nuodan,' desert wheatgrass and 'shandan' psathyrostachys.

Sub-regions include:

(1) Sub-region of *Elymus sibiricus, E.dahuricus, Leymus chinensis* in South-central Inner Mongolia—Cool and

(1) 大兴安岭羊草、苜蓿、沙打旺亚区。

(2) 三江平原苜蓿、无芒雀麦、山野豌豆亚区。

(3) 松嫩平原羊草、苜蓿、沙打旺亚区。

(4) 松辽平原苜蓿、无芒雀麦亚区。

(5) 东部长白山山区苜蓿、胡枝子、无芒雀麦亚区。

(6) 辽西低山丘陵沙打旺、苜蓿、羊草亚区。

10.4.2 内蒙古高原苜蓿、沙打旺、老芒麦、蒙古岩黄芪栽培区

本区地处内蒙古高原，包括河北坝上地区。主要特点为：无霜期短，热量明显不足，冬季严寒有暴风雪，春季有大风沙为害，干旱少雨，全年降水量50~450 mm，但水热同期，全年70%左右降水量集中在7、8、9三个月份。主要经营游牧、渔猎业，河套平原是商品粮基地，宁夏黄河灌区亦是商品粮基地。适宜的品种有耐寒性强的‘草原1号’‘草原2号’‘润布勒’苜蓿；适应沙地栽培丰产性较好的‘中草1号’塔落岩黄芪，耐旱性较强的蒙古冰草、‘诺丹’及沙生冰草，‘山丹’新麦草等。亚区有：

(1) 内蒙古中南部老芒麦、披碱草、羊草亚区——温凉半干旱，年降水量300~400 mm，海拔1 100 m以上，冬寒夏凉，羊草是该亚区的主要草种。

(2) 内蒙古东南部苜蓿、沙打旺、羊草亚区——温暖半干旱草原地带，年降水量350~400 mm，海拔500~800 m，该区实行灌溉农业，粮草轮作。

(3) 河套-土默特平原苜蓿、羊草亚区——热量资源丰富，地势平坦，灌溉农

semi-arid, annual rainfall is 300~400 mm. Altitude is above 1 100 m and the climate is cold in winter and cool in summer. *Leymus chinensis* is the dominate species in this sub-region.

(2) Sub-region of alfalfa, adsurgens, *L. chinensis* in Southeastern Inner Mongolia—Warm semi-arid steppe region, annual rainfall is 350~400 mm, altitude is 500~800 m. Irrigated agriculture region, where the rotation system of crops and forage exist.

(3) Sub-region of Alfalfa, *Leymus chinensis* in Hetao-Tumote Plains—Warm temperature, flat terrain, irrigated agriculture region, where crops and forage are arranged in a rotational system.

(4) Sub-region of Elymus, adsurgens, Korshinskii in north-central Inner Mongolia—Arid area, snowstorms, annual rainfall less than 300 mm.

(5) Sub-region of Korshinskii, Mongolia Hedysarum, adsurgens in Ordos region—Pastoral areas, Ordos Plateau, arid, sandy loam, serious soil erosion, steep slopes and deep groove, rich in heat resources.

(6) Sub-region of Haloxylon, Calligonum in Western Inner Mongolia—Desert, annual rainfall is less than 150~200 mm, sometimes even less than 50 mm. Evaporation is higher than 2 300 mm. The typical climate is dry and warm.

(7) Sub-region of Alfalfa, adsurgens, Korshinskii, twigs Hedysarum in Ningxia, Gasnsu and Hexi Corridor—Arid desert climate, in the north and south of Hexi Corridor dominated by livestock, irrigated area in the middle is mainly crop production. Alfalfa, Melilotus and spring vetch are suitable for planning in this sub-region.

10.4.3 Alfalfa, adsurgens, smooth bromegrass, tall fescue cultivation area in Huang Huai Hai Plain

This region includes Beijing, Tianjin, Hebei, Shandong, Jiangsu, eastern Henan and northern of Anhui province. This area called the North China Plain, is a typical alluvial plain, impacted by the Yellow River, Huaihe River and Haihe River. Yanshan, Taihang Mountains are located in the north and west of the North China Plain, respectively. Northern of China Plain has a flat terrain, and superior soil conditions of

区，粮草轮作。

(4) 内蒙古中北部披碱草、沙打旺、柠条亚区——干旱，多雪灾，降雨量不足 300 mm。

(5) 鄂尔多斯柠条、蒙古岩黄芪、沙打旺亚区——牧区，鄂尔多斯高原，干旱、沙质土壤，水土流失严重，坡陡沟深，热量资源丰富。

(6) 内蒙古西部梭梭、沙拐枣亚区——荒漠地带，年降水量 150~200 mm 以下，甚至不足 50 mm，蒸发量高达 2 300 mm 以上，干燥、温热是典型的气候特点。

(7) 宁甘河西走廊苜蓿、沙打旺、柠条、细枝岩黄芪亚区——干旱荒漠气候，河西走廊南北以牧为主，中部灌区以农为主，适宜种植苜蓿、草木樨、春苕子等。

10.4.3 黄淮海苜蓿、沙打旺、无芒雀麦、苇状羊茅栽培区

本区包括北京、天津、河北、山东、苏北、豫东和皖北，是由黄淮海三大水系冲击而成的华北平原，北部、西部有燕山、太行山隆起。平原地势平坦，水土条件较优越，气候属暖温带，无霜期达 140~220 d，年降水量达 500~850 mm，但季节分配不均匀，春季雨少，夏季雨水集中，形成春旱夏涝，对生产不利。沿海地区多盐碱地，适宜开发草业。主要的栽培草种有苜蓿、沙打旺、无芒雀麦、苇状羊茅及葛藤、二色胡枝子、鸭茅、长穗冰草等。适宜的品种有耐盐性较强的‘中苜一号’苜蓿、‘沧州’苜蓿、‘林肯’无芒雀麦等。利用冬闲田可种植‘冬牧 70’黑麦、‘中饲 237’小黑麦等。亚区有：

soil and good water availability. The climate is representative of the warm temperate zone. Total annual frost-free period is 140~220 d. Annual rainfall is 500~850 mm. However, less rainfall occurs in spring but concentrated in summer. Uneven distribution of precipitation is unfavorable for crop production. Saline soils are common in coastal lines and are suitable for forage production. The main cultivated species include alfalfa, adsurgens, smooth bromegrass, tall fescue, kudzu, lespedeza bicolor, orchardgrass and Agropyron elongatum. The salt-tolerant varieties include *Medicago sativa* L.cv. 'zhongmu No.1', 'cangzhou' alfalfa and 'lincoln' smooth bromegrass. Winter rye .cv. 'dongmu 70' or winter triticale 'zhongsi 237' can be planted during the winter-fallowed period.

Sub-regions include:

(1) Sub-region of alfalfa, adsurgens, kudzu, smooth bromegrass in Northwest Mountain, which include Yanshan, Taihang Mountains, continental monsoon climate. Crown vetch, tall fescue, smooth bromegrass and Robinia pseudoacacia are suitable for this region.

(2) Sub-region of Alfalfa, adsurgens, smooth bromegrass in North China Plain.

Foothills plains: Alluvial plains, dominated by crops, areas unadapted for cropping might benefit from forage establishment.

Heilonggang low plains: Geology and soils influenced by the ancient Yellow River and Haihe, partial arid zone with saline soils, alfalfa is the good choice.

Coastal plain: Located in the Bohai seacoast, poor drainage may frequently lead to a short-term seasonal ponding. Soil salinity is serious. There are opportunities to integrate crop, livestock and fisheries. Natural grassland management could be strengthened by establishment of artificial grasslands and implementing forage/crop rotation. Suitable forages are Alfalfa, adsurgens, smooth bromegrass, crown vetch and tall fescue.

(3) Sub-region of Alfalfa, adsurgens, smooth bromegrass in Huang Huai Hai Plain. The subregion includes North of Jiangsu, north of Anhui, northwest of Henan province, and western plains in Shandong Province.

(1) 北部西部山地苜蓿、沙打旺、葛藤、无芒雀麦亚区——燕山、太行山，大陆性季风气候，适宜草种还有小冠花、苇状羊茅、无芒雀麦、刺槐。

(2) 华北平原苜蓿、沙打旺、无芒雀麦亚区。

山麓平原：两山前形成的冲积平原，以农为主，充分利用间隙地种草。

黑龙港低平原：由古黄河及海河水冲击而成，偏旱区，盐渍化，适宜种苜蓿。

滨海平原：位于渤海沿岸，排水不畅，短期季节性积水，盐渍化严重，贯彻农牧渔结合原则，加强天然草地管理，建立人工草场，实行粮草轮作，适宜栽培的草种有苜蓿、沙打旺、无芒雀麦、小冠花、苇状羊茅。

(3) 黄淮海平原苜蓿、沙打旺、无芒雀麦亚区——包括苏北、淮北、豫北西部，山东省鲁西平原。

(4) 鲁中南山地丘陵沙打旺、苇状羊茅、小冠花亚区——半湿润气候区。

(5) 胶东低山丘陵苜蓿、百脉根、黑麦草亚区——海洋性湿润气候，空气潮湿，云雾多，日照较少，冬暖夏凉，昼夜温差小、雨量多而匀，年降水量 750~900 mm，还可种植三叶草、鸭茅、羊茅。

10.4.4 黄土高原苜蓿、沙打旺、小冠花、无芒雀麦栽培区

本区包括山西、河南西部、陕西中北部、甘肃中东部、宁夏南部和青海东部。海拔 1 000~1 500 m，土层厚达几十米至几百米，水土流失十分严重，地貌支离破碎。气候温和干燥，年降水量在 350~700 mm，但地区间

(4) Sub-region of Adsurgens, tall fescue and crown vetch Mountains, hills in Central and southern Shandong-Semi-humid climate zone.

(5) Sub-region of Alfalfa, lotus, ryegrass in Shandong hilly areas—Moist maritime climate, wet, foggy, less sunshine, warm in winter and cool in summer, day and night temperature difference is small. There is abundant and uniform rainfall. Rainfall is 750~900 mm. Clover, orchard grass and tall fescue are suitable.

10.4.4 Loess Plateau alfalfa, adsurgens, crown vetch, smooth bromegrass cultivation area

This area includes Shanxi, western Henan, central and northern Shaanxi, central and eastern Gansu, southern Ningxia and eastern Qinghai. Altitude is 1000~1500 m. The soil layer is tens of meters to several hundred meters thick. Soil erosion is very serious. Climate is mild and dry. Precipitation is 350 to700 mm. However, rainfall distribution is uneven. There are large areas of farmland, but with low grain productivity, agriculture combined with animal husbandry usually historically. Suitable forages include: alfalfa, adsurgens, crown vetch, smooth bromegrass, tall fescue, orchardgrass, sainfoin and *Agropyron cristatum* (L.) Gaertn. The suitable cultivars include 'jinnan' alfalfa, 'pianguan' alfalfa, 'guanzhong' alfalfa, 'shanbei' alfalfa, 'longdong' alfalfa, hybrid alfalfa 'gannong No.1', hybrid alfalfa 'gannong No.2', 'yellow river No.2', Precocious Adsurgens, Gemstones crown vetch, Binjifute crown vetch, Carlton smooth bromegrass, Gansu sainfoin, and some annual forages like narrow leaf vetch '333/A', sonchifolia and red amaranth.

Sub-regions include:

(1) Sub-region of alfalfa, adsurgens, crown vetch, smooth bromegrass, tall fescue in hills and mountains of eastern Shanxi and western Henan—southeast of the Loess Plateau, warm semi-arid climate. Drought is the most serious problem.

(2) Sub-region of alfalfa, crown vetch, smooth bromegrass, halymenia dentata, tall fescue in Fen River and Wei

分布不均匀。该区农田面积大，但粮食产量低，历史上农牧结合为主。主要栽培草种有苜蓿、沙打旺、小冠花、无芒雀麦、苇状羊茅、鸭茅、红豆草、扁穗冰草等。适宜品种有‘晋南’苜蓿、‘偏关’苜蓿、‘关中’苜蓿、‘陕北’苜蓿、‘陇东’苜蓿等地方品种，及‘甘农1号’‘甘农2号’杂交苜蓿、早熟沙打旺、‘宝石’和‘宾吉夫特’小冠花、‘卡尔顿’无芒雀麦、‘甘肃’红豆草等，也可种植‘333/A’狭叶野豌豆、苦荬菜、红苋等一年生牧草。主要亚区有：

(1) 晋东豫西丘陵山地苜蓿、沙打旺、小冠花、无芒雀麦、苇状羊茅亚区——黄土高原东南部，温暖半干旱气候，干旱是最突出的问题。

(2) 汾渭河谷苜蓿、小冠花、无芒雀麦、鸡脚草、苇状羊茅亚区——山西晋中、晋南，陕西关中平原，农业发达，还可种植沙打旺、红豆草。

(3) 晋、陕、甘、宁高原丘陵沟壑苜蓿、沙打旺、红豆草、小冠花、无芒雀麦、扁穗冰草亚区——山西北部、西部，陕北，甘肃东部，宁南地区，暖温带半湿润气候，历史上以牧业为主。

(4) 陇中青东丘陵沟壑苜蓿、沙打旺、红豆草、扁穗冰草、无芒雀麦亚区——农牧结合区，干旱、土地贫瘠是农业的限制因素。

10.4.5 长江中下游白三叶、黑麦草、苇状羊茅、雀稗栽培区

包括江西、浙江、上海及湖南、湖北、江苏、安徽的大部及河南小部。本区位于中亚热带和北亚热带，气候温暖湿润，冬冷夏热，四季分明，水热资源丰富，气候具有明显的过渡性质，

River Valley—Central and south of Shanxi province, Guanzhong Plain in Shaanxi province. Grain production is fully developed. Adsurgens, crown vetch and sainfoin are well adapted as well.

(3) Sub-region of alfalfa, adsurgens, sainfoin, crown vetch, smooth bromegrass, *Agropyron cristatum* (L.) Gaertn in Shanxi, Shaanxi, Gansu, Ningxia hilly plateau—The north and west in Shanxi, Northern Shaanxi, Eastern Gansu and Southern Ningxia belong to warm temperate semi-humid climate. This region historically is dominated by livestock production.

(4) Sub-region of alfalfa, Adsurgens, sainfoin, *Agropyron cristatum* (L.) Gaertn, smooth bromegrass in the central regions of Gansu and the gully areas of eastern Qinghai—Cropping combined with livestock. Arid and barren lands are limiting factors for agricultural development.

10.4.5 Middle and lower part of Yangtze River white clover, ryegrass, tall fescue, *Paspalum* cultivation area

This region includes Jiangxi, Zhejiang, Shanghai and most of Hunan, Hubei, Jiangsu, Anhui and a small part of Henan Provinces. This region is located in the central and northern subtropical zone. Climate is warm and humid, hot in summer and cold in winter. There are four distinct seasons, resources of water and heat is abundant. Climate has an obvious transitional character. Temperate forages cannot grow well in summer and it is hard for the tropical forages to over wintering. Acidic soil is yellow brown, red and yellow, with pH ranging from 4.0 to 6.5. Deficiency of phosphorus and potassium in Soil leads to the low fertility.

Grain production is well developed and the the multiple crop index is high in this area. Grassland resources are mainly grass mountains and slopes intertidal zone. White clover, perennial ryegrass, tall fescue, Paspalum urillei, Paspalum wettsteinii, bermudagrass, orchardgrass and red clover are the main cultivated forages in this region. Suitable cultivars include white clover 'emu No.1', festulolium 'nannong No.1', tall fescue Fawn, 'yancheng' meadow fescue, bermudagrass 'anza No.1', Pennisetum americanum 'ningza No.3', Sor-

温带牧草不容易过夏，热带牧草又不容易越冬，土壤为黄棕壤、红壤和黄壤，多呈酸性，pH4.0~6.5，缺磷少钾，土壤肥力较低。

该区农业生产发达，复种指数高，草地资源主要是草山草坡、滩涂地。主要栽培草种有白三叶、多年生黑麦草、苇状羊茅、小花毛花雀稗、宽叶雀稗、狗牙根、鸭茅、红三叶等。适宜的品种有‘鄂牧1号’白三叶、‘南农1号’羊茅黑麦草、‘法恩’苇状羊茅、‘盐城’牛尾草、‘岸杂1号’狗牙根、‘宁杂3号’美洲狼尾草、‘皖草2号’高粱－苏丹草杂交种、‘冬牧70’黑麦、‘盐城’多花黑麦草、‘上农’四倍体多花黑麦草等。亚区有:

(1) 苏浙皖鄂豫平原丘陵白三叶、苇状羊茅、苜蓿亚区——亚热带向暖温带的过渡带，温暖湿润，雨量充沛，多集中于夏季，年降水量1 000 mm以上。温带牧草难越夏，热带牧草难越冬，但过渡区各种均可生长。

(2) 湘赣丘陵山地白三叶、‘岸杂1号’狗牙根、苇状羊茅、紫花苜蓿、雀稗亚区——亚热带气候，温和湿润，年降水量1 300~1 700 mm，水热条件优越，温热带牧草均可播种。海拔500 m以上的山地降雨多气温低，可选择三叶草、鸡脚草、多年生黑麦草。

(3) 浙皖丘陵山地白三叶、苇状羊茅、多年生黑麦草、鸡脚草、红三叶亚区——中亚热带气候，四季分明，水土流失严重，少雨易旱、多雨易涝。

10.4.6 华南宽叶雀稗、卡松古鲁狗尾草、大翼豆、银合欢栽培区

包括海南、广东、广西、福建及云南南部。本区是我国水热资源最充足的地区，北回归线穿越大部分地区，光照

ghum-sudangrass hybrid varieties ‘wancao No.2’, rye ‘dongmu 70’, ‘yancheng’ Italian ryegrass, Italian ryegrass ‘ganxuan No.1’ and ‘shangnong’ tetraploid Italian ryegrass.

Sub-regions include:

(1) Sub-region of white clover, tall fescue, alfalfa in Jiangsu, Zhejiang, Anhui, Hubei, Henan hilly plains—Subtropical and warm temperate transition zone, warm and humid. Rainfall is abundant and concentrated during summer. Precipitation exceeds 1 000 mm annually.

(2) Sub-region of white clover, bermudagrass “Anza No.1”, tall fescue, alfalfa, paspalum in Hunan and Jiangxi hills and mountains—Subtropical climate, mild and humid, rainfall is 1 300~1 700 mm. Climatic conditions are excellent, both temperate and tropical forages can be planted. Mountains where the altitude is higher than 500 m have abundant precipitation and the moderate temperatures. Clover, Halymenia dentate and perennial ryegrass are adapted in this area.

(3) Sub-region of white clover, tall fescue, perennial ryegrass, Halymenia dentata, red clover in Zhejiang and Anhui hills and mountains—Central Asian tropical climate, with four distinct seasons in a year. Soil erosion is a serious challenge. Less rainfall leads to drought and heavy rainfall results in floods.

10.4.6 Latifolia paspalum, Kasonggulu foxtail, large wing beans, leucaena cultivation area in southern of China

This region includes Hainan, Guangdong, Guangxi, Fujian and southern Yunnan provinces. This area has the most favorable climate in China. The tropic of Cancer crosses most areas and make abundant strongly sunshine. Ocean monsoons lead to abundant and reliable rainfall. Most areas have a long summer without winter. It is a warm, rainy tropical and subtropical climate. The soil is mountainous red soil, lateritic red soil, latosol. The pH is range from 4.5 to 5.5. Nitrogen content is low in this region, and the phosphorus is often deficient.

Tropical forages are suitable for this area.

Sub-regions include:

强烈，海洋性季风使得雨量充沛，绝大部分地区呈现长夏无冬、温热多雨的热带、南亚热带气候。土壤为山地红壤、赤红壤、砖红壤，pH 多在 4.5~5.5 之间，氮含量低，磷普遍缺乏。

本区最适宜栽培热带牧草，亚区有：

(1) 闽、粤、桂南部丘陵平原大翼豆、银合欢、格拉木柱花草、卡松古鲁狗尾草、宽叶雀稗、象草亚区——南亚热带和热带气候，长夏无冬。年均温 20℃以上。

(2) 闽、粤、桂北部低山丘陵银合欢、银叶山蚂蝗、绿叶山蚂蝗、宽叶雀稗、小花毛花雀稗亚区——中亚热带气候，丘陵山地为主。

(3) 滇南低山丘陵大翼豆、格拉木柱花草、宽叶雀稗、象草亚区——南亚热带，地形复杂，西南边缘为湿热多雨的热带气候。

(4) 台湾山地平原银合欢、山蚂蝗、柱花草、毛花雀稗、象草亚区——热带、亚热带，水热充沛为我国之首，降雨量 2 000 mm 以上，农业发达。

10.4.7 西南白三叶、黑麦草、红三叶、苇状羊茅栽培区

本区包括陕西南部、甘肃东南部、四川、云南大部、贵州全部、湖北、湖南西部，地处亚热带，全区 95% 的面积是丘陵山地和高原。气候为亚热带湿润气候，冬季气候温和，生长期较长，雨量充沛，年降水量 1 000mm 以上，冬无严寒，夏无酷暑。主要栽培草种是喜温和湿润气候的温带牧草，如白三叶、红三叶、多年生黑麦草、多花黑麦草、鸭茅、苇状羊茅等。适宜的品种有 ‘胡依阿’白三叶、‘贵州’白三叶、‘川引拉地诺’白三叶、‘巴东’红三叶、‘亚溪’

(1) Sub-region of *Macroptilium lathyroides* (L.) Urban, *Leucaena glauca* (L.) Benth., *Stybsanthes guianensis* (Aubl) Sw., *Setaria* anceps Stapf cv., *Paspalum wettsteinii* Hackel, *Pennisetum purpureum* Schum in Fujian, Guangdong, Guangxi southern hilly plains—South subtropical and tropical climates with long summers but without winter. The average annual temperature is above 20 degrees.

(2) Sub-region of *Leucaena glauca* (L.) Benth., silver leaf beggarweed, green leaf beggarweed, *Paspalum wettsteinii* Hackel, Paspalum urillei in northern Fujian, Guangdong, Guangxi hilly—subtropical climate, the terrain is hilly and mountainous.

(3) Sub-region of *Macroptilium lathyroides* (L.) Urban, *Stybsanthes guianensis* (Aubl) Sw. *Paspalum wettsteinii* Hackel, *Pennisetum purpureum* Schum in southern Yunnan hilly—South subtropical climates with complex terrain. The southwest portion is a humid tropical rainy climate.

(4) Sub-region of *Leucaena glauca* (L.) Benth., beggarweed, *Stybsanthes guianensis* (Aubl) Sw., *Paspalum dilatatum* Poir., *Pennisetum purpureum* Schum in Taiwan mountain plains—Tropical and subtropical. It is the most hot and humid region in China. More than 2 000 mm of annual rainfall and grain production is well developed.

10.4.7 Southwestern white clover, ryegrass, red clover, tall fescue production region

This is a subtropical area, includes southern Shaanxi, southeastern Gansu, Sichuan, most of Yunnan, Guizhou, Hubei and western Hunan. Hills, mountains and plateaus comprise 95% of the area. The climate is humid subtropical with mild winters, a long growing season and abundant rainfall. Annual rainfall is more than 1 000 mm. Summer and winter temperatures are moderate, no extreme temperature. The main cultivated forage species which thrive is in the mild and humid climate includes whiter clover, red clover, perennial ryegrass, Italian ryegrass, orchardgrass and tall fescue. Suitable cultivars include ‘ huia’ white clover, ‘guizhou’ white clover, ‘sichuan latino’ white clover, ‘patong’ red clover, ‘yaxi’ red clover, ‘mountain min’ red clover, ‘narok’ setaria, ‘moun-

红三叶、‘岷山’红三叶、‘纳罗克’非洲狗尾草、‘岷山猫’尾草、‘涪陵’十字马唐、‘古蔺’鸭茅、‘阿伯德’多花黑麦草等。亚区有：

(1) 四川盆地丘陵平原白三叶、黑麦草、苇状羊茅、扁穗牛鞭草、聚合草亚区——中亚热带，春旱、夏热、秋雨、冬暖，年降水量 900~1 200 mm。

(2) 川陕甘秦巴山地白三叶、红三叶、苜蓿、黑麦草、鸭茅亚区——陕南、陇东、湖北西北部及四川盆地北部边缘——暖温带向北亚热带过渡，90% 为山地，适宜于建立人工草场放牧。

(3) 川鄂湘黔边境山地白三叶、红三叶、黑麦草、鸡脚草亚区——中亚热带湿润气候，冬无严寒，雨量充沛，日照不足，畜牧业较发达，以养殖牛、羊、猪为主。

10.4.8 青藏高原老芒麦、垂穗披碱草、中华羊茅、苜蓿栽培区

本区包括西藏全部、青海大部、甘肃的甘南、四川西部、云南西北部，是我国面积最大、地势最高、气候最冷的高原，号称世界屋脊，为大陆性高原气候，冬寒夏凉，日照长，雨水少，太阳辐射强。气候寒冷干燥，无霜期短，生态环境严酷。

该区以牧业为主。主要栽培草种有老芒麦、垂穗披碱草、中华羊茅、无芒雀麦、苜蓿、沙打旺、扁蓿豆、红豆草等。适宜品种有‘川草 1 号’‘川草 2 号’老芒麦、‘甘南’垂穗披碱草、早熟沙打旺等。亚区有：

(1) 藏南高原河谷苜蓿、红豆草、无芒雀麦亚区——位于西藏西南部，北部为一江两河（雅鲁藏布江、拉萨河、年楚河），地势平缓，气候温凉，西藏粮食的

tain min' phleum pretense, 'fuling' *Digitaria cruciata*, 'gulin' orchardgrass, 'abed' Italian ryegrass.

Sub-regions include:

(1) Sub-region of white clover, ryegrass, tall fescue, Hemarthria compressa, Symphytum in the Hilly plains of Sichuan Basin—Central subtropical, dry in spring, hot in summer, rainy in autumn, warm in winter. Annual precipitation is 900~1 200 mm.

(2) Sub-region of white clover, red clover, alfalfa, ryegrass, orchard grass in Sichuan, Shaanxi, Gansu mountains, which covers Southern of Shaanxi, eastern of Gansu, northwestern Hubei and the northern edge of the Sichuan Basin, a transition zone from warm temperate to northern subtropical. 90% of this sub-region is mountainous, but very suitable for establishment of artificial grasslands.

(3) Sub-region of white clover, red clover, ryegrass, orchard grass in Sichuan, Hubei, Hunan and Guizhou. Mountainous—Central humid subtropical climate, winters are not cold and rainfall is abundant, but with less sunshine. Livestock production is well developed, especially cattle, sheep and pigs.

10.4.8 Qinghai-Tibet Plateau *Elymus sibiricus*, *Elymus nutans*, China fescue, alfalfa production region

This region includes all of Tibet, most of Qinghai, south of Gansu, west of Sichuan and northwest of Yunnan. This plateau is the largest, highest and coldest climate in China and is also known as the roof of the world. It has a continental plateau climate. Winters are cold and summers are cool. There are long sunshine hours, limited rainfall and strong solar radiation. The frost-free period is short and the environment is very harsh.

The area dominated by livestock production. Primary forage species are Siberian wildrye, Elymus nutans, Festuca sinensis, awnless brome, alfalfa, Astragalus adsurgens, Medicago ruthenica and Sainfoin. Suitable varieties are Siberian wildrye 'sichuan grass No. 1' and Siberian wildrye 'sichuan grass No. 2', 'Gannan' Elymus nutans and erect milkvetch.

Sub-regions include:

主产区，主要牲畜是藏羊和牦牛，牧区以低、中型牧草为主。

(2) 藏东川西河谷山地老芒麦、无芒雀麦、苜蓿、红豆草、白三叶亚区——农林牧垂直分布层次分明，可谓“一山有两季，十里二层天”。

(3) 藏北青南垂穗披碱草、老芒麦、中华羊茅、冷地早熟禾亚区——青藏高原的主体，占全区面积的一半，是一个地高天寒（年均温 0 ℃以下，年降水量 100~250 mm）、夏短冬长、草场辽阔的纯牧区。人烟稀少，靠天养畜。常见的栽培豆科牧草难以越冬。

(4) 环湖甘南老芒麦、垂穗披碱草、中华羊茅、无芒雀麦亚区——青藏高原东北边缘，地势平缓，雨量较多，是高原上最好的草甸草场。燕麦面积大，多年生牧草种类较单一，豆科牧草难越冬，农牧业经营粗放，草地补播任务大。

(5) 柴达木盆地沙打旺、苜蓿亚区——青海省西北部，是我国海拔最高的盆地，属干燥大陆性气候，海拔高、雨量少，年降水量 17.6~210 mm，年蒸发量 2 088~3 298 mm，沙打旺最适于本区种植，苜蓿需要灌溉条件。

10.4.9 新疆苜蓿、无芒雀麦、老芒麦、木地肤栽培区

新疆位于我国西北部，地处欧亚大陆，远距海洋，四周高山环绕，天山横列中部将全疆分为南疆和北疆自然条件有明显差异的两部分，气候干燥而温暖，全疆年均降水量 150 mm，北疆各地为 150~200 mm，南疆只有 20 mm，山区、迎风坡降水量较多。新疆是我国第二大牧区，畜牧业发达，有农牧结合和实行季节轮牧的传统。利用水源较好的地方发展了著名的绿洲农业。

(1) Sub-region of alfalfa, sainfoin, awnless brome in valleys of the southern Tibet on plateau—located in southwestern Tibet. The northern area has three rivers (Brahmaputra, Lhasa and Nianchu rivers). Flat terrain and cool climate characterize this region. This is the principle grain producing area in Tibet. The main livestock species are tibetan sheep and yak, raised in pastoral areas with low to moderate forage production.

(2) Sub-region of Siberian wildrye, awnless brome, alfalfa, white clover and Sainfoin in the valley of eastern Tibet and western Sichuan. Livestock and crop production reflect a hierarchy. There is an old saying for this region: "There are two seasons in a mountain, and two different weather within ten miles".

(3) Sub-region of Elymus nutans, siberian wildrye, Festuca sinensis, Poa ciymophila in northern Tibet and southern Qinghai. This is the main area of Qinghai-Tibet Plateau which is half of the subregion. It is completely a pastoral area, with high altitude, cold and dry weather (annual average temperature is below 0 ℃ and rainfall is 100~250 mm), short summer and long winter. It is sparsely populated and livestock rely on the climate. Use of cultivated legumes is difficult because of frequent winterkill.

(4) Sub-region of the southern siberian wildrye, Elymus nutans, Festuca sinensis, awnless brome subregion lies in the northeastern margin of the Qinghai-Tibet Plateau. Terrain is flat and the climate is cool in this region. The best meadow on the plateau occurs in this subregion. Oat is widely grown. Perennial forages species are less and it is difficult for legumes over wintering. Crop and livestock management is extensive. It is very important to oversowing.

(5) Sub-region of Astragalus adsurgens, alfalfa sub district in Qaidam Basin is located in northwest Qinghai and it is the highest basin in China. It has a continental dry climate at high altitude and low rainfall. Annual precipitation is 17.6~210 mm, but evaporation is 2 088~3 298 mm. Astragalus adsurgens is the species most suitable for cultivation. Alfalfa requires proper irrigation.

农区种植苜蓿有良好的基础，不仅增加牲畜饲草，并可肥田养土。主要栽培草种有苜蓿、无芒雀麦、老芒麦、木地肤、沙枣、红豆草等。适宜的品种有‘新疆’大叶苜蓿、‘北疆’苜蓿、‘新牧 1 号’杂花苜蓿、‘新牧 2 号’紫花苜蓿、‘阿勒泰’杂花苜蓿、‘新雀 1 号’无芒雀麦、‘奇台’无芒雀麦、‘紫泥泉’新麦草、‘巩乃斯’木地肤、‘伊犁蒿’等。亚区有：

(1) 北疆苜蓿、木地肤、无芒雀麦、老芒麦亚区——天山以北（包括准噶尔盆地、阿尔泰山南坡、博乐等），气候较湿润，年降水量 150~260 mm，木地肤飞播区。

(2) 南疆苜蓿、沙枣亚区——四周高山，中央低陷的巨大盆地。天山阻断西来湿气，气候干燥，年降水量 20~40 mm，无霜期 200~220 d。塔里木盆地边缘、哈密等地是大叶苜蓿的分布区。

10.4.9 Xinjiang alfalfa, smooth brome grass, siberian wildrye, *Kochia prostrate* cultivation region

Xinjiang is located in northwestern China, far from the ocean and surrounded by high mountains. The Middle Tianshan Mountains divide Xinjiang into two parts, southern and northern Xinjiang where differences in natural conditions are apparent. Climate is dry and warm. Average annual precipitation in Xinjiang is 150 mm, with about 150~200 mm in the north and only 20 mm in the south, where precipitation is influenced by the location which is windward of the mountains. Xinjiang is China's second major pastoral area where livestock production is well developed. There are traditions influence crop and livestock production, their combination and allocation of grazing. Optimizing use of local water sources lead to development of oasis agriculture. Planting alfalfa in crop fields provides a reliable forage resource. It increases livestock forage and also improves raise soil fertility. The main forage species are alfalfa, smooth brome grass, siberian wildrye, Kochia prostrata, Elaeagnus angustifolia and sainfoin. Suitable varieties are 'xinjiang' *Medicago sativa* L., 'beijiang' alfalfa, Medicago varia 'xinmu No.1', alfalfa 'xinmu No.2', 'aletai' Medicago varia, awnless brome 'xinque No.1', 'qitai' awnless brome, *Psathyrostachys juncea* (Fisch.) Nevski, 'gongnaisi' Kochia prostrate and Artemisia 'yili'.

Sub-regions include:

(1) Sub-region of alfalfa, Kochia prostrata, awnless brome, siberian wildrye in northern Xinjiang, north of the Tianshan Mountains (including Junggar Basin, the south slope of Altai Mountains and Bole.). The climate is warm. Annual precipitation is 150~260 mm. Kochia prostrate is aerial seeded in this subregion.

(2) Sub-region of alalfa, Elaeagnus angustifolia in southern Xinjiang, surrounded by mountains and central the sunken Great Basin. Tianshan mountains block moisture movement from the west. The climate is desert. Annual precipitation is 20~40 mm. The frost-free season is 200 ~ 220 d. Broad leaves alfalfa is distributed at the edge of the Tarim Basin and Hami.

主要参考文献

BARNES R F, NELSON C J, MOORO K J, et al. Forages. 6th ed. Anles, Iowa: blackwell publishing, 2001.
陈宝书．牧草饲料作物栽培学．北京：中国农业出版社，2001.
陈冬季．论西域战争岩画的文化意义．新疆社科论坛，1992(3)：60-70.
陈默君，贾慎修．中国饲用植物．北京：中国农业出版社，2002.
董宽虎，沈益新．饲草生产学．北京：中国农业出版社，2003.
谷安琳，王宗礼．中国北方草地植物彩色图谱．北京：中国农业科学技术出版社，2009.
韩建国，毛培胜．牧草种子学．北京：中国农业大学出版社，2011.
韩建国．草地学．3 版．北京：中国农业出版社，2009.
韩建国．牧草种子学．北京：中国农业大学出版社，1997.
洪绂曾．中国草业史．北京：中国农业出版社，2011.
胡金良．植物学．北京：中国农业大学出版社，2012.
贾慎修．中国饲用植物志．北京：中国农业出版社，1987.
李毓堂．草产业和牧区畜牧业改革发展 30 年．草业科学，2009，26(1)：3-7.
全国牧草品种审定委员会．中国牧草登记品种集．北京：中国农业大学出版社，中国农业出版社，1999—2006.
任继周．草地农业生态系统通论．合肥：安徽教育出版社，2004.
任继周．草业大辞典．北京：中国农业出版社，2008.
师尚礼．草类植物种子学．北京：科学出版社，2011.
王成章．饲料生产学．郑州：河南科学技术出版社，1998.
王栋．牧草学各论．任继周修订．新一版．南京：江苏科学技术出版社，1989.
王明利．中国牧草产业经济．北京：中国农业出版社，2012.
王贤．牧草栽培学．北京：中国环境科学出版社，2006.
邢旗．内蒙古草原常见植物图鉴．呼和浩特：内蒙古文化出版社， 2008.
颜启传．种子学．北京：中国农业出版社，2001.
贠旭疆．中国主要优良栽培草种图鉴．北京：中国农业出版社，2008.
张子仪．中国饲料学．北京：中国农业出版社，2000.
周寿荣．草地生态学．北京：中国农业出版社，1996.
周寿荣．饲料生产手册．成都：四川科学技术出版社，2004.

附录 1　常见栽培牧草饲料作物英拉汉名称

英文名称	拉丁文名称	中文名称	别　名
Alfalfa lucerne	*Medicago sativa* L.	紫花苜蓿	苜蓿、紫苜蓿
Sickle alfalfa Yellow sickle medick	*Medicago falcata* L.	黄花苜蓿	野苜蓿、镰荚苜蓿
Varigated alfalfa	*Medicago varia* Martyn	杂花苜蓿	杂种苜蓿
Sainfoin	*Onobrychis viciifolia* Scop.	红豆草	驴食豆、驴食草
White clover, White trefoil	*Trifolium repens* L.	白三叶	白车轴草、荷兰翘摇
Red clover, Red trefoil	*Trifolium pratense* L.	红三叶	红车轴草、红荷兰翘摇、红椒草
Russian fenugreek	*Medicago ruthenica* (L.) Trautv.	扁蓿豆	花苜蓿、野苜蓿
White sweetclover	*Melilotus albus* Desr.	白花草木樨	白香草木樨
Yellow sweetclover	*Melllotus offlcinalis* (L.) Desr.	黄花草木樨	草木樨、香马料、黄香草木樨
Caribbean stylo, Pencil flower, Mother Segal Tebeneque	*Stylosanthes hamata* (L.) Taub.	有钩柱花草	有钩笔花豆
Erect milkvetch	*Astragalus adsurgens* Pall.	沙打旺	直立黄芪、麻豆秧
Chinese milkvetch	*Astragalus sinicus* L.	紫云英	翘摇、红花菜、米布袋
Common vetch, Fodder vetch	*Vicia sativa* L.	箭筈豌豆	大巢菜、春巢菜、野豌豆、救荒野豌豆
Hairy vetch, Winter vetch	*Vicia villosa* Roth.	毛苕子	冬巢菜
Glabrous villose vetch	*Vicia villosa* Roth var. glabrescens Koch.	光叶紫花苕	光叶紫花苕子、细毛苕子
Korshinsk peashrub	*Caragana korshinskii* Kom.	柠条锦鸡儿	柠条、白柠条、毛条
Crownvetch	*Coronilla varia* L.	多变小冠花	小冠花、绣球小冠花
Bushclover	*Lespedeza bicolor* Turcz.	胡枝子	二色胡枝子、扫条
Leucaena	*Leucaena leucocepllala* L. De.Wit.	银合欢	萨尔瓦多银合欢、新银合欢

英文名称	拉丁文名称	中文名称	别　名
Birdsfoot trefoil	*Lotus corniculatus* L	百脉根	五叶草、牛角花、鸟趾草
Siratro	*Macroptllium atropur pureum* (DC)Urban.	大翼豆	紫菜豆、紫花大翼豆
Wheatgrass, Crested wheatgrass	*Agropyron cristatum* L.	扁穗冰草	野麦子
Desert wheatgrass	*Agropyron desertorum* (Fisch.) Schult.	沙生冰草	
Smooth bromegrass	*Bromus inermis* Leyss.	无芒雀麦	
Perennial ryegrass	*Lolium perenne* L.	多年生黑麦草	宿根黑麦草、黑麦草、英国黑麦草
Italian ryegrass	*Lolium multiflorum* L.	多花黑麦草	意大利黑麦草、一年生黑麦草
Chinese leymus	*Leymus chinensis* (Trin.) Tzvel.	羊草	碱草
Dahuria wildryegrass	*Elymus dahuricus* Turcz.	披碱草	
Drooping wildryegrass	*Elymus nutans* Griseb.	垂穗披碱草	
Sibirian wildryegrass	*Elymus sibiricus* L.	老芒麦	西伯利亚披碱草
Tall fescue	*Festuca arundinacea* Schreb.	苇状羊茅	苇状狐茅、高羊茅
Bermuda grass	*Cynodon dactylon*(L.)Pers.	狗牙根	
Bahiagrass	*Paspalum notatum* Flugge.	巴哈雀稗	百喜草、巴喜亚雀稗
Pearl Millet	*Pennisetum americanum* (L.) LeekexP. purpureum Schum.	杂交狼尾草	
Sudan grass	*Sorghum sudanense*(Piper)Stapf.	苏丹草	野高粱
Reed canarygrass	*Phalaris arundinacea* L.	虉草	草芦、丝带草、金色草苇
Timothy	*Phleum pratense* L.	猫尾草	梯牧草
Kentucky bluegrass	*Poa pratensis* L.	草地早熟禾	六月禾、蓝草
Maizecorn	*Zea mays* L.	玉米	玉蜀黍、包谷、珍珠米、苞芦
Barley	*Hordeum vulgare* L.	大麦	
Rye	*Secale cereale* L.	黑麦	粗麦、洋麦

英文名称	拉丁文名称	中文名称	别　名
SorghumBroom-corn	*Sorghum bicolor* (L.) Moench	高粱	蜀黍
Triticale	*Triticale* Wittmack	小黑麦	
Common chicory	*Cichorium intybus* L.	菊苣	
Samaradaisy	*Pterocypsela indica* (L.) Shih.	苦荬莱	翅果菊、苦麻菜、山莴苣、野莴苣、八月老
Cup plant	*Silphium perfoliatum* L.	串叶松香草	菊花草、杯草、法国香槟草
Common comfrey	*Symphytum pezegrinum* L.	聚合草	紫菜根、紫草、爱国草、友谊草、肥羊草
Turnip	*Brassica rapa* L.	芜菁	
Pumpkin	*Cucurbita moschata* (Duch. ex Lam.) Duch.ex Poiret.	南瓜	中国南瓜、倭瓜、番瓜、饭瓜、番南瓜、北瓜

附录 2　生词表

Introduction

alfalfa power 苜蓿粉
arable land 可耕地
bale *n.* 包；捆
browse *n.* 嫩草；枝条
breeding *n.* 繁殖；饲养；育种
crude fiber (CF) 粗纤维
crude protein(CP) 粗蛋白质
ecological *adj.* 生态的；生态学的
edible *adj.* 可食用的
forage *n.* 饲料；草料
feed *n.* 饲料 *v.* 饲喂
ferment *vt.* 使发酵 *vi.* 发酵 *n.* 发酵；酵素发酵
grain *n.* 粮食
grass-crop rotation 草田轮作
grass agro-ecological zones 草地农业生态区域
greenhouse gas 温室气体
harvest *n.* 收获
hay *n.* 干草
herbage *n.* 草本植物；牧草
husbandry *n.* 畜牧业
leaf *n.* 叶
legume *n.* 豆类；豆科植物
livestock *n.* 牲畜；家畜
loess plateau 黄土高原
mast *n.* 树木果实或种子
melamine *n.* 三聚氰胺
mitigate *vt.* 使缓和；使减轻
multiple functions 多功能
nitrogen *n.* 氮
nonwoody *adj.* 非木本的
per unit area 单位面积
root *n.* 根
root nodules 根瘤菌
ruminant *n.* 反刍动物
stem *n.* 茎
seed *n.* 种子
soil organic matter 土壤有机质
species *n.* 物种；种类　*adj.* 物种上的
soil fertility 土壤肥力
total output 总产出

Chapter 1　Classification of Forages

annual *n.* 一年生植物　*adj.* 年度的；每年的
axillary bud 腋芽
biennial *n.* 两年生植物　*adj.* 持续两年的
binomial *adj.* 二项式的；两种名称的
bottom grass 下繁草
bunch grasses 疏丛型禾草
compositae *n.* 菊科；菊科植物
diversity *n.* 多样性
erect *adj.* 直立的　*vt.* 使竖立
grasses *n.* 禾本科
grazing *n.* 放牧；牧草
lateral *n.* 侧部　*adj.* 侧面的；横向的
lifespan *n.* 寿命
long-lived perennials 长寿多年生
morphology *n.* 形态学；形态论
perennial *n.* 多年生植物　*adj.* 多年生的
polemoniaceae *n.* 旋花科
regrowth habit 再生习性
rhizomatous *adj.* 地下茎的；生地下茎的
rhizomatous legumes 根蘖型豆科牧草

rhizome-bunch grasses 根茎疏丛型禾草
root crown 根冠；根颈
rosette forming forages 莲座状草
short-lived perennials 短寿多年生
sod forming grasses 密丛型禾草
solanaceae *n.* 茄科
stoloniferous forages 匍匐型牧草
tap-rooted legumes 轴根型豆科牧草
taxonomic *adj.* 分类的
tendrils *n.* 卷须；卷须状物（tendril 的复数形式）
tiller *n.* 分蘖 *vi.* 分蘖
top grass 上繁草
umbelliferae *n.* 伞形科；伞形花序植物

Chapter 2 Growth and Reproduction Characteristic of Forages

amino acid(AA) 氨基酸
anther *n.* 花药；花粉囊
auricle *n.* 耳廓；叶耳
assimilation *n.* 同化；吸收；同化作用
auxin *n.* 植物生长素；植物激素
biological *adj.* 生物的；生物学的
blade *n.* 叶片；刀片
blossom *n.* 花；开花期
boot stage 孕穗期
botanical *adj.* 植物学的
caryopsis *n.* 颖果
catalysis *n.* 催化作用
cell division 细胞分裂
chloroplast *n.* 叶绿体
coleoptile *n.* 胚芽鞘
coleorhiza *n.* 胚根鞘
collar *n.* 叶枕
corolla *n.* 花冠
cotyledon *n.* 子叶
cross-pollination 异花授粉
cuticle *n.* 角质层；表皮
cytokinin *n.* 细胞分裂素
defoliation *n.* 落叶，去叶；脱叶
development *n.* 发展；开发；发育
dicots *n.* 双子叶植物（等于 dicotyledons，dicot 的复数）
differentiation *n.* 变异；分化；区别
dough stage 蜡熟期
elongate *adj.* 伸长的；延长的 *vi.* 拉长；延长；伸长
embryo *n.* 胚芽；胚胎；初期 *adj.* 胚胎的；初期的
embryo sac 胚囊
emergence *n.* 出现，浮现；发生；露头
endosperm *n.* 胚乳
entomophilous *adj.* 虫媒的
enzymes *n.* 酶（enzyme 的复数）
epicotyl *n.* 上胚轴
epidermis *n.* 上皮；表皮
epigaeous *adj.* 出土的
extravaginal *adj.*（生长于）鞘外的
fertilization *n.* 施肥；受精
fibrous roots 须根
flag leaf 旗叶
floret *n.* 小花
fragrance *n.* 香味；芬芳
germ *n.* 胚芽
germination *n.* 发芽；发生
growth *n.* 增长；发展；生长
heading stage 抽穗期；出穗期
heritance *n.* 遗产；遗传
hydration *n.* 水合作用
hydrolysis *n.* 水解作用
hypocotyl *n.* 下胚轴
hypogeal *adj.* 地下的；留土的
imbibition *n.* 吸入；吸取
inclined *adj.* 趋向于……的；倾斜的
indefinite inflorescence 无限花序
inflorescence *n.* 花；花序；开花
internode *n.* 节间

intravaginal *adj*. 叶鞘内的
jointing (elongation) stage 拔节期
karyotype *n*. 染色体组型；核型
keel *n*. 龙骨
lamina *n*. 叶片
lateral root 侧根
leaf area index(LAI) 叶面积指数
leaf axil 叶腋
leaflet *n*. 小叶
leaf margin 叶缘
ligule *n*. 舌叶；叶舌
maturation *n*. 成熟
mesophyll *n*. 叶肉
metamorphic leaf 变态叶
midrib *n*. 中脉
milk stage 乳熟期
moisture *n*. 水分；湿度
monocots *n*. 单子叶
mother plant 母株
nectar *n*. 花蜜
node *n*. 节；节点
often cross-pollinated 常异花授粉
overwinter *vi*. 过冬；越冬
oxygen *n*. 氧气
palmate *adj*. 掌状的
panicle *n*. 圆锥花序
perpendicular *n*. 垂线　*adj*. 垂直的；正交的；直立的；陡峭的
petiole *n*. 叶柄；柄部
petiolule *n*. 小叶柄
pinnate *adj*. 羽状的
phenological period 物候期
photosynthesis *n*. 光合作用
physiological metabolism 生理代谢
pistil *n*. 雌蕊
plumule *n*. 胚芽
pod *n*. 豆荚；荚果
pollen grain 花粉粒
pollinate *vt*. 对……授粉
poor soil 贫瘠土壤
primordium *n*. 原基
prostrate *adj*. 俯卧的；平卧的
protoplast *n*. 原生质体
raceme *n*. 总状花序
radical *n*. 胚根
respiration *n*. 呼吸；呼吸作用
rhizome *n*. 植物的根状茎
ripening period 成熟期
root system 根系
rosette *n*. 莲座丛
seed coat 种皮
seedling *n*. 幼苗
self-pollinated *adj*. 自花授粉的
sheath *n*. 鞘；护套；叶鞘
sigmoid curve S 形曲线
sperm *n*. 精子
spikelet *n*. 小穗；小穗状花
stamen *n*. 雄蕊
starch *n*. 淀粉
stigma *n*. 柱头
stolon *n*. 匍匐茎
stoma *n*. 气孔；叶孔
sympetal flower 合瓣花
taproot *n*. 直根；主根
three fundamental points 三基点（最高、最适、最低）
thin-walled cell 薄壁细胞
trifoliolate *adj*. 有三小叶的
trifoliolate leaf 三出复叶
variety *n*. 品种；种类
varying temperature 变温
vascular bundle 维管束
vegetative *adj*. 植物的；营养的
vertically *adv*. 垂直地
vigorous *adj*. 有活力的；精力充沛的
wind-pollinated 风媒的

Chapter 3 Environmental Aspects of Forage Growth and Development

alkaline *adj*. 碱性的
biomass *n*. 生物量
calcium *n*. 钙
cell *n*. 细胞
cell turgor 细胞膨压
clay soil 黏质土壤
day-neutral plant 日中型植物
dry matter (DM) 干物质；固形物
evaporation *n*. 蒸发
granular *adj*. 颗粒的；粒状的
humidity *n*. 湿度；湿气
individual development 个体发育
irrigation *n*. 灌溉
lime *n*. 石灰
loam soil 壤质土
long-day plant 长日照植物
macronutrient *n*. 大量营养素
magnesium *n*. 镁
metabolism *n*. 新陈代谢
microclimate *n*. 小气候
micronutrient *n*. 微量营养素
parent material 母质
phasic development 阶段发育
phosphorus *n*. 磷
photoperiod *n*. 光周期
photophase *n*. 光照阶段
potassium *n*. 钾
precipitation *n*. 降水
seedling emergence 出苗
short day plant 短日照植物
soil bulk density 土壤容重
soil moisture 土壤湿度；墒情
soil texture 土壤质地
sowing *n*. 播种
springiness *n*. 春性
sulfur *n*. 硫黄
tillage *n*. 耕作；耕种
tissue *n*. 组织；纸巾
transpiration *n*. 蒸发；散发；蒸腾作用
vernalisation *n*. 春化现象
wavelength *n*. 波长
winterness *n*. 冬性
yield *n*. 产量

Chapter 4 Artificial Grassland Establishment and Management

animal husbandry 畜牧业；畜牧学
artificial grassland 人工草地
awn *n*. 芒
band application 带状施肥
band seeding 带状播种；施肥条播
broadcast *n*. 广播；撒播 *adj*. 广播的
companion crop (cover crop, nurse crop) 保护作物
companion seeding 保护播种
corn stover 去穗整株玉米
cultivar *n*. 培育植物；栽培变种
dormancy *n*. 休眠；冬眠
drill seeding 条播
forage-crop rotation 草田轮作
frost seeding 顶凌播种
glume *n*. 颖；颖片
gramineous *adj*. 禾本科的；草的；似草的
hard seed 硬实种子
inoculation *n*. 接种
leguminous *adj*. 豆科的
livestock *n*. 牲畜；家畜
manure *n*. 肥料；粪肥
monoculture *n*. 单作；单一栽培
mixture seeding 混播
no-till 免耕
paddock *n*. 围场；小牧场
pasture *n*. 草地；牧场

postripeness *n.* 后熟
primary tillage 基本耕作
pure live seed 纯活种子；种子用价
reduced tillage 少耕栽培
rhizobium *n.* 根瘤菌
rotational grazing 循环放牧；轮牧
seeding rates 播种密度；播种量
soil erosion 土壤侵蚀；土壤流失
sprigging *n.* 插钉；幼苗移殖
stand *n.* 一片生长的植物
symbiotic relationship 共生关系
variable temperature 变温

Chapter 5 Legumes

3-nitro-1-propanol 3- 硝基 -1- 丙醇
accumulated temperature 积温
acid *adj.* 酸性的
alkaloids *n.* 生物碱
alternative *adj.* 互生的
ammonium nitrate 硝酸铵
amorpha *n.* 紫穗槐属
apex *n.* 先端
aphid *n.* 蚜虫
artemisia desterorum spreng 沙蒿
asexual reproduction 无性繁殖
astragalus *n.* 黄芪属
base fertilizer 底肥
beak *n.* 喙
beetle *n.* 金龟子
bladder-shaped 膀胱状的
bloat *n.* 臌胀病
bloom *n.* 开花
boron *n.* 硼
branching stage 分枝期
bud stage 孕蕾期
bumblebee *n.* 大黄蜂
calcium superphosphate 过磷酸钙
calligonum *n.* 沙拐枣
canopy height 草层高度
capitulum *n.* 头状花序
carbendazim *n.* 多菌灵
carbonate *n.* 碳酸盐
caucasus sainfoin 高加索红豆草
Chinese lespedeza 截叶胡枝子
Chinese milkvetch 紫云英
cicer milkvetch 鹰嘴紫云英
clavate *adj.* 棒状的
common vetch 箭筈豌豆
compound fodder 配合饲料
concave *adj.* 凹的
conception rate 受孕率
convex *n.* 凸面体　*adj.* 凸面的
cortex *n.* 皮质；树皮
coumarin *n.* 香豆素
deciduous *adj.* 落叶的
desertification *n.* （土壤）荒漠化；沙漠化（等于 desertization）
diarrhea *n.* 腹泻；痢疾
dibble seeding 点播
dodder *n.* 菟丝子
bud stage 现蕾期
elliptic *adj.* 椭圆形的
emarginated *adj.* 顶端微凹的
epidemic *n.* 流行病
erect milkvetch 沙打旺
estadiol *n.* 香豆雌醇
estrogen *n.* 雌激素
estrogen-coumadin 雌性激素
evaporation coefficient 蒸腾系数
even-pinnately 偶数羽状的
evergreen *adj.* 常绿的
fall dormancy 秋眠性
fasciculate *adj.* 丛生的
forage powder 草粉
frost-free period 无霜期
fungicide *n.* 杀菌剂

genera *n.* 属
germination rate 发芽率
glandular spot 腺点
glycerinum 甘油
green hay 青干草
green manure 绿肥
green-up 返青
hairy vetch 毛苕子
haylage *n.* 半干青贮
heliophilous *adj.* 阳性的
herbaceous *adj.* 草本的
horse tamarind 银合欢
hydrocyanic acid 氢氰酸
hydrogen cyanide 氢氰酸
hydrogen glycosides 氢甙
indehiscent *adj.* 果皮不裂的
intercrop *n.* 间作
isoflavones *n.* 异黄酮
keratinize *vt.* 使角质化 *vi.* 角质化
kidney-shaped 肾形的
lanceolate *adj.* 披针形的
landscaping plant 景观植物
leafcutter bees 切叶蜂
lignify *vt.* 木质化
little-leaf peashrub 小叶锦鸡儿
manganese *n.* 锰
meadow fescue 牛尾草
melilotus *n.* 草木樨属
membranous *adj.* 膜状的
middle peashrub 中间锦鸡儿
mimosine *n.* 含羞草素
mixed stands 混播草地
mold *n.* 发霉；霉菌
molybdenum *n.* 钼
monogastric *adj.* 单胃的
mould *n.* 霉 *vi.* 发霉
nectariferous plant 蜜源植物
non-dormant 非休眠
obcordate *adj.* 倒心形的
obround *adj.* 长圆形的
odd-pinnate *adj.* 奇数羽状的
organic fertilizer 有机肥
ornamental plant 观赏植物
ovary *n.* 子房
papilio *n.* 凤蝶
papilionaceae *n.* 蝶形花科
papilionaceous *adj.* 蝶形的
parasitic plant 寄生植物
peduncle *n.* 花梗
pesticide *n.* 杀虫剂
petal *n.* 花瓣
phosphate diammonium 磷二铵
phosphate rock 磷矿粉
plant ash 草木灰
plant heaving 冻拔现象
plough *n.* 犁；耕
powdery mildew 白粉病
propagate *vt.* 繁殖
psammophytes *n.* 沙生植物
pubescence *n.* 柔毛
rangeland *n.* 天然牧场
reticular veins 网状纹
rhombus *n.* 菱形的
root hair 根毛
root rot 根腐病
saponin *n.* 皂素
scarification *n.* 松土；划破
sclerotium disease 菌核病
seed viability 种子生活力
shoot *n.* 地上部分
shrubbery *n.* 灌木林
shrub *n.* 灌木
silage *n.* 青贮
solitary *n.* 单生的
spherical *adj.* 球状的
stubble *n.* 残茬
subshrub *n.* 半灌木
subterraneous *adj.* 地下的
subtriangular *adj.* 近似三角形的
subtropical zone 亚热带

tendril *n.* 详细　*n.* 卷须
thorn *n.* 小尖刺
thrips *n.* 蓟马
trampling resistance 耐践踏
transverse *adj.* 横向的
truncate *vt.* 把……截短　*adj.* 截形的
tubular *adj.* 筒状的
tuzet *n.* 退菌特
twig *n.* 嫩枝
umbel *n.* 伞形花序
vexil *n.* 旗瓣
waterlog *n.* 积水
winter hardy 抗寒的
β- glycoside β- 葡萄糖糖苷
β-nitro propionic acid β- 硝基丙酸

Chapter 6　Grasses

adaptability *n.* 适应性；可变性；适合性
adaxial surface 近轴表面
Aneurolepidium 赖草属
annual ryegrass 一年生黑麦草；多花黑麦草
anthesis *n.* 开花；开花期
arid *adj.* 干旱的
aristate *adj.* 有刺的；有芒的
barley *n.* 大麦
bermuda grass 百慕大草；狗牙根
bisexual *adj.* 两性的；雌雄同体的
broadleaf paspalum 宽叶雀稗
bulb canarygrass 球茎虉草
cespitose *adj.* 丛生的（等于 caespitose）
chestnut soil 栗钙土
Chinese wildrye 羊草
ciliolate *adj.* 有纤毛的
clonal *adj.* 无性系的；无性繁殖系的
cob *n.* 玉米穗轴
cold resistance 耐寒性
conical *adj.* 圆锥的
cool season grass 冷季型草
crested wheatgrass 冰草
culm *n.* 茎
cultivation *n.* 培养；栽培
dallis grass 毛花雀稗
desert wheatgrass 沙生冰草
detrimental *adj.* 不利的；有害的
drooping wildryegrass 垂穗披碱草
drought tolerance 耐旱性
ecological amplitude 生态幅度
field capacity 田间持水量
fusiform *adj.* 梭形的；纺锤形
glabrous *adj.* 无毛的；光滑的
harrowing *n.* 耙地
heat resistance 耐热性
intertillage *n.* 中耕；间作
involute *adj.* 内卷的
Kentucky bluegrass 早熟禾；肯塔基蓝草
leafy *adj.* 多叶的；叶茂盛的
lemma *n.* 外稃
meadow *n.* 草地；牧场
meadow fescue 牛尾草；草地羊茅
mixed sowing 牧草混播
napiergrass *n.* 象草；紫狼尾草
oblate *adj.* 扁圆的；扁平的
orchardgrass *n.* 果园草；鸭茅
palatability *n.* 适口性；风味
pale turquoise 淡粉蓝色
palea *n.* 内稃；托苞
pearl millet 狼尾草；御谷
pedicel *n.* 花梗
perennial ryegrass 黑麦草；多年生黑麦草
procumbent *adj.* 匍匐的；平伏的；伏卧的
prominent *adj.* 突出的；显著的
pubescent *adj.* 青春期的；被短柔毛的
rachilla *n.* 小穗轴；小花轴

raking *n*. 用耙子扒垄；*v*. 耙松
reed canarygrass 虉草
rescue brome 扁穗雀麦
rough *adj*. 粗糙的；粗略的
row spacing 行距
saline *n*. 盐湖；*adj*. 盐的
scabrous *adj*. 粗糙的；难解决的
seedbed *n*. 苗床；苗圃
semiarid *adj*. 半干旱的
serration *n*. 锯齿状
sessile *adj*. 无柄的
setaceous *adj*. 长有刚毛的
Siberian wildryegrass 老芒麦
smooth bromegrass 无芒雀麦
sorghum *n*. 高粱
sterile *adj*. 不育的；无菌的；贫瘠的
sudan grass 苏丹草
tall fescue 高羊茅；苇状羊茅
temperate zone 温带
terete *adj*. 圆柱状的；圆筒形的
thermophilic *adj*. 适温的；喜温的
thousand seed weight 千粒重
timothy *n*. 梯牧草
tiny *adj*. 微小的；很少的
trampling tolerance 耐践踏性
tufted *adj*. 成簇状的
vegetative growth 营养生长
ventral *n*. 腹鳍　*adj*. 腹侧的
vestigial *adj*. 退化的；残余的；发育不全的
viridescent *adj*. 淡绿色的
xerophyte *n*. 旱生植物

Chapter 7　Forbs

achene *n*. 瘦果
amaranthaceae *n*. 苋科
aromatic *adj*. 芳香的
bract *n*. 苞；苞片
carbohydrate *n*. 糖类
carotene *n*. 胡萝卜素
ceratoides *n*. 驼绒藜属；驼绒藜
chenopodiaceae *n*. 藜科
chicory *n*. 菊苣
comfrey *n*. 紫草科植物；聚合草
corolla *n*. 花冠
cuneate *adj*. 楔形的
cup plant *n*. 菊花草；串叶松香草
digestibility *n*. 消化性；可消化性
diterpene *n*. 双萜，二萜
feeding value 饲用价值
fleshy *adj*. 肉的；肉质的
forb *n*. 非禾本草本植物
India lettuce 苦荬菜
inulin *n*. 菊粉
juicy *adj*. 多汁的
kochia *n*. 木地肤
lysine *n*. 赖氨酸
methionine *n*. 甲硫氨酸
monoecious *adj*. 雌雄同株的
nitrogen free extract (NFE) 无氮浸出物
palatable *adj*. 美味的；可口的
perianth *n*. 花被
polysaccharide *n*. 多糖；多聚糖
poultry *n*. 家禽
propagation coefficient 繁殖系数
residual *n*. 剩余；残茬
succulent *n*. 肉质植物；多汁植物
swine *n*. 猪
tryptophan *n*. 色氨酸

Chapter 8 Cereal Forage Crops

cereal *n.* 谷物；谷类植物
corn *n.* 玉米
constipation *n.* 便秘
continuous cropping *n.* 连作
cutworm *n.* 切根虫；地老虎
ear *n.* 耳朵；穗
fibrous root system 须根系
foxtail millet 谷子；粟
grub *n.* 蛆；幼虫
head smut 黑穗病
hydrocyanic acid (HCN) 氢氰酸
husk *n.* 苞叶
maize *n.* 玉米
mole cricket 蝼蛄
oat *n.* 燕麦
rye *n.* 黑麦
silk *n.*（玉米）吐丝期
soybean *n.* 大豆
straw *n.* 稻草；秸秆
subspecies *n.* 亚种
sugar beets 甜菜
tannin *n.* 丹宁酸；鞣酸
tassel *n.* 天花；穗状雄花
wean *vt.* 使断奶

Chapter 9 Forages of Tuberous Root, Tuber and Melon Vegetables

aggregate fruit 聚合果
brassica *n.* 芸苔属植物
brown patch 褐斑病
carrot *n.* 胡萝卜
cucurbitaceae *n.* 葫芦科
field management 田间管理
fodder beet *n.* 饲用甜菜
leaf spot disease 叶斑病
marginal land 边际土地
mosaic disease 花叶病
mustard oil 芥子油
nursery *n.* 苗圃；温床
radish *n.* 萝卜
root rot 根腐病
silique *n.* 长角果
sweet potato *n.* 甘薯
solanaceae *n.* 茄科
tuberous *adj.* 块茎状的
turnip *n.* 萝卜；芜菁甘蓝
powdery mildew 白粉病
pumpkin *n.* 南瓜
silique *n.* 长角果
utricle *n.* 胞果

Chapter 10 Forages Distribution and Regional Planning

binominal *n.* 双重命名法
dominant species 优势种；建群种
geography *n.* 地理；地形
legend *n.* 图例
Mediterranean basin 地中海盆地
regional planning 区划
subregion *n.* 亚区
tetraploid *adj.* 四倍体的
topography *n.* 地势；地形学

▶附录 3　牧草生长及利用总结性补充

Attached table 1. Morphological descriptions for growth stages of forage grasses and legumes
附表1. 豆科和禾本科牧草各生育时期的形态学描述

Grasses 禾本科	
***First growth* 第一茬生长**	
Vegetative 苗期	Leaves only; stems not elongated 只长叶，茎未伸长
Stem elongation 拔节期	Stems elongated 茎伸长
Boot 孕穗期	Inflorescence enclosed in flag leaf sheath and not showing 花序被包在旗叶叶鞘内，未露头
Heading 抽穗期	Inflorescence emerging or emerged from flag leaf sheath, but not shedding pollen 花序露头或从旗叶叶鞘中抽出，但未散落花粉
Anthesis 开花期	Flowing stage; anthers shedding pollen 开花期，花粉囊开始散落花粉
Milk stage 乳熟期	Seed immature, endosperm milk 种子未成熟，胚乳乳熟
Dough stage 蜡熟期	Well-developed seed; endosperm doughy 种子发育完成，胚乳成面团状
Ripe seed 成熟期	Seed ripe; leaves green to yellow brown 种子成熟，叶片由绿色变成黄褐色
Postripe seed 果后营养期	Seed postripe; some dead leaves; some heads shattered 种子后熟，部分叶片死去，一些花序落粒
Stem-cured 茎叶卷曲	Leaves cured on stem; seed mostly cast 叶片卷曲包在茎秆上，大多数种子掉落
***Regrowth* 再生**	
Vegetative 苗期	Leaves only; stems not elongated 只有叶，茎不伸长
Jointing 拔节期	Green leaves and elongated stems 叶绿，且茎伸长
Late growth 后期生长	Leaves and stems weathered 叶与茎风干
Legumes 豆科	
***Spring and summer growth* 春夏生长**	
Vegetative (or prebud) 苗期	No buds visible 没有，花蕾
Bud 现蕾期	Buds visible, but no flowers 有花蕾，但没开花
First flower 初花期	First flowers appear on plants 第一批花出现
Bloom (flower) 开花期	Plants flowering 植物开花
Pod (or green seed) 结荚期	Green seedpods developing 绿色豆荚发育
Ripe seed 成熟期	Mostly mature brown seedpods with lower leaves dead and some leaf loss 大多数成熟豆荚变成褐色，下部叶片开始死亡或凋落
***Fall recovery growth* 秋后再生**	Vegetative or with floral development 苗期或小花发育

Attached table 2. Forage plant families and species that sometimes contain toxic compounds in sufficiently high concentrations to harm animals consuming them
附表2. 植株体内含有足够浓度的有毒抗营养化合物时会危害采食动物的饲用植物

Forage family 科 别	Common name 牧草名称	Antiquality compound 抗营养化合物
Poaceae(grasses) 禾本科	Forage sorghum 饲用高粱	Ergot alkaloids (ergoline) 麦角生物碱（麦角林）
	Tall fescue 高羊茅	Tremorgens (lolitrem B) 震颤素（黑麦草神经毒素）
	Perennial ryegrass 多年生黑麦草	Oxalates, saponins 草酸盐类、皂甙
	Tropical grasses 热带禾草	Cyanogens (HCN) 氰类（氢氰酸）
Fabaceae(legumes) 豆科	Alfalfa 苜蓿	Saponins,phytoestrogens,bloating agents 皂苷、植物雌激素、膨胀剂
	White clover 白三叶	Cyanogens, phtyoestrogens, bloating agents 氰类、植物雌激素、膨胀剂
	Red clover 红三叶	Slaframine, phtyoestrogens, bloating agents 根霉菌胺、植物雌激素、膨胀剂
	Alsike clover 杂三叶	Photosensitization agents 光敏素
	Sweet clover 草木樨	Coumarin (dicumarol) 香豆素（双香豆素）
	Subterranean clover 地三叶	Phtyoestrogens 植物雌激素
	Crownvetch 小冠花	Glycosides 糖苷类
Brassicaceae 十字花科	Turnip 芜菁	Brassica anemia factor 芸薹属贫血因子
	Rape 油菜	Glucosinolates 硫苷

Attached table 3. Adaptation and management characteristics of perennial cool-season grasses adapted to northern areas

附表3. 适宜于北方地区的主要冷季型禾草适应性和管理特性

Common name 牧草名称	Soil adaptation 土壤适应性	Easy to establish 建植难易	Winter hardiness 耐寒性	Drought tolerance 耐旱性	Tolerance of wet soils 耐湿性	Primary use 主要用途	Forage quality 饲草品质	Grazing tolerance 耐牧性	Competitiveness in mixtures 混播竞争性	Regrowth potential 再生潜力
Smooth bromegrass 无芒雀麦	Well-drained,deep,fertile soils 排水良好的深层肥沃土壤	Slow to establish	Excellent	Excellent	Poor	Hay or silage	Very good	Low	Fair	Fair
Perennial ryegrass 多年生黑麦草	Most heavy, fertile soils 黏重肥沃土壤	Establishes rapidly	Fair	Poor	Good	Pasture	Excellent	Very good	Poor	Very poor
Annual ruegrass 一年生黑麦草	Most heavy, fertile soils 黏重肥沃土壤	Establishes rapidly	Fair	Poor	Good	Pasture	Excellent	Good	Poor	Very poor
Tall fescue 高羊茅	Tolerant of both wet and dry soils 耐湿耐干燥土壤	Easy	Good	Very good	Good	Pasture or hay	Poor to good (endophyte dependent)	Good	Very good	Good
Orchardgrass 鸭茅	Well-drained loams and silt loams 排水良好的壤土、沙壤土	Easy	Good	Good	Fair	Hay or silage	Good	Fair	Very good	Good
Reed canarygrass 虉草	Moist or wet lowlands to dry upland fertile soils 潮湿低洼到干旱高燥的肥沃土壤	Problematic	Very good	Very good	Excellent	Hay or silage	Poor to good (alkaloid dependent)	Good	Good	Good
Timothy 梯牧草	Well to somewhat poorly drained, fine textured soils 排水稍差结构良好的土壤	Easy	Excellent	Poor	Fair	Hay or silage	Very good	Low	Fair	Poor
Crested wheatgrass 冰草	Fertile, light sandy loams to heavy clays 肥沃的轻沙壤土到黏重土壤	Easy	Excellent	Excellent	Poor	Grazing or hay	Good	Excellent	Excellent	Fair
Kentucky bluegrass 早熟禾	Fine-textured, well-drained to somewhat poorly drained soils 结构和排水良好或一定程度的不良土壤	Easy	Excellent	Poor	Good	Pasture	Excellent	Very good	Very good	Poor

注：Very poor–很差；Poor–较差；Fair–一般；Good–较好；Very good–很好；Excellent–非常好。

Attached table 4. Adaptation characteristics, yield, and bloat-causing potential of forage legumes
附表4. 几种主要豆科牧草的适应性、产量和致臌胀病风险

Legume 豆科	Tolerance to low soil fetility 耐贫瘠	Tolerance to soil acidity 耐酸性土壤	Tolerance to drought 耐旱性	Tolerance to poor drainage 耐排水不良	Cold hardiness 耐寒性	Yield potential 产量潜力	Bolat potential 臌胀病风险
Alfalfa 紫花苜蓿	Poor	Sensitive	Excellent	Poor	Excellent	Very high	Yes
Red colver 红三叶	Fair	Moderate	Good	Fair	Very good	High	Yes
White clover 白三叶	Fair	Moderate	Poor	Good	Very good	Medium	Yes
Aslike clover 杂三叶	Fair	Moderate	Fair	Excellent	Very good	Medium	Yes
Birdsfoot trefoil 百脉根	Good	Good	Good	Good	Good	Good	No
Crownvetch 小冠花	Good	Moderate	Good	Fair	Good	Medium	No
Sainfoin 红豆草	Good	Low	Excellent	Poor	Very good	Medium	No
Cicer milkvetch 鹰嘴紫云英	Good	Moderate	Very good	Good	Excellent	High	No
Sweetclover 草木樨	Good	Sensitive	Excellent	Poor	Excellent	High	Yes
Annual lespedeza 一年生胡枝子	Excellent	Tolerant	Good	Good	None	Low	No

注：Very poor–很差；Poor–较差；Fair–一般；Good–较好；Very good–很好；Excellent–非常好；Moderate–适度的；Medium–中等。

Attached table 5. Seasonal or conditional disorders caused by climate, soil, plant, or animal factors or by a combination of these factors

附表5. 因气候、土壤、植物或动物因素，或以上因子的结合导致的季节性或条件性的失调

Disorder 紊乱症	Description 病因描述	Symptoms 症状	Plants 主要牧草	Animals 影响动物	Prevention 预防措施	Treatment 治疗方法
Grass tetany (hypomagnesemia) 禾草抽搐症低镁症	Inadequate blood serum（血清） magnesium	Stiff gait（步态僵硬）, staggering, twitching muscles（肌肉震颤）, convulsions（惊厥抽搐）	Cool-season grasses in early spring	Brood cows, ewes, often heavy milkers in early lactation	Split K and N fertilizer applications, provide Mg supplements	Intravenous injections（静脉注射） of calcium-gluconate solution（葡萄糖钙溶液）, subdermal（皮下） injections of magnesium
Frothy bloat (rumen tympany) 臌胀病	Inability to eructate（嗝出） rumen gases	Distended（膨胀的） rumen visible first on the left side	Alfalfa, red and white clover, lush grass pastures, including wheat	Cattle, sheep	Using antibloating agents, use grass/legume mixtures, supplement ionophore monesin, strip-graze	Provide hay or more-mature grass forage
Nitrate poisoning 硝酸盐中毒	Toxic accumulation of nitrate converted to nitrates that bind hemoglobin（血红蛋白）	Labored breathing（呼吸困难）, abdominal pain（腹痛）	Sudangrass, oats, rape, wheat, corn, on high-N soils or under drought conditions	All livestock	Split N applications when high annual rates are used	Remove from toxic plants, dilute with other forage
Prussic acid poisoning〔hydrocyanic acid（HCN） poisoning〕 氢氰酸中毒	Young, white, frosted, or stunted plants; glycosides（苷类） degrade to release HCN	Muscle tremors（震颤）, rapid breathing, convulsions, asphyxiation（窒息）	Sorghums, white clover, vetch seed	Cattle	Avoid stuned, frosted plants; split N fertilizer application; test samples; prevent selective grazing	Remove from causative plants, dilute with other forage
Phtyoestrogens 雌激素	Plant compounds that mimic estrogens and cause reproductive problems	Poor reproductive performance. In serve cases, changes are visible	Certain species of forage legumes	Sheep are more susceptible than cattle	Replace problem species or avoid use during susceptible periods	Subclinical（亚临床症状） effects are alleviated by removal from problem forages; Permanent infertility can result
Primary photosensitization 光敏反应	Blood compounds reacting with ultraviolet light（紫外线） on the skin, producing free radicals（自由基） that react with dermal tissue（皮肤组织） proteins	Dermatitis（皮炎） affecting light skin	Hypercium spp., buckwheat, spring parsley, lush pastures	Cattle, horses, goats, sheep	Remove from food source	Dependent on toxicant
Facial eczema 湿疹	Secondary potosensitization caused by the hepatoxic mycotoxin sporidesmin（肝毒性霉菌毒素）	Severe dermatitis of light-skinned areas	Ryegrass pastures following warm, wet weather	Sheep, cattle	Provide zinc supplements and iron salts, mange around affect pastures	Provide shade

Attached table 6. Forage species or cultivar-related livestock disorders

附表6. 与牧草种或品种有关的牲畜病症

Disorder 紊乱症	Description 病因描述	Symptoms 症状	Plants 主要牧草	Animals 影响动物	Prevention 预防措施	Treatment 治疗方法	Future 未来方向
Tall fescue toxicosis 羊茅中毒症〔summer slump, fat necrosis(坏死), fescue foot〕	Hyperthermia(elevated body temperature) hypoprolactemia(low serum prolactin levels)	Weight loss; dull, rough coat; excessive salivation	Tall fescue	Cattle	Check your field before grazing	Removal from contaminant	Use only fungus freeseed stock for pasture
Ryegrass staggers 黑麦草蹒跚症	Hyperthermia, hypoprolactemia	Weight loss; dull, rough coat; excessive salivation	Perennial ryegrass	Sheep, cattle, horses	Change fields	No treatment	Use endophyte-free seed
Reed canarygrass alkaloids 虉草生物碱	Alkaloids negatively affect forage consumption	Reduced intake, muscle tremors	Certain reed canarygrass cultivars	Cattle, sheep	Avoid high-alkaloids cultivars	Removal from the affected forage	Low-alkaloid varieties are available
Sweetclover poisoning 草木樨中毒	Internal blood loss	Swellings under skin, pale membranes, weakness	Spoiled sweetclover hay	Cattle, horses, sheep	Avoid feeding moldy sweetclover hay	Removal from contaminated feed, vitamin K and blood injections	
Tannins 单宁中毒	High levels of tannins reduce intake, DM digestibility, and weight gains		Mostly legumes and woody browse species	Sheep,cattle, goats	Avoid high-tannin forage		Use low-tannin cultivars where available

Note: The above 6 attached tables are all from Robert F. Barnes etc., Forages (6th Edition), Blackwell publishing, 2001.

Attached fig. 1 Anatomy of a legume

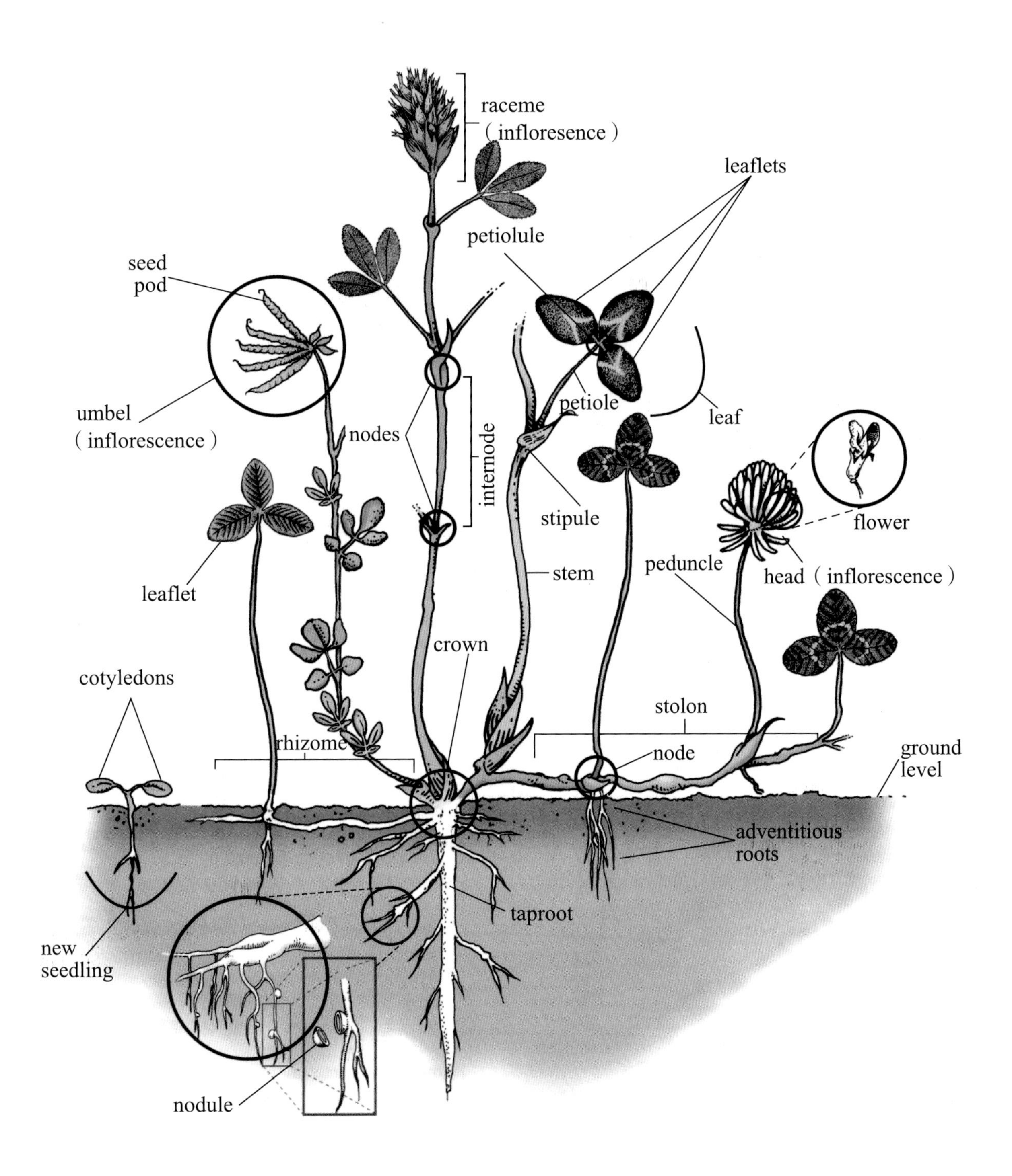

Attached fig. 2 Anatomy of a grass

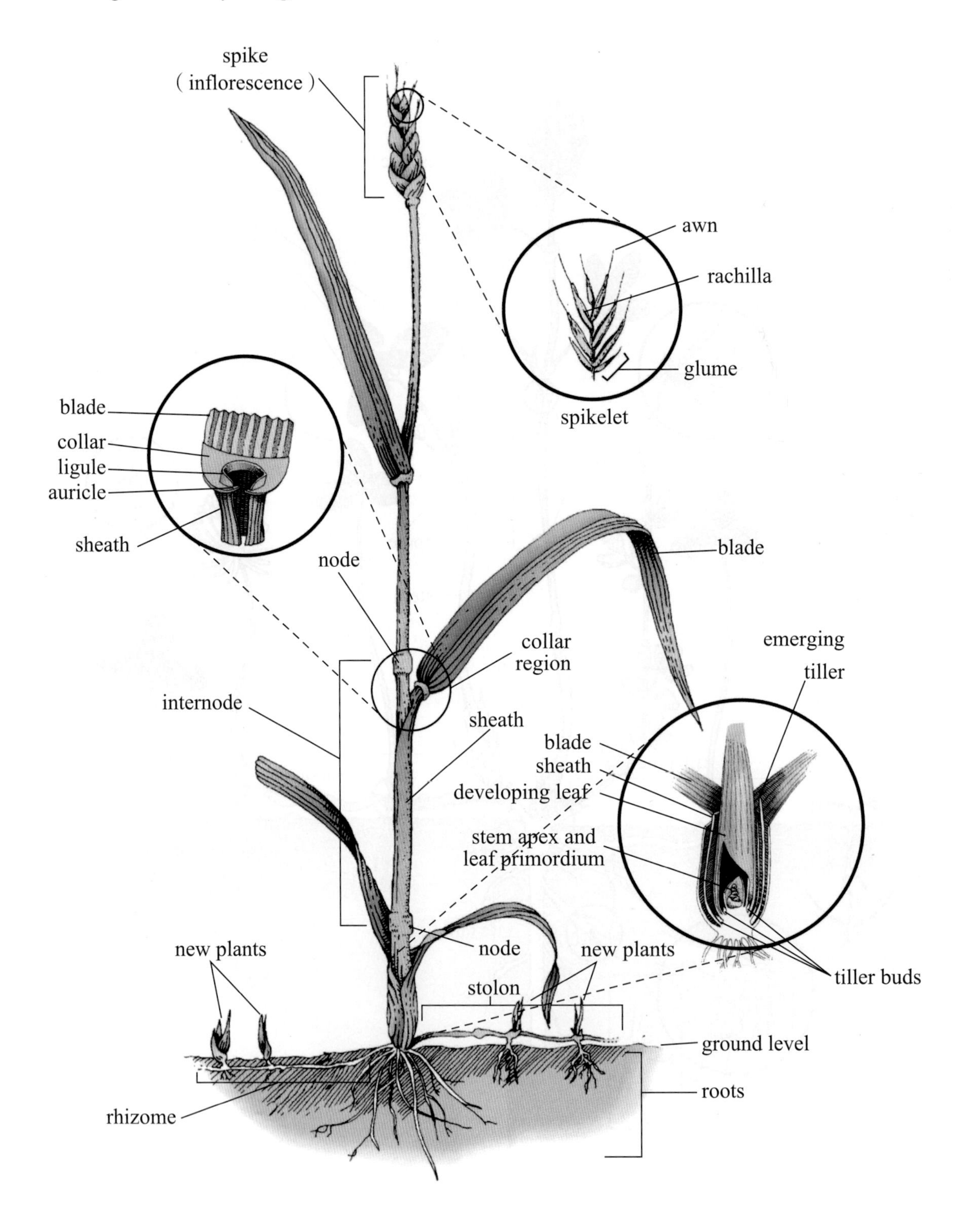

后 记

《牧草栽培学双语辑要》即将付梓出版了，从组织策划、撰写前言始，掐指算来，虽然历时三年有余，但不尽如人意之处甚多。因此，对编写的目的和编写过程再赘述几点。

高等教育走到今天面临许多挑战，教师作为主体，除承担教育教学任务外，更要承担科学研究、社会服务和文化传承的繁重任务。在战斗间隙，组织这一群年轻人编写教材，特别是双语教材，不忘初心，重视教育教学，过程大于结果。

在教育日益国际化的大背景下，学生的国际交流学习已成为常态，对于我校草业科学或相近专业的学生而言，出国学习或外教来校授课已不是什么新鲜事。因此，实现中文专业名词、英文专业名词和实物的准确对应是编写本教材的初衷之一。尽管还有许多不完善之处，好在他们还年轻，在未来的实践过程中，相信会日臻完善。

需要说明的是，本书初稿的绪论、第一章、第十章和附录由龙明秀博士编撰；第二章、第三章和六章由何学青博士编撰；第四章由许岳飞博士编撰；第五章由杨培志博士编撰；第七章、第八章和第九章由何树斌博士编撰。在统稿和反复修订过程中，龙明秀博士付出了大量心血，充分担当了主编的重要责任，我只是做了些润色和整体考量的工作。

在本书编写过程中参考借鉴了不少国内外的教材、专著和其他资料，在此谨向各位编著者深表谢意。感谢副校长罗军教授对本学科国际交流的长期支持，感谢教务处和学院领导的不懈支持，感谢曹社会教授和美国南达科他州立大学 Roger N. Gates 教授对教材初稿的修改润色，特别感谢高等教育出版社李光跃同志在出版过程中付出的艰辛努力。

呼天明

2018 年 8 月 30 日

于西北农林科技大学